The person charging th'
sp   ble for   turn.

MUSIC, SPEECH
&
HIGH FIDELITY

MUSIC, SPEECH

&

HIGH FIDELITY

A Descriptive Acoustics Worktext

BRIGHAM YOUNG UNIVERSITY

WILLIAM J. STRONG
GEORGE R. PLITNIK

ISBN 0-8425-0797-3

78 2100 35007

Music

Contents

Preface

The present *Descriptive Acoustics Worktext* has grown out of our attempts to create resource materials for student use in descriptive acoustics courses. Our original intent was to create an exercise book which could be used in conjunction with several good books that were already available. However, after experimenting with this approach for several years, we found it less than satisfactory. This worktext represents our attempt to provide reading materials and associated exercises that deal with topics we have found to be of interest in descriptive acoustics courses.

Most books available at the descriptive level in acoustics concentrate their emphasis on one or two selected areas of acoustics to the partial or complete exclusion of others. This may be desirable when depth of treatment, at the necessary expense of a broad overview, is the aim. However, we feel that once some of the basic conceptual tools have been developed it is not only possible, but highly desirable, to treat several areas of acoustics at the descriptive level. In particular, many of the same ideas that are applicable in musical acoustics are applicable in speech acoustics and vice versa. We have attempted in this worktext to develop the relevant concepts of physics and to apply these concepts to acoustical aspects of music, speech, high fidelity, listening environments, and hearing. The approach here is, of necessity, broad brush in nature, and many of the more subtle points are only alluded to or not even mentioned. In keeping with our aim that the worktext be primarily descriptive, we have had to sacrifice absolute scientific accuracy in some instances in order to present the material in a simple manner. We hope the student will make use of the resource materials listed at the end of each unit to explore areas of interest in greater depth and also to enlarge upon topics too sparsely developed in the current worktext.

We have attempted to emphasize the application of physical principles in explaining and describing many diverse phenomena of acoustics. We expect that the student will give serious consideration to the various questions, problems, and exercises provided at the end of each unit. Most of the exercises are self-contained and can be performed with the material supplied in the text. However, many of the exercises can be made more meaningful and exciting by using laboratory instruments to collect the data rather than using the data that are supplied. The data in the book are from "real world" situations and should be meaningful from that point of view, but we firmly believe that there is no substitute for collecting one's own data when possible.

We hope that the student and instructor alike will find the listing of resource materials useful. The resource materials include books, journal articles, audio tapes, films, and demonstrations. Many of these items will be available to the student in the school library or learning resource center. Books that provide useful reference material for several sections are listed in a bibliography at the end of this text and are referred to by chapter number at the end of the appropriate sections. Journal articles and books with restricted content domain are referenced in full at the end of each section. The audiotape and film listings provide enough information for ordering. In some instances the demonstrations listed are only suggestive; in other cases some detail or a reference is provided.

More than enough material for a one-semester course is included in this worktext. After Chapters 1 and 2 (the physical principles) have been completed the other chapters can be used more or less independently. It is possible to structure a course involving only hearing and music (Chapters 3 and 6), hearing and speech (Chapters 3 and 5), hearing and hi-fi (Chapters 3, 4, and 7), or other combinations which may be desirable.

Undoubtedly much useful refining of this worktext can still be realized, and we will be pleased to receive comments and criticism from interested readers and users. We regard it as an evolving effort and so will be happy to include additional features and corrections in such future editions as might be warranted.

We acknowledge our indebtedness to many workers in the fields of acoustics presented in this book. Many of our figures have been taken or adapted from journal articles and books as noted in the figure captions; others have been supplied by researchers from their unpublished material. We acknowledge our debt to researchers and journals for this material.

The books on musical acoustics by John Backus and Arthur Benade and *The Speech Chain* by Peter Denes and Elliott Pinson have particularly stimulated our thinking in the writing of this book. Members of the Education Committee of the Acoustical Society of America have also encouraged us in this endeavor.

We are indebted to John Backus and Arthur Benade for assistance in developing impedance measuring devices and for the impedance curves with which they have provided us. We are indebted to many other colleagues for fruitful discussions, useful information, and other assistance. In particular, we must mention Irvin Bassett, Duane Dudley, Harvey Fletcher, Carleen Hutchins, Justin Olsen, Paul Palmer, Thomas Rossing, and Ingo Titze. The following graduate students have also given their assistance: Norman Kinnaugh, Alan Melby, Ronald Millett, and Steve Stewart.

We appreciate the encouragement and support given us by the Brigham Young University Department of Physics and Astronomy. We are indebted to Gordon Johnston for typing the manuscript, Karen Jeffs and Stacy McKim for assisting with the typing, and Nancy Snyder for donating her time to type rough draft copies. We also owe a debt to the staff of Brigham Young University Press.

Finally, we wish to thank our wives, Charlene and Gail, who gave us practical help in rewriting, proofreading, and other necessary tasks related to the book and who always offered their encouragement and enthusiasm for our work.

Chapter 1

Fundamentals of Physics and Vibration

1.A. SCIENCE AND SCIENTIFIC METHOD

What Is Science

According to Webster's dictionary, science is "systematized knowledge"; more specifically, physical science is an organized body of knowledge about the physical universe, by which we seek to establish general relationships connecting a number of particular facts. Through the process of *generalization* a system of scientific knowledge is organized; in fact, science could be described as a vast collection of generalizations. If you ask a physicist what the weight of a moose would be on top of a tower 6400 km high, he will answer, "Its weight will be about one-fourth what it is on the earth's surface." How did he know this? No one has ever built such a tower; nor is it likely that one ever will be built. He knows the answer to your question from a generalization that states that an object's weight decreases in proportion to the inverse square of its distance from the earth's center. (The radius of the earth is about 6400 km, and if a moose is on a 6400 km-high tower the moose will be twice as far from the earth's center as when it is on the earth's surface.)

Although we can consider science to be a collection of generalizations, the generalizations of science—unlike such common generalizations as "all redheads are hot tempered"—are motivated by careful, methodical observation along with insight and imagination. A scientist first collects data by making carefully controlled measurements. Then he tries to establish relationships among the data by the process of generalization. Once a generalization is obtained, the scientist then uses it as the basis for further experiments. Science, then, is not a static process of making observations and collecting data, but a dynamic, on-going process which is constantly changing and expanding as new observations are made and new generalizations are formed.

Distinguishing Characteristics of Science

To what extent do the various sciences have common characteristics which distinguish them from other disciplines, such as the humanities? The answer, of course, is not as simple as the question—something we learn to expect in any field of study. All we can do is mention a few points for consideration, stressing four characteristics which, when combined, make the sciences unique: generalization, classification, quantification, and experimentation.

Generalization is the reduction of particular situations to more general terms. For example, a person may begin to sneeze and cough on repeated occasions when petting the family cat and therefore draw the general conclusion that he/she is allergic to cats. The sciences make use of generalization in varying degrees. We can picture

1

a chain of natural sciences in this order: physics, chemistry, geology, biology. The current extent to which generalization has been found possible and fruitful decreases as we go along this chain. However, virtually all other fields of human knowledge also employ generalization in varying degrees.

A second feature is *classification,* or the sorting of objects or phenomena into different classes or categories. Classification catalogs information into a more easily accessible form and has been a mark of science since the days of the Greeks. It is particularly important for biology and geology.

A third characteristic is *quantification,* which is the attaching of numerical values to generalizations. While this could be done in any field (e.g., attaching numerical "weights" to various authors in English literature), if quantification is to be meaningful it must be used to express unique relationships in mathematical form (as, for example, in an algebraic equation).

Generalization, classification, and quantification, however, are still not sufficient to unequivocably set the sciences apart from other fields. (Business accounting, for example, may well have all these characteristics.) Science, however, possesses another important characteristic—*experimentation.* Experimentation is a procedure whereby generalizations are tested.

The importance of experimentation to science cannot be overemphasized. Often in the past, when people have tried to have a science without experimentation they have caused more harm than good. For example, the ancient Greeks liked to theorize about nature and they did contribute greatly to the development of science. The famous Greek, Aristotle (who lived in the fourth century B.C.), however, preferred theorizing to experimenting. He concluded that the heavier an object is the faster the object will fall toward the earth. This seems like a reasonable generalization, for we have all observed that if a stone and a feather are dropped simultaneously the stone reaches the ground first. But if Aristotle had performed several carefully controlled experiments, he might have noticed (as was finally observed by Galileo about 1600 A.D.) that all objects, regardless of weight, fall toward the earth with the same acceleration. Any differences are due to air resistance. Aristotle was lax in his experimentation, but scientists in the following centuries were also wrong by accepting Aristotle's work with no further experimentation.

Scientific Method

The so-called "scientific method" is not one, specific, detailed method of procedure used by all scientists, but rather it is any systematic procedure used by scientists in their work. There are probably as many different methods used by scientists as there are scientists; nevertheless, the general way in which most scientists tackle a problem can be considered a scientific method.

First of all, when a scientist attempts to start from a specific observation and arrive at a generalization, he is using *inductive* reasoning. When he then attempts to use the generalization to search for evidence of it in scientific facts, he is using *deductive* reasoning. Both of these types of reasoning are important in science, and each, for different reasons, must ultimately refer back to experimentation. In induction we must constantly devise new experiments to ascertain that we have not overextended ourselves with our generalization. In deduction we must rely on experimentation to verify the validity of our initial assumptions, because the conclusions we draw will be valid only if the initial assumption is valid. Abraham Lincoln once asked, "If you call a tail a leg, how many legs does a dog have?" If we take the first clause as a postulate, the deductive process tells us that the dog has *five* legs. But it remains an unalterable fact that a dog has only *four* legs. Nothing is wrong with the deductive process here; it is only that we have a faulty postulate. As Lincoln observed, "Calling a tail a leg does not make it a leg."

In capsule form, the scientific method *usually* consists of the following steps:

1. The *problem* to be solved is stated as precisely as possible.

2

2. *Published material* is searched to find all the known information which is pertinent to the particular problem.
3. *Experimentation* is then conducted to observe and collect data applicable to the problem.
4. The data are then *organized* (by means of tabulation, graphs, etc.), and an attempt is made to find consistent relationships (or correlations) among the data. If correlations are poor, it may be necessary to repeat one or more of the previous three steps.
5. A *generalization* from the correlation is then formulated to provide an answer to the problem.
6. *Predictions* are then deduced from the generalization.
7. Additional experiments are then performed to see if the predictions are *verified*. If the predictions are not verified, modifications may be made to the generalization and another attempt made to achieve experimental verification of further predictions.
8. Finally, the findings are recorded and *published* so that others may learn the results. Publishing accounts for much of the success in the ongoing processes of science and makes the second step possible.

As an example of the application of the above steps—and omitting, for simplicity, steps 2 and 8—assume that you have different pieces of brass tubing that are identical to each other except for length. Assume further that each piece of tubing is fitted with an identical clarinet mouthpiece. The *problem* is to find the relationship between the length of tubing and the playing pitch. When you *experiment* by blowing on the different lengths of tube, you observe that there are different pitches associated with each. You further discover through *organization* of your data that the longer lengths of tube produce the lower pitches. You make the *generalization* that the pitch is inversely proportional to the length of the tube. You then make some *predictions* about the pitches that will be produced by some of the tubes that you have not yet blown on. You blow on these tubes and measure the pitches and compare the measured values with the predicted values in an attempt to *verify* your generalization. You find that your generalization is approximately verified, but that you get more accurate results if you include the length of the mouthpiece along with the length of the tube. This leads to some modifications of your original generalization and may be followed by new attempts at experimental verification of additional predictions.

Application of the scientific method to the understanding of physical phenomena should enable you after some practice to do the following: (1) discover new knowledge, (2) structure knowledge so that it is always available, (3) apply knowledge to new situations, (4) test the validity of new knowledge, and (5) judge the value of new knowledge. (Test yourself as you progress by checking to see if you can do the above with new phenomena as you encounter them.)

Laws, Hypotheses, and Theories

A physical scientist generally proceeds on the assumption that the phenomena he observes in his laboratory are governed by natural laws. His task then is to discover these physical laws and to express and explain them in a lucid manner. If his expression of the laws is accurate, relationships among any experimental data he gathers can be accommodated within the existing framework of these laws.

The experimental observations or data measurements that a scientist makes constitute the *facts* of science. Certain controls and conditions are usually associated with the measurements of data, and these in effect become a part of the facts. Furthermore, scientific facts or measurements must be able to be duplicated by different observers in different parts of the world at different times, or at least be equally accessible to all comparably equipped scientists in the case of nonrepeatable phenomena.

When a scientist attempts to take a particular body of facts or experimental observations and state general relationships between them, he creates a *law*. A law generally refers to the generalizations of science; it implies definite relationships between natural phenomena under certain pescribed conditions. When using a scientific law, we must be aware of the limitations and conditions under which the law applies.Some laws appear to be universal, but sooner or later a situation may arise in which the laws will be found to be inaccurate or too limited. However, within the limits prescribed when a law was derived, we expect the law to hold because it is based on the previously observed behavior of matter. When a law is supplemented by new information, it does not always mean that the old law must be discarded. Typically, new information can be expressed in the form of additional conditions on the law, and so long as we remember the limitations of a law and do not exceed them, the old form of the law may be very useful and in some cases easier to apply than a new and more general law.

A *hypothesis* is generally considered a tentative explanation of the observed facts. It implies an insufficiency of experimental evidence and is often intended only as a tentative assumption to be treated and modified.

As a hypothesis is tested and refined it may evolve into a *theory*. A theory is a conceptual scheme or model that provides an explanation of the natural phenomena or the laws of science. A theory implies a greater range of experimental evidence and more refinement than a hypothesis does. Theories are more general than hypotheses or empirical laws as empirical laws can often be deduced and explained within the framework of a theory. To be most useful in science a theory should be fruitful in making predictions that can be tested experimentally. Ideally a theory should correlate many separate and seemingly unrelated facts into a logical, easily grasped structure. A good theory will make it possible to predict specific new observable phenomena and explain all previously observed phenomena. A theory that is not fruitful is not a useful theory because it does not lead to any new knowledge. (Notice that the criterion of a "useful theory" depends not on whether it is true or not, but rather only on the consequences of the theory.)

Obviously scientists do not expect a theory to go through the ages unchanged. Theories are always subject to change, and an important part of science is trying to *disprove* existing theories. A theory can never be proven but only disproven, and the scientist who formulates a theory will then immediately try to disprove it, for he knows that if he does not disprove it somebody else probably will. No number of observations or experiments will prove a theory to be true. Rather, we speak of observations and experiments as *verifying* theories. The most beautiful of theories, however, is subject to failure if it cannot explain the facts. (We should note in passing that the word *theory* is often misused by the layman. It is often used to refer to an educated guess or a tentative answer to a trivial problem. This is not the way scientists use the word, as a theory has a great deal of careful experimental and conceptual work going into its makeup.)

The Threefold Nature of Science

The activities with which science concerns itself may be regarded as an interrelated threefold process, as shown in Fig. 1.A-1. First, science utilizes experimentation (circle A); that is, scientists are involved in observing nature and gathering data. Second, it is theoretical (circle B); science produces laws and theories which generalize, predict, interpret, and otherwise make the data more manageable and more easily understandable. Third, science can be *applied* (circle C) to transform our environment. Whereas theoretical, or "pure," science is concerned with the formulation of new laws and theories, "applied" science seeks to find useful applications of these laws and theories in the matters of everyday living.

The three components of science are inseparable. Each is given meaning only in its relation to the others, as shown in Fig. 1.A-1. Furthermore, each circle is connected with each of the others by two oppositely directed arrows, which indicates that the process of science can and does go in either direction. The arrows between

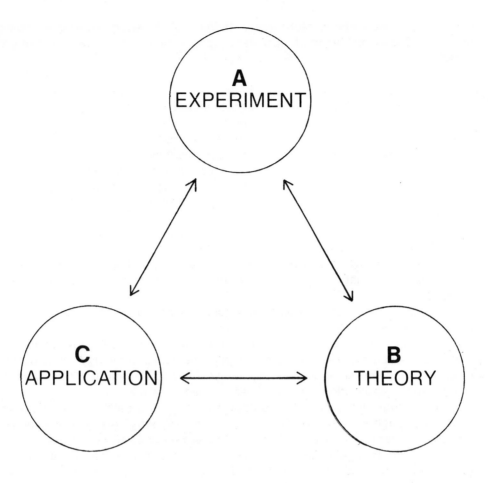

Fig. 1.A-1. The threefold nature of science.

circles A and B, for instance, show that experimentation leads to theory and that theory often influences the type of experiment performed and what one expects to observe from the experiment. The double arrows between circles B and C indicate that theory is important in assaying the possibility, as well as the probable direction, of some useful application. Conversely, an enrichment of theory is often the result of a struggle with new problems arising from an attempt to apply the theory. Finally, circles A and C are also connected, showing that an experiment may directly suggest an application, or that technological developments may suggest new areas for gathering data.

In summary, it cannot be stressed too strongly that science is an indivisible unity of these three aspects. Science functions most efficiently when all three components are thriving and properly balanced.

Exercises
1. In induction a scientist may observe that a simple relationship exists among his data. Hooke, for example, noted that when weights were hung on a wire the length of the wire increased in proportion to the weight. As a "law" this observation so far fits only these particular measurements on this particular piece of wire. How should you argue by induction to generalize the law for—
    a. any pull on the particular wire?
    b. any metal wire?
    c. wire of any material (i.e., rubber)?
    Do you expect the law to hold for all of these cases? How would you check the validity of the law in each case?

2. Each of the two sets of steps given below could be considered a "scientific method." Based on the reading material, place each set of steps in the proper order:

Set A	Set B
analysis (form hypothesis)	accumulation of data
observation	measuring instruments
prediction	desire to understand
recording (measure)	useful application
verification	classification and organization of data
	testing and modification of theories
	development of theories

3. Is the assigning of numbers to the works of different authors (as done in a library) an example of quantification? Why or why not?

4. You observe that when you hold a lighted match to a piece of paper the paper begins to burn. What are the facts and what are the theories involved in this observation?

5. What is the difference between a statutory law and a natural law? Can natural laws ever be violated?

6. Are the steps of the scientific method peculiar to science? How could similar steps be applied to music, speech, religion, medicine, sociology? How would the means for accomplishing each step, particularly the measurement step, compare in the various applications with the means used in science?

7. What is the logical procedure for producing a scientific theory? Are theories constructed through inductive or deductive processes? Why? Are theories verified by inductive or deductive means? Explain. Why are theories never "proved"?

8. Is the seventh step (after proper ordering) of set B in problem 2 really a part of the scientific method? Do science and technology have any interdependence? What?

9. Explain how the science of high fidelity involves both experiment and theory. Are the applications an important aspect of high fidelity? Give several examples.

10. How can the science of musical acoustics be theoretical and oriented toward application? Is experimentation an important aspect of musical acoustics? Give examples of the interrelationships between circles A and B, B and C, and C and A (Fig. 1.A-1) in this field of study.

11. Explain how scientific study of human speech involves experiment, theory, and application. Give examples of the interrelationship for speech and science.

12. In several facets of life, including music and speech, symbols are used to represent "real things" or other symbols. Symbols allow us to economize. In the table of exercise 13, write symbols that represent the same "real things" as the word symbols in the "measured quantity" column. Why are standard symbols necessary? (See discussion of symbols in Appendix 3.)

13. In accomplishing the "measuring step" in science, it is necessary to use certain standard units. Why are standard units necessary? (See Appendix 3 for a dicussion of standard units.) In the following table give standard units to express the measured quantity. (See Appendix 3.)

Measured Quantity	Symbol	Standard Unit
length		
mass		
time		
displacement		
velocity		
acceleration		

Measured Quantity	Symbol	Standard Unit
angle		
force (weight)		
pressure		

14. Application of Scientific Method. (The concepts of frequency and frequency analysis will be introduced later. The instructor may explain air volume and frequency as abstract entities at this point.)

   a. Take a soft drink bottle or other narrow-necked bottle and blow across the top when it is filled with various amounts of water. What do you *observe* in regard to the frequency (pitch)?

   b. A soft drink bottle was filled with various amounts of water, producing the different air volumes in the bottle *recorded* in the table below. A microphone was used to pick up the sounds produced by blowing the bottle and to pass them to a frequency analyzer which made it possible to *measure* frequencies recorded in the table.

   c. *Analyze* the results of examples 1 and 2 and make a hypothesis about how frequency and air volume are related for the particular bottle. Express the result symbolically by determining which of the following formulas expresses the correct relation: $f = (constant)(V)$; $f = constant/V$; $f = constant/SQRT(V)$. Determine the value of the constant.

   d. Use your hypothesis to *predict* the frequencies that should result from the air volumes in examples 3 and 4. Then compare your predicted frequencies to the measured frequencies. (Keep in mind that the predicted and measured frequencies will not be identical because of measuring errors and because the hypothesis may not be exact.)

Example	Air Volume (cm³)	Measured Frequency (Hz)	Predicted Frequency (Hz)
1	320	187	——
2	235	223	——
3	150	272	
4	70	414	

Further Reading
Ashford, chapter 1
Backus, Introduction
Ballif and Dibble, chapter 1
Krauskopf and Beiser, chapter 3
Roederer, chapter 1

Audiovisual
1. *Science and Superstition* (11 min, B&W, 1947, CF)
2. *Scientific Method* (12 min, color, 1954, EBE)
3. *How a Scientist Works* (17 min, color, 1960, EHF)
4. *What Is Science?* (11 min, color, 1947, CF)

Demonstrations
1. Pitch versus string length for a monochord
2. Musical instrument spectrum analysis
3. Soft drink bottle spectrum analysis

1.B FUNDAMENTAL PHYSICAL QUANTITIES AND LAWS

Operational Definitions

Suppose you ask four people to measure a given piece of string. They each perform your request and report back with the following information:

Person	#1	#2	#3	#4
Measurement	0.25 m	25.40 cm	10.00 in	0.83 ft

Immediately you notice that they did not all perform the measurement in the same way, nor did they all get the same answer. Nevertheless, you are struck by the fact that they all give an answer in some unit of length (meters, feet, etc.). Furthermore, each obtained his/her respective answer in the same manner, by comparing the piece of string to a standard (although different in each case) measuring device. The important point is not that they all used different standards of length, but the fact that they each had the same concept of length, namely that it is something obtained by measurement.

To these people, and to us, length is defined by the operation of making a measurement, and this is what is meant by an *operational definition.* Does an operational definition tell us what length "really" is? No, it only tells us how to measure it, and for physical science that is all that is necessary. Where physics is concerned, any concept which is not ultimately defined by operations is meaningless, and any question which cannot be answered by making a measurement (even if we don't have the technical ability at present to perform the task) is meaningless. A question like "Does a table cease to exist when not being observed?"—while it may be an interesting topic of discussion for a philosophy class—is totally meaningless for physics.

Four Fundamental Physical Quantities

All the terms which are used in science must be defined carefully in order to avoid possible confusion and semantic misunderstanding. As you will see, once a term is defined it is used to define still other terms, and in this manner the scientific jargon is formulated. Somewhere, however, there must be a starting point; that is, there will have to be several basic quantities which are used to start the hierarchy of definitions. For the physics which we will use, four fundamental quantities are needed. Although the choice of which physical quantities are taken as fundamental is somewhat arbitrary, the four which are usually chosen are length, time, mass, and charge. Since these quantities are used to define all the other physical quantities which will follow, they can only be defined operationally; that is, we cannot tell you what length is apart from measurement since length is "defined" by the act of making a measurement. Likewise, time, mass, and charge have meaning in physics only in terms of their measurement.

Length is the first fundamental quantity to be considered and can be defined as the spatial distance between two points. The measurement is accomplished by comparing the unknown distance to some standard length. The international standard of length is the *standard meter,* which is defined as the distance between two lines engraved on gold plugs near the ends of a platinum-iridium bar kept at the International Bureau of Weights and Measures at Sevres, near Paris. (Since 1960 the standard has been defined by comparison to one particular color of light emitted by a certain isotope of Krypton.) A convenient unit of length for many acoustical measurements is the centimeter, which is defined as one one-hundredth of a meter.

Time is the second fundamental quantity to be considered. As with length, we have an intuitive feeling for time even though we would probably have great difficulty in defining it. We may think of time in terms of the duration of events, and the measurement of time as the comparing of the duration of an unknown event with that of a standard event. If we define a solar day as the time for the earth to make one complete rotation on its axis (with respect to the sun), we have a standard event which can be used to measure time. Careful measurement, however, has shown that the solar day varies slightly in time during the course of the year; so a mean solar day was defined. This is just the "time length" of a solar day averaged over a year. The mean solar day can then be subdivided into twenty-four hours, each hour into sixty minutes, and each minute into sixty seconds. Then there are $24 \times 60 \times 60 = 86,400$ seconds in a mean solar day, and the standard second can be defined as 1/86,400 of a mean solar day. (Recent astronomical measurements have shown that the rotating earth can no longer be regarded as a satisfactory clock because there are slight irregularities in its rotation. Just as the meter can be defined more accurately in terms of a certain color of light, time can be measured more accurately by using the vibration properties of atoms or molecules.)

Mass is the third fundamental quantity considered. The concept of mass is not as intuitively obvious as length and time, but we can try to develop some feeling for this concept by considering he following experiment. Suppose you are walking barefoot along a country lane and you see an old tin can. You kick the can as hard as you are able and it sails down the street. Farther down the lane you encounter a similar tin can, which you also kick. This time, however, someone has previously filled the can with lead. When you kick the can it barely moves, but you experience considerable pain in yor large toe—an experiment unwise to actually conduct, because of the danger of toe-main poisoning. One way to describe the difference between the two cans is to say that the second can (because of all the lead) had a greater mass than the first can. The mass of an object is related to the difficulty of changing the state of motion of the object, which ultimately depends upon the amount of matter present. To determine the mass of any object it is only necessary to compare the unknown to a standard mass by use of a balance. The international standard of mass is the *standard kilogram,* defined by a cylinder of platinum-iridium kept with the standard meter at the International Bureau of Weights and Measures. The gram, defined as one one-thousandth of a kilogram, is often used to express mass.

All objects are composed of particles possessing *electric charge,* which is the fourth fundamental quantity. Friction can "rub off" some of these particles so that the effect of isolated charges can be demonstrated, even if the charge itself is not directly observable. For example, an ordinary hard rubber comb, when drawn briskly through your hair, becomes charged and will attract small pieces of paper. Further experiments show that when two materials are charged by being rubbed together the charge of each material is fundamentally different. In order to distinguish between these two types of charge one is labeled negative (the charge found on the rubber comb) while the other is called positive. (These labels are arbitrary, albeit conventional. The charges could just as easily have been called red and green.) It is now known that electrons all possess an identical negative charge, hence a negatively charged comb is a comb with an excess of electrons. Like the other fundamental quantities, charge is defined operationally by comparison to a standard, the *coulomb.* A coulomb may be defined as the total charge of 6.24×10^{18} electrons.

Derived Quantities

The following quantities are defined in terms of the four fundamental quantities. The dimensions associated with each quantity signify the manner in which it was obtained from the fundamental quantities.

When an object changes position, it is said to undergo a *displacement,* and the magnitude of that displacement is the length or distance it has moved. Distance alone, however is not sufficient to determine displacement. Displacement must also include the direction in which the object is moved. In many practical applications displacement is measured from some natural or rest position, as, for example, when considering the displacement of a point on a violin string. To completely specify the displacement, we must specify not only the distance the string is moved, but also the direction of motion from the rest position.

Area is obtained by multiplying two perpendicular lengths (such as length and width) and consequently has units of square meters (or square centimeters, etc.).

Volume arises from the multiplication of three perpendicular lengths and has units of cubic meters (written m³) or cubic centimeters (cm³).

Average speed is defined as the average time rate at which distance is traveled. That is,

$$\text{avg speed} = \frac{\text{distance traveled}}{\text{elapsed time}} = \frac{\text{distance}}{\text{time}}$$

As an example, suppose it takes you one-half hour to travel from one town to another town twenty kilometers away. Your average speed for the trip is calculated as follows:

$$\text{avg speed} = \frac{20 \text{ kilometers}}{0.5 \text{ hr}} = 40 \text{ km/hr}$$

The average speed, however, does not specify how the speed varied during your trip. For instance, part of the trip was on a freeway where you traveled at 60 km/hr and part of the trip was through town where your speed varied from 25 km/hr on some roads to zero km/hr while you waited for traffic. To take into account these variations of speed, another concept, that of *instantaneous speed,* is developed. The instantaneous speed is merely your speed at any instant of time. This is the speed which shows on your speedometer at any moment.

Average velocity is defined as the time rate of change of *displacement*. That is,

$$\text{avg velocity} = \frac{\text{change of displacement}}{\text{elapsed time}} = \frac{\text{displacement}}{\text{time}}$$

Velocity, then, includes speed (the magnitude of the velocity) and direction. To accurately state a velocity you must give the speed (20 km/hr) and the direction (due north). As an example, suppose you travel 20 kilometers due west in a half an hour. Your average velocity would be

$$\text{avg velocity} = \frac{20 \text{ km due west}}{0.5 \text{ hr}} = 40 \text{ km/hr } due \ west.$$

Suppose you then return to your starting point by a different route, but your average speed is again 40 km/hr. Now, your average speed for the entire trip is 40 km/hr; but your average velocity for the trip is zero km/hr because the net displacement is zero, since you returned to your starting point.

Instantaneous velocity is your velocity at any instant of time, just as instantaneous speed is your speed at any instant of time. In dicussing acoustical systems, instantaneous velocity is more often of interest than any average velocity.

An object experiences an *acceleration* when its velocity changes, i.e., it speeds up or slows down. More precisely,

$$\text{acceleration} = \frac{\text{change in velocity}}{\text{elapsed time}} = \frac{\text{velocity}}{\text{time}}$$

Since velocity includes both speed and direction, an acceleration can be obtained by either a change of speed or a change of direction, or a change of both speed and direction.

Electric current is defined as a flow of charge with time, that is

$$\text{current} = \frac{\text{charge}}{\text{time}}$$

The unit of current is then the coulomb/sec, which is defined as an *ampere*. In solids it is the electrons which are free to move, so an electric current can be considered as a flow of electrons. The motion of the electrons does not produce a current; rather the moving electrons are the current. To maintain a continuous current requires a continuous supply of electrons, on one end of a wire and a place for the electrons to go on the other end. These two requirements can be met by a battery, and the flow of electrons from one end of the battery through the wire to the other terminal of the battery is analogous to the flow of water through a pipe—a system which is easier to visualize. To obtain a continuous flow of water through a pipe, a pump and a large supply must be provided at one end, and an opening for the water to escape at the other end. The resultant flow is measured in gallons/sec, just as electric current is measured in coulombs/sec. The total water available is the number of gallons in the reservoir, just as the total charge available would be the number of coulombs.

Laws of Motion

The concept of *force,* for which we all have some intuitive feeling, is defined by three laws first formulated by Isaac Newton and known collectively as Newton's Laws of Motion.

The *first law* states that a body in a state of uniform motion (or at rest) will remain in that state of uniform motion (or rest) unless acted upon by an outside force. The first law is a qualitative definition of force, for while it does not tell us what a force is, it does tell us when a force acts. If we observe an object at rest, we know there is no net force acting on the object. If you take this book out to the edge of the solar system and heave it as hard as you can off into space (as you would perhaps like to do), it will travel forever in a straight line at a constant speed (unless, of course, influenced by another object).

Newton's first law tells us when a force acts, while Newton's *second law* tells us what happens when a force acts. "When an object is acted upon by an outside force, the object is accelerated. The acceleration is directly proportional to the magnitude of the force (and in the same direction) and inversely proportional to the mass." What happens, then, when a force acts is that there is a change in the body's motion—that is, the object is *accelerated.* This acceleration is proportional to the force and is in the same direction. We can represent this symbolically as

$$\text{acceleration} = \frac{\text{force}}{\text{mass}}, \text{ or } a = \frac{F}{m}.$$

Rearranging the above equation, we can write $F = ma$; so Newton's second law is a quantitative definition of force—for the law tells us that force is the product of mass and acceleration. The unit of force is a newton, which is defined as the force which causes a mass of 1.0 kg to accelerate at a rate of 1.0 m/sec/sec.

As an example of the second law, suppose that you exert your maximum force on two different objects—a baby carriage and an automobile. Which one receives the greater acceleration? Obviously the baby carriage, because of its much smaller mass. Dividing the small mass of the baby carriage into the maximum force, a fairly large acceleration would be obtained. On the other hand, dividing the large mass of the automobile into the same maximum force results in a much smaller acceleration. (When dividing numbers, the larger the denominator, the smaller the dividend.)

Newton's *third law* is probably the best known and least understood of the three laws. It is stated: "For every action (force) there is an equal and opposite reaction (force)." This law tells us that forces do not occur singly in nature, but in pairs. When two objects interact and object A exerts a force on object B, then object B exerts an equal and opposite force on object A. However, only one force in the pair acts on each object, and it is this force that we must consider when applying the second law of motion to determine whether an object will be accelerated. For instance, when a violin string is bowed, the bow exerts a force on the string and the string exerts an equal and opposite force on the bow; but for determining the acceleration of the string, we consider only the force exerted on the string by the bow and not the force exerted on the bow by the string.

Electric and Magnetic Forces

Experiments with charged objects demonstrate that the objects either attract or re-pel each other. The generalized *electric force law* is: Two like charges exert a repul-sive force on each other while two unlike charges exert an attractive force. The inter-action takes place through a field of influence (called an electric field). Every charge produces an electric field which exerts a force on any additional charges in the vi-cinity of this charge. The electric field surrounding a positive charge is illustrated in Fig. 1.B-2a.

"One step closer, Bong, and I will shoot!"

Fig. 1.B-1. An example of Newton's Third Law. What happens when the man shoots?

Although magnetic forces were first studied by means of bar magnets, it is now known that magnetic forces are due, ultimately, to electric charges in motion (cur-rent). A fundamental *law of magnetism* states that any electric current produces a magnetic field. The magnetic field surrounding a current-carrying wire is illustrated in Fig. 1.B-2b.

(a)

(b)

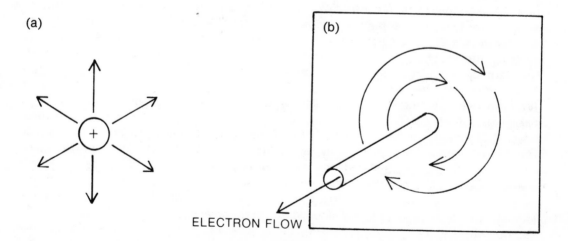

ELECTRON FLOW

Fig. 1.B-2. a. The electric field surrounding a positive charge.
b. The magnetic field around a current-carrying wire.

The *magnetic force law* states that (with one exception) a charged particle experiences a force when moving in an external magnetic field. (The exception occurs whenever a charge moves along the magnetic field: in this case there is no force.) The direction of this force is perpendicular to both the magnetic field and the motion of the particle.

If an electric current produces a magnetic field, is it possible for a magnetic field to produce an electric current? Experiments with magnetic fields demonstrate that it is indeed possible, but only when the magnetic field changes in strength. (A changing magnetic field can be achieved by moving a permanent magnet or by changing the strength of a current.) This phenomenon is summarized by *Faraday's Law,* which states that a changing magnetic field will generate an electric current in a nearby loop of wire. The above laws of magnetism are essential to the workings of several types of loudspeakers and microphones.

Additional Derived Quantities

Pressure is defined as the force per unit area acting on the surface of an object. As an example, consider a balloon filled with air. The pressure is caused by the force of the air molecules hitting the inside surface of the balloon.

A force may also produce *work* whenever a force causes the displacement of a body. The work done by the force is the product of the force (in the direction of the displacement) multiplied by the displacement caused by the force. Symbolically,

$$\text{work} = \text{force} \times \text{displacement} = F \cdot d \quad (F \text{ in direction of } d)$$

According to this definition, if you push and push on a wall and the wall does not move, no work is done, since the force results in no displacement of the wall.

Mechanical energy is the ability or capacity to do work. If a body has mechanical energy, it can exert a force on a second body and cause that object to be displaced. Since a force times a displacement caused by the force is work, the first object did work on the second. When we do work on an object we add to that body an amount of energy equal to the work done on it. Mechanical energy occurs in one of two possible forms, as potential energy or as kinetic energy. An object has *potential energy* when it is able to work by virtue of its position or state. For example, if a spring is stretched or compressed from its natural state, it acquires a potential energy equal to the work done in causing this change. On the other hand, if work is performed on a body in order to give the object a certain speed, the moving object now has *kinetic energy* by virtue of that motion. If you attempt to stop a moving apple by holding out your hand it will exert a force on your hand and cause your hand to move backwards; hence it is able, by virtue of its motion, to do work on your hand. In addition to mechanical energy, energy occurs in many other forms. For our present purposes it is necessary to consider only two additional forms: heat (internal energy) and electrical energy.

Heat is a form of energy which is due to the kinetic energy of molecular motion. *Electrical energy* arises from the work done during the mutual attraction or repulsion of electric charges. The unit of energy (or work) in the metric system is the *joule.*

Electrical potential is related to the work done on a charge when moving in the force field of other charges. Imagine that you have a backyard waterfall and pump system arranged so that the falling water is used to turn a paddle wheel. The pump drives the water to the top of the falls, where the water has a certain potential energy. When the water drops to the base of the falls, its potential energy is converted into the work involved in turning the paddle wheel. The work obtainable from a gallon of water during its fall is equal to this decrease in potential energy. Now consider a negative charge on the negative terminal of a battery. When a wire is connected across the battery terminals the electron moves because of the attraction of the positive charge and the repulsion of the negative charge. In making this journey, the electron loses electrical potential energy (the ability to do work due to position) but it does do work equal to the potential energy lost. The potential energy lost by one coulomb of charge in traveling from the minus pole to the positive pole is called the *potential difference,* or voltage. The potential difference is analogous to the difference in elevation for the waterfall. We measure the difference of elevation in meters; potential difference is measured in volts, that is:

$$\text{potential difference (volts)} = \frac{\text{work on charge (joules)}}{\text{charge (coulombs)}}$$

Power is defined as the time rate of doing work, or the amount of energy supplied per unit time, and has the unit of *watts*. Symbolically,

$$\text{power (watts)} = \frac{\text{work done or energy supplied (joules)}}{\text{time (sec)}} = \frac{E}{t}.$$

As an example, suppose you must exert a force of 400 newtons to climb up to the fourth floor of a certain building (10 meters up). One time you dash up the stairs in 10 seconds. Another day you walk up the stairs in 100 seconds. After one particularly rough day you crawl up the stairs in 1000 seconds. What was different in each of these cases? The work done by you was the same in each case because your weight and the distance traveled was identical. The work done (or energy supplied) by you would be E = F·d = 400 newtons × 10 m = 4000 joules. But the time it took you to do this amount of work was different in each case, so the power you had to supply was also different. The power exerted in the first case was P = E/t = 4000 joules/10 sec = 400 joules/sec = 400 watts. For the second case, P = 4000 joules/100 sec = 40 watts, and for the third case, P = 4000 joules/1000 sec = 4 watts. Hence, the longer it takes you to travel up the stairs, the less the power required. The total work or energy in each case was the same because when less power was supplied it was being applied for a longer time.

Electrical power is the rate at which an electric current performs work. It can be shown that for electrical systems

$$\text{Power (watts)} = \text{current} \times \text{voltage}.$$

Conservation of Energy

The *law of conservation of energy* states that energy can be neither created nor destroyed, but it can be changed from one of its many different forms to another. For example, when friction (mechanical resistance to motion) is present in a mechanical system, some of the mechanical energy is transformed into heat and the system becomes warmer. Electrical energy is a particularly useful form of energy because it can be converted easily to many other types of energy—as, for example, in a light bulb, an electric motor, or an electric oven.

The total mechanical energy (the sum of the kinetic and potential energy) itself is conserved when no friction is present. As an example of this law, consider a vibrating pendulum with no frictional losses. As the pendulum goes from position A through position B to position C and back again, there is a continual transfer of energy from all potential (at A and C because the ball is not moving there) to all kinetic (at B where the speed is greatest). At the in-between points the energy is a combination of potential and kinetic, but at all points the sum of the potential and kinetic energy is a constant. In any real system, however, frictional forces are always present and consequently part of the

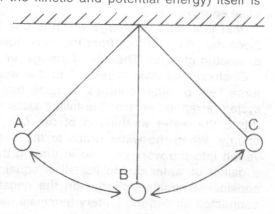

Fig. 1.B-3. Vibrating pendulum.

energy of the system is being transformed into heat. In the above examples the friction present is due primarily to the motion of the string and the bob through the air. If we combine the heat generated with the mechanical energy of the system we can still state that the total energy remains constant. The useful energy, however, is decreasing because the mechanical energy is being converted into heat via friction.

An electrical concept which is closely related to mechanical resistance (or friction) is that of electrical resistance. The units of resistance are volts/amperes = ohms.

As a manifestation of the law of conservation of energy, consider a smoothly flowing fluid, such as water, in a pipe. When the water arrives at a constriction in the pipe, as shown in Fig. 1.B-4, obviously it must move faster to avoid piling up. (This effect would be analogous to six lanes of bumper-to-bumper traffic moving at 30 km/hr and merging into three lanes of traffic. If the traffic is to keep moving, so that a delay is avoided, each car must speed up to 60 km/hr at the constriction. Although this method of merging traffic would be unsafe, it would avoid further congestion.

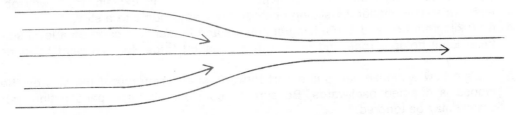

Fig. 1.B-4. Constricted water pipe.

When the flowing fluid reaches a smooth constriction it increases its speed and the smooth flow continues with no congestion. But if the fluid gains speed, it also gains kinetic energy. Where does this energy come from? In the eighteenth century a Swiss scientist, Daniel Bernoulli, reasoned that the kinetic energy was acquired at the expense of a lowered pressure. A lower pressure can be thought of as a reduction of potential energy. One form of *Bernoulli's Law* states that where kinetic energy is large, in a flowing fluid, potential energy is small (and vice versa) so that the total energy remains constant. Put another way, when the speed of a fluid, such as water or air, increases, the fluid pressure must decrease. Fig. 1.B-5 illustrates this law for a pipe with a constriction.

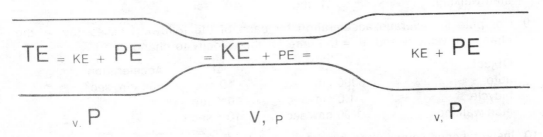

Fig. 1.B-5. Bernoulli's Law.

The concept of energy conservation is very important in its applications to speech production, musical systems, control of the energy level in an auditorium, and so on. Speech production and production of musical sounds require the performer to supply energy to the vocal mechanism or to the musical instrument. Part of the energy supplied by the performer is converted into acoustical energy that is emitted by the instrument as useful musical energy; part of the energy is lost internally in the instrument due to frictional heating. We are often concerned with supplying the appropriate amount of sound energy loss, or absorption, in buildings so that the sound does not reverberate too long and so that it is not too loud.

Exercises

1. Which of the following statements and questions are operationally meaningless?
   a. Are there natural laws which man can never hope to discover?
   b. All matter and space are permeated with an undetectable substance.
   c. The entire universe is expanding so that everything in it has all its linear dimensions double every month.
   d. It is impossible to prove logically the validity of the rules of logical reasoning.
   e. Is the sensation which I experience when I see green the same as that which you experience when you see green?

2. Describe experiments which would demonstrate that there are two and only two types of electricity.

3. According to Newton's first law, once an object is started in motion it will continue in straight line motion forever if no forces act on it. Explain, then, why objects we start in motion by sliding or throwing always come to a stop.

4. A 50-kilogram boy and a 100-kilogram man stand in identical carts on a level surface. Each holds a rope and pulls with a constant force. Explain what happens and why.

5. Explain how a moose can pull a cart forward when the force of the cart on the moose is directed backwards. Be sure to consider all force pairs acting, but gravity may be ignored.

6. When you drive a car along a road at a constant speed, the engine is continuously doing work on the car, and yet the car does not gain any energy. Explain.

7. When you raise a book above a table you do work on the book. When the book is at rest one meter above the table top, in what form is the energy and from where did it come? When you drop the book, in what form is the energy the instant before the book hits the table? When the book hits the table and is at rest, what happened to the energy?

8. Compute the average velocity for each of the following cases. (Δd = change in displacement and Δt = time to traverse the given distance.)

Object	Δd	Δt	Velocity
auto	60,000 cm	100 sec	600 cm/sec
bicycle	1.0 cm	10^{-2} sec	
man walking	−.1 cm	10^{-4} sec	

9. Compute the average acceleration for each of the following cases. (Δv = the change in velocity and Δt = the time for the velocity to change.)

Object	Δv	Δt	Acceleration
auto	6000 cm/sec	10 sec	600 cm/sec^2
bicycle	−1.0 cm/sec	10^{-2} sec	
man walking	1.00 cm/sec	10^{-3} sec	

10. Instantaneous velocity can be determined by taking a very small change in displacement (represented as Δd) and dividing it by the very small elapsed time (Δt) to give $v = \Delta d/\Delta t$. If a vibrating string moves a distance of .001 cm in .0001 sec, calculate the instantaneous velocity.

11. The relation between instantaneous velocity, displacement, and time can be written $v = \Delta d/\Delta t$. If $\Delta d/\Delta t = 0.50$ cm and $\Delta t = 0.01$ sec, find the istantaneous velocity. If $v = 100$ cm/sec and $\Delta t = .05$ sec, find Δd. If $v = 100$ cm/sec and $\Delta d = 0.20$ cm, find Δt.

12. The instantaneous acceleration of an object can be written $a = \Delta v/\Delta t$. When $\Delta v = 0.10$ cm/sec and $\Delta t = 0.05$ sec, find a. When $a = 6.0$ cm/sec/sec and $\Delta t =$

0.05 sec, find Δv. When $a = 5.0$ cm/sec/sec and $\Delta v = 0.10$ cm/sec, find Δt.

13. Imagine that you are driving your car on a perfectly straight highway. Calculate your acceleration for each of the following situations:

    a. You increase your speed from 20 km/hr to 40 km/hr in 10 sec.

    b. You decrease your speed from 40 km/hr to 20 km/hr in 5 sec.

    c. You remain at a constant speed of 30 km/hr for 20 sec.

14. Explain (in terms of electrons) why producing electricity by friction (static electricity) always yields equal amounts of positive and negative charge.

15. Why must an automobile continuously burn fuel just to continue moving at a constant speed?

16. Why do automobile brakes get hot when the car is stopping?

17. Tell for each of the following examples what type of energy the electrical energy is being changed to: a light bulb, an electric motor, an oven.

18. Why does an electric motor require more current when it is started than when it is running continuously?

19. When you switch on a lamp with a 60-watt light bulb (the operating voltage is 120 volts),

    a. what current exists in the bulb?

    b. what is the resistance of the bulb?

    c. how much power is consumed by the bulb?

20. All atoms contain moving electrons. What connection do you suspect between this fact and the fact that all atoms exhibit magnetic properties?

21. An oboe is found to be only 3% efficient in converting the player's input energy into useful acoustic energy that is radiated as sound. What happens to the rest of the energy? Where is it lost?

22. Two excellent musicians (an oboist and a trumpeter), with comparable lung capacity, have a contest to see who can blow his/her instrument for the longer time. Since both performers are able to sustain a continuous tone for the same time period, the result of the contest is a tie. However, people who listened to the sound output observed that the trumpet tone was consistently louder than the oboe tone. Does this observation offer any clues as to the relative efficiencies of the two instruments? Explain.

Further Reading

Ashford, chapters 3,4

Backus, chapter 1

Ballif and Dibble, chapters 2, 3, 4, 5, 7.

Krauskopf and Beiser, chapters 1, 2, 4, 6, 10, 11, 12

Sears, Zemansky, and Young, chapter 1

Audiovisual

1. *Laws of Motion* (12 min, color, 1952, EBE)
2. *Energy and Power* (30 min, color, 1957, EBE)
3. *Electricity* (13 min, color, 1971, WD)
4. *What Is Electricity?* (13 min, B&W, 1953, EBE)
5. *What Is Electric Current?* (14 min, color, 1961, EBE)
6. *Ohm's Law* (30 min, color, 1957, EBE)
7. *Newton's First and Second Laws* (Loop #80-273, EFL)
8. *Newton's Third Law* (Loop #80-274, EFL)
9. *Energy Conversion* (Loop #80-3437, EFL)
10. *Conservation of Energy* (Loop #80-276, EFL)
11. *Force on a Current in a Magnetic Field* (Loop #80-169, EFL)
12. *Faraday's Law of Induction* (Loop #80-4179, EFL)

Demonstrations

1. Energy conservation

    a. simple pendulum

    b. mass on a string

2. Electrical attraction: Hold a charged rubber rod near a thin stream of water which is running smoothly from a faucet

3. Faraday's Law
    a. permanent magnet moving in coil of wire
    b. coil with attached flashlight bulb

1.C. PROPERTIES OF SIMPLE VIBRATORS

Simple Harmonic Motion

Generally speaking, any material object which is capable of vibrating has a rest position where it remains when not vibrating. Such a position is called the equilibrium, or rest, position, because all the forces acting on the object add to zero. Any disturbance of the object will cause it to move away from this position of equilibrium to some new position in which the forces on the object no longer add to zero. The object then experiences a "restoring force" which pulls it back toward its original position. This restoring force is due to the internal stiffness (elasticity) of the material or (in the case of strings) to an externally applied tension. As the restoring force pulls the object back toward its rest position, the object eventually returns to the rest position, but in doing so it acquires some speed which causes it to overshoot its rest position and travel some distance on the other side of the equilibrium position before coming to rest. The object now experiences a new force which again pulls it back toward the rest position; again it overshoots the position and the motion then continuously repeats itself. Motion of this general type is called an oscillation or vibration. It is characterized by the fact that the farther the object moves from the equilibrium position the greater the restoring force which attemps to pull the object back. If the restoring force is *exactly* proportional to the distance from the rest position, we have a special type of oscillation known as *simple harmonic motion*. Simple harmonic motion (abbreviated SHM) is characterized by a periodic repetition of the motion.

As an illustration of SHM, consider a mass which is permanently attached to a spring. The spring is placed in a horizontal position on a frictionless table top (e.g., an air table), one end being fastened to the table. Fig. 1.C-1 is a top view of this arrangement and shows that the mass is free to slide on the table. We can start the mass oscillating by first displacing it (which stretches the spring) directly away from the table support and then releasing the mass. Suppose that we now mount a movie camera directly above the oscillating mass and film the motion at a rate of 10 frames per sec. The movie film provides us with a detailed record of the location of the mass from its rest position at time intervals of 0.1 sec. To obtain the displacement of the mass from its rest position at each time interval we need only

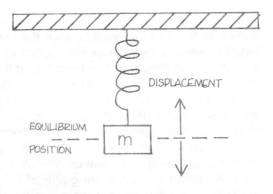

Fig. 1.C-1. Mass and Spring Vibrator.

measure the appropriate distance on each frame of the film. Some of these measured data are recorded in the displacement column of Table 1.C-1. The initial frame (labeled #0) was arbitrarily chosen from one of the frames where the mass was passing through its rest position. The displacement data as a function of time can now be used to construct the upper graph of Fig. 1.C-2. The displacement has both positive and negative values because it is measured either "above" or "below" the rest position. After the data are plotted the individual points are connected by a smooth curve (the solid line in the graph). This graph now clearly shows how the displacement changes with time for the vibrating mass. The resultant smooth curve is called a sinusoid. Any system which vibrates with SHM has a sinusoidal displacement curve.

The column of velocities in Table 1.C-1 was computed from the frames of the movie film to give the *instantaneous* velocity of the mass. A positive velocity represents

Table 1.C-1. Values of displacement, velocity, and acceleration
for the mass and spring vibrator.

Frame #	Time (sec)	d (cm)	v (cm/sec)	acc (cm/sec/sec)
0	0	.00	6.3	0
1	.1	.59	5.1	−23.
2	.2	.95	1.95	−38.
3	.3	.95	−1.95	−38.
4	.4	.59	−5.1	−23.
5	.5	.00	−6.3	0
6	.6	−.59	−5.1	23.
7	.7	−.95	−1.95	38.
8	.8	−.95	1.95	38.
9	.9	−.59	5.1	23.
10	1.0	.00	6.3	0
11	1.1	.59	5.1	−23.
12	1.2	.95	1.95	−38.
13	1.3	.95	−1.95	−38.
14	1.4	.59	−5.1	−23.
15	1.5	.00	−6.3	0
16	1.6	−.59	−5.1	23.
17	1.7	−.95	−1.95	38.
18	1.8	−.95	1.95	38.
19	1.9	−.59	5.1	23.
20	2.0	.00	6.3	0

motion of the mass *upward* (even if the mass is below the equilibrium position) and a negative velocity represents the motion of the mass *downward*. A plot of the velocity versus time is shown in the middle graph of Fig. 1.C-2. You will note that the curve of velocity versus time is also a sinusoid, but it is shifted with respect to the displacement curve. When the displacement is zero, the velocity is a maximum, i.e., as the object passes through the equilibrium position (displacement zero) it is traveling at its maximum speed. On the other hand, when the displacement is maximum (at the position of greatest extent) the velocity is zero (the object stops momentarily). Finally the curve of acceleration versus time (also computed from the movie frame data) is plotted in the bottom graph of Fig. 1.C-2. Again, you will note that the resulting curve is a sinusoid, but the curve is shifted with respect to both the displacement and velocity curves. When the displacement is a positive maximum the acceleration is a negative maximum. This means that when the object has reached its position of greatest extent the acceleration is greatest in the opposite direction because the restoring force is greatest. (Recall, by Newton's second law, that force and acceleration are proportional.) When the displacement is zero, however, the restoring force (or acceleration) is also zero. In summary, then, for an object undergoing SHM the curves representing its displacement, velocity, and acceleration as functions of time are sinusoids.

Definitions for SHM

Now that simple harmonic motion has been defined, several additional terms can be defined for an object which is undergoing SHM. These terms are general definitions which apply whenever an object undergoes SHM.

The *displacement amplitude* is the maximum displacement in either direction from the rest position. The total displacement from maximum in one direction to maximum in the other direction is then twice the amplitude. Amplitude may also be used to refer to the maximum velocity or the maximum acceleration. In general, the amplitude

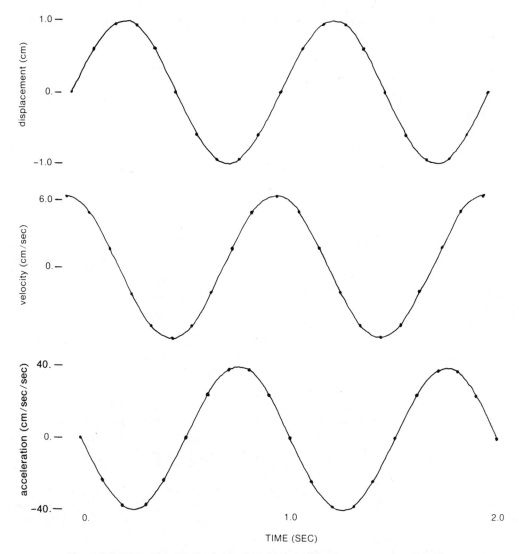

Fig. 1.C-2. Dislacement, velocity, and acceleration for the mass and spring vibrator.

of any sinusoid is the distance from the rest position to the highest point on the sine curve, as shown in Fig. 1.C-3 where the letter A represents amplitude.

One cycle of a vibration represents one complete excursion of the mass from the rest position over to one extremity, back through the rest position to the other extremity, and back again to the rest position. In terms of the sinusoid this represents one complete wave. It is not mandatory, however, that a cycle be counted from the equilibrium position. Any point of the sinusoid is a valid starting point, but one cycle always means that you have returned to an identical point on the following sine curve. This concept is illustrated in Fig. 1.C-3.

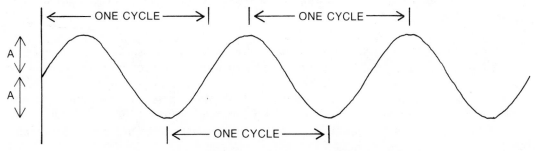

Fig. 1.C-3. The amplitude and several ways of defining a "cycle".

21

The *period* of a vibration is the time (usually expressed in seconds) required for one cycle or for one complete vibration.

Frequency is the number of cycles (or vibrations) completed in one second. For this reason the units of frequency are generally given as cycles per second. The units of cycles per second are termed *Hertz* (Hz) in honor of one of the great scientists of wave study. Since the frequency is the number of vibrations per second and the period is the time per vibration, these two quantities are not independent but are always inversely related to each other. For example, if a vibrating mass on a spring has a frequency of 10 Hz, its period would be 1/10 sec, or 0.1 sec. If the frequency of vibration is decreased to 2 Hz, the period is increased to 0.5 sec.

Phase is the fraction of a cycle which has been covered from some arbitrarily chosen reference point. The phase can be expressed as fractions of a cycle. Considering the graphs of displacement, velocity, and acceleration (Fig. 1.C-2), let the phase of the displacement wave be 0.0 cycle. Since the velocity wave starts one quarter cycle earlier than the displacement wave, the velocity wave has a phase of 0.25 cycle with respect to the displacement wave. Since the acceleration wave starts one half cycle earlier than the displacement wave, the phase of the acceleration curve is 0.5 cycle. Since when one is positive, the other is negative, and vice versa, we can say that the displacement and acceleration are exactly *out of phase* with respect to each other.

An expression can be written for the displacement of a simple vibrator in terms of a sinusoid as $d = A \sin(t/T)$, where d is the instantaneous displacement of the vibrator from equilibrium, A is the amplitude of vibration, T is the period of the vibration, and t is the amount of time that has elapsed since we started the timing. The ratio (t/T) forms a dimensionless argument for the sine function enumerated in Appendix 5. If we use the relationship that $f = 1/T$, where f is the frequency, we can write the above equation in an equivalent but often more convenient form, $d = A \sin(ft)$. Similar expressions can be written for velocity and acceleration.

The Effect of Mass and Stiffness on SHM

Let us now reconsider the previous example of a mass attached to a spring and set vibrating with SHM. If we measure the period of vibration we find it to be 1.0 second. If additional mass is now added to the first mass, the same restoring force is acting on a larger mass. By Newton's second law, $a = F/m$, we see that the larger mass will now experience a smaller acceleration. The smaller acceleration means that the time of vibration will be increased because it will now take longer for the mass to make the back-and-forth trip. Increasing the mass, then, results in an increased period for the vibration. When the period of a SHM is increased, say from 1.0 second to 2.0 seconds, the frequency of the SHM is correspondingly decreased from 1.0 Hz to 0.5 Hz.

Suppose we go back to the original mass but increase the stiffness of the spring. The stiffer spring will provide a greater restoring force and consequently a greater acceleration. But an increased acceleration means the vibration will be completed in a smaller time interval; so the period is decreased. Increasing the stiffness of a simple harmonic vibrator, then, results in a decreased period or an increased frequency. The frequency with which a simple mass-spring system vibrates is called its *natural frequency*.

Resonance

One of the most important phenomena encountered in the study of sound is that of resonance. Before we can explain the phenomenon of resonance, however, it is necessary to understand the difference between free and forced vibrations. *Free vibration* means that a vibrator is started and then allowed to vibrate without further disturbances. The object vibrates at its own natural frequency, which remains constant even though the amplitude decreases continuously. An example of a free vibrator is a tuning fork struck and left to itself.

In *forced vibration* a vibrator is acted on by a continuous force. If the force has a

22

certain frequency of vibration the vibrator to which it is applied will vibrate at this same frequency, and generally with a small amplitude. For instance, if a sounding tuning fork is held over the opening of a bottle a feeble sound of the same pitch as the fork will be heard from the air in the bottle. The vibration of the air is forced by the periodic sound waves emitted by the tuning fork. Blowing across the mouth of the bottle will result in a much louder sound of different frequency, as this corresponds to the free vibration of the air in the bottle.

Now suppose that we use a tuning fork whose frequency can be changed continuously. We take the fork and hold it over the bottle again, but this time we vary the frequency of the fork and listen to the sound emitted by the bottle. We find that there is one frequency where the emitted sound is particularly loud. In measuring this frequency we discover that it is about the same as that emitted by the blown bottle. This situation, in which the vibrator is driven at its natural frequency, is known as *resonance.* When a system resonates, the vibrator has its largest amplitude of vibration. Stated in another form: the amplitude of a forced vibration increases as the frequency of the vibration approaches the natural frequency of the system.

As an example of the principle of resonance, consider the ringing of a heavy church bell. A big bell is usually too heavy to be set ringing by a single pull on the rope. Instead, the bell ringer pulls the rope as hard as possible and then releases it. The bell then begins to swing, which causes the rope to move up and down. After the bell has performed a complete swing the rope is back to its original position but it is moving downward. If the ringer now pulls down on the rope again the amplitude of the bell's swing will be increased with only slight additional effort. With repetition of the process time after time the amplitude of the swing increases until the bell is ringing with as much vigor as is desired. Physically, the force exerted by the ringer is a periodic force with the same period as that of the free oscillations of the bell; so the amplitude of the vibration builds to a large value.

Let us return for a moment to the experiment with the bottle and the variable frequency tuning fork. Recall that the response from the bottle was greatest when the fork had the same frequency as the air in the bottle. If we perform this experiment a little more carefully we notice that other frequencies close to the resonant frequency also elicit a fairly large response from the bottle. As a matter of fact, as the frequency is slowly varied from its resonance value, the response of the bottle falls off rather slowly and we have to make a fairly large change in the frequency before the bottle's response decreases appreciably. We describe this situation by saying that the resonance is a *broad resonance.* Compare this broad resonance with the situation exhibited by two identical tuning forks mounted on resonance boxes. If the open end of each box is faced toward the other, then striking one fork will cause the other fork to vibrate by resonance. If the frequency of one fork is now changed slightly by placing a small piece of clay on one prong, very little energy will be transferred by resonance, even though the actual frequency change of the fork has been very small. In this case, the resonance is highly selective, or a *narrow resonance.*

The relative amount of resistance in a vibrating system determines whether the system will have narrow resonance or broad resonance. Again we consider the mass on a spring and assume that it is being driven by a small forcing device of variable frequency. If we plot the amplitude of vibration as a function of frequency, we get a curve like the one shown by the solid line in Fig. 1.C-4. If we now let the mass vibrate on a rough table top, friction will have a considerable effect on the amplitude of vibration. Plotting the amplitude for different driving frequencies, as was done before, we get the dashed curve shown in Fig. 1.C-4. Note that with the increased resistance the originally narrow response curve has been appreciably broadened. Note also that the amplitude achieved at resonance is much smaller for the larger resistance than for the original case. Systems with a narrow resonance respond strongly to a limited, narrow range of frequencies; systems with a broad resonance respond slightly to a greater range of frequencies.

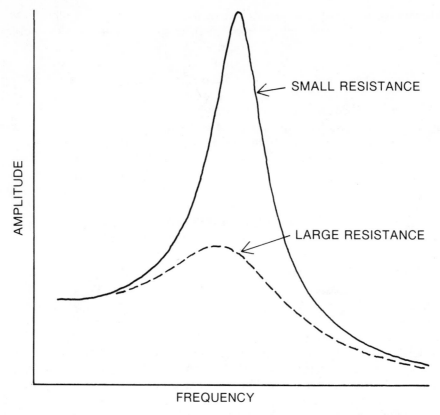

Fig. 1.C-4. Vibration amplitude plotted versus driving frequency for a mass and spring vibrator. Note that the greater the resistance the lower the height and the greater the width of the amplitude curve.

We have so far considered the effect of resistance on forced systems only. Free vibrating systems exhibit an analogous behavior when friction is present. If, for example, the freely vibrating mass and spring of Fig. 1.C-1 is placed on a table where there is a small amount of friction, the vibration amplitude will decrease very slowly (as illustrated in the upper part of Fig. 1.C-5). When the system is subjected to a very large resistance (such as would result from vibrating the mass on sandpaper), however, the amplitude decreases quite rapidly, as can be seen in the lower part of Fig. 1.C-5.

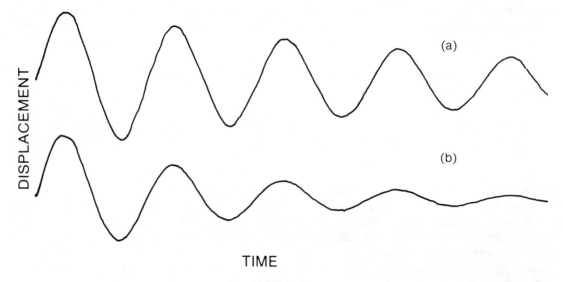

Fig. 1.C—5. Displacement vs. time for free vibrators: (a) with small resistance, (b) with large resistance.

We can summarize the effects of resistance as follows:

(1) A large resistance causes free vibrations to decay rapidly and produces a broad resonance (or response curve) when acting on a forced vibrator.

(2) A small resistance results in a long-lasting free vibration and a narrow resonance for a forced vibration.

Exercises

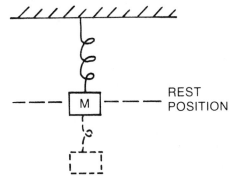

REST POSITION

1. If a suspended mass is pulled down by stretching the spring as shown, what direction will it move when released? What provides the force to accelerate it? When the mass reaches the rest position, will it stop or keep moving? What role does inertia play in the behavior of the mass? What forces are acting on the mass when it is in the rest position?

2. For SHM, what is the phase of the acceleration curve relative to the velocity curve?

3. Consider a mass vibrating on a spring. Ignoring friction, at what position of the mass is the kinetic energy greatest? At what position is the potential energy greatest? How does the sum of the kinetic and potential energies compare at these various points? What happens to the total energy if friction is present?

4. Two identical tuning forks are struck identical blows and caused to vibrate. One fork is held in the air, the other has its base pressed against a table. Which tuning fork will sound louder? Why? Which fork will vibrate for a longer time? Why?

5. Name some musical instruments which are free vibrators. Name some musical instruments which are forced vibrators. Are the vocal cords free or forced?

6. What is the effect of resistance on an ocillatory system? What causes the resistance?

7. Make a rough graph of the displacement-versus-time curve for a free vibrator when resistance is present to a moderate degree.

8. List several examples of resonance.

9. Will a narrow or a broad resonance allow the greater displacement of a forced oscillator? Which type of resonance is more frequently selective?

10. Describe the mass, the restoring force, and the resistance for each of the following three simple vibrators: (a) a ball on a massless spring, (b) a ball on a stretched massless tring, (c) an air-filled bottle with a narrow neck.

    The relation $f \simeq 0.16 \sqrt{s/m}$, where m is the mass of the vibrator and s is a measure of its stiffness, gives the approximate frequency of vibration for these vibrators. An air-filled soft-drink bottle can be viewed as a simple vibrator with the air in the bottle providing the stiffness and the air that moves in the neck of the bottle providing the mass.

11. A mass of 100 gm is hung on a spring set into motion. A movie camera with a framing rate of 20 frames per second is used to photograph the mass. In the table below, the values for the displacement of the mass were obtained by reading each frame of the film. How were the velocities determined? How were the accelerations determined? Does a constant force act on the mass? What produces the force?

t (1/20 sec)	d (cm)	v (cm/sec)	a (cm/sec/sec)
0	.00	6.3	0
1	.31	6.0	−12.2
2	.59	5.1	−23.3
3	.81	3.7	−32.0
4	.95	1.95	−37.5
5	1.00	0	−39.2
6	.95	−1.95	−37.5
7	.81	−3.7	−32.0
8	.59	−5.1	−23.3
9	.31	−6.0	−12.2
10	.00	−6.3	0
11	− .31	−6.0	12.2
12	− .59	−5.1	23.3
13	− .81	−3.7	32.0
14	− .95	−1.95	37.5
15	−1.00	0	39.2
16	− .95	1.95	37.5
17	− .81	3.7	32.0
18	− .59	5.1	23.3
19	− .31	6.0	12.2
20	.00	6.3	0

12. Graph the values for displacement, velocity, and acceleration from the table of problem 11. (Notice the relationships among these three quantities and that each can be represented with a sinusoid.)

*13. Measurement of displacement, velocity, and acceleration of a moving vibrator.
   a. Attach a mass to the end of a suspended spring. Place a Polaroid camera and a stroboscopic light about 1 meter away from the mass. Set the strobe to flashing about 20 times per second. Set the mass in motion. Open the camera shutter for about 2 seconds and pan the camera. Then close the shutter and develop the print.
   b. Construct a table having the form of table 1.C-1.
   c. Fill in the table by measuring the displacements on the print and then calculating the velocities and accelerations. If you do not have a usable print, the diagram provided below may be employed for these measurements. (This picture was obtained by simulating the motion of the mass.)
   d. At which of the tabulated times was the velocity greatest? Smallest? How can you tell?
   e. At which of the tabulated times was the acceleration greatest? Smallest?

*14. Simple vibrators and forced vibration.
   a. Suspend a mass from a spring and set the system in oscillation. Measure the period. (You could, for example, time 10 oscillations and this period divided by 10 would be the period for one oscillation.) Give the value of the period in seconds and fractions of a second. Give the value for frequency in Hz.
   b. Increase the mass appreciably (e.g., double it). Repeat the above. What happens to the period (increase, decrease, unchanged)? What happens to the frequency? Give the vaiue of T and f.
   c. Now suspend the original mass on a stiffer spring. Repeat the above. What happens to the period? What happens to the frequency? Give values for T and f.
   d. Attach a driving apparatus (your finger if nothing better) to the mass and spring. Adjust the driving frequency until the maximum amplitude of oscillation is obtained. (This is called the *resonance condition*.) Measure T and f

26

Exercise 13

27

under this condition. Are they approximately the same period and frequency as those obtained for the natural oscillation? Why?

    e. Give some other examples of simple vibrators.

Further Reading

Backus, chapters 2, 4
Benade, chapter 10
Crawford, chapters 1, 3
Denes, chapter 3
French, chapters 3, 4
Sears, Zemansky, and Young, chapter 11

Audiovisual

1. *Vibrations* (14 min, color, 1961, EBE)
2. *Velocity and Acceleration in Simple Harmonic Motion* (Loop #80-225, EFL)
3. *Tacoma Narrows Bridge Collapse* (Loop #80-2181, EFL)
4. *Simple Harmonic Motion* (Loop #80-3098, EFL)
5. *Damped Oscillators* (Loop #80-166, EFL)

Demonstrations

1. Assorted mass and spring vibrators
2. Driven mass and spring vibrator
3. Heavy and light damping of vibrators
4. Pop bottle resonators

1.D. PROPERTIES OF COMPOUND VIBRATORS

Natural Modes

Most vibrating systems, including musical instruments and the voice, are not simple oscillators, nor can they be described solely in terms of the definitions obtained for the simple vibrating systems. On the other hand, analysis has shown that almost any complicated vibrating system can be represented as a combination of many different vibrator elements. In section 1.C we considered an example of a mass attached to a spring and vibrating on a frictionless table top. Let us now consider a more complicated version of this situation: two masses attached with springs, as shown in Fig. 1.D-1. This compound vibrator has all the same ingredients of a simple vibrator—mass, restoring force, resistance. Note that now, however, instead of having one mass and a restoring force, we have two masses and restoring forces. What is the simplest manner in which this system can vibrate? Both masses can move "up" and "down" together—a system very similar to the single mass on a spring. It may occur to you, however, that there are other ways in which this system can vibrate in the "up" and "down" direction. The next simplest type of motion would be for the masses to move opposite to each other (as one mass moves "up" the other moves "down"). This type of vibration occurs at a higher frequency than when the masses move together. The two simple types of motion which a two-mass system may exhibit are called the *natural modes* of the vibrator. If the system is started vibrating in one natural mode it will continue to vibrate in that mode. Any other complicated type of vibration can be constructed from a combination of the two natural modes for this oscillating system. Note that the one-mass system has one natural mode and the two-mass system has two natural modes.

Consider now a set of vibrators consisting of blocks attached to a string, as shown in Fig. 1.D-2. The one-block system has one natural mode, which consists of the block vibrating up and down. The two natural modes of the two-block system are shown in row 2 of Fig. 1.D-2. Note that in the first mode the masses always move together, while in the second mode the blocks move opposite to each other. A three-mass system would have three natural modes, as shown in row 3. The lowest-frequency mode (first natural mode) again has all the masses moving together as a unit. The higher frequency second mode has the masses moving as two units. (For this mode the center mass does not move at all.) Finally, in the highest frequency mode (third natural mode) the masses move as three units, each moving in the direction opposite that of its nearest neighbor. Note that in the three-mass system the frequency of the second mode is higher than that of the first mode, and the frequency of the third mode is higher than that of either the first or second mode.

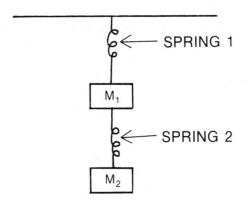

Fig. 1.D-1. Two mass and two spring vibrator.

Let us now generalize our observations:

1. Multi-mass systems can oscillate in different ways.
2. There are as many natural modes as there are masses in a system.

29

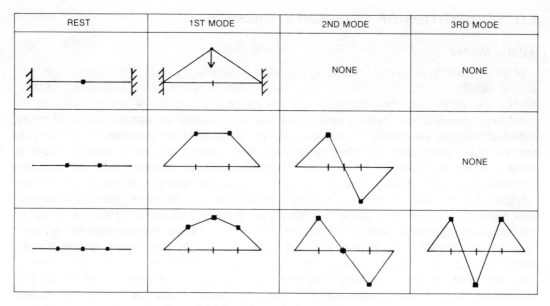

REST	1ST MODE	2ND MODE	3RD MODE
		NONE	NONE
			NONE

Fig. 1.D-2. Natural modes of blocks on a string.

3. The higher modes have higher frequencies than the lower modes.

Suppose now that the number of blocks on the string is increased from two to ten. There will now be ten natural modes. Let us consider the first three natural modes and compare them to the first three natural modes for the three-block system. (See Fig. 1.D-3.)

Note the similarities of the first, second, and third modes. Also note that when we increase the number of blocks the string shape of vibration smooths out; that is, the sharp angles present in the three-block system are replaced by smoother curves. Imagine now what would happen if we increase the number of blocks to 500, or 5000, or even to an infinite number. The latter case would produce a perfectly smooth curve and, as a matter of fact, we would have a piece of string vibrating. The form of the curve would then be a sine wave.

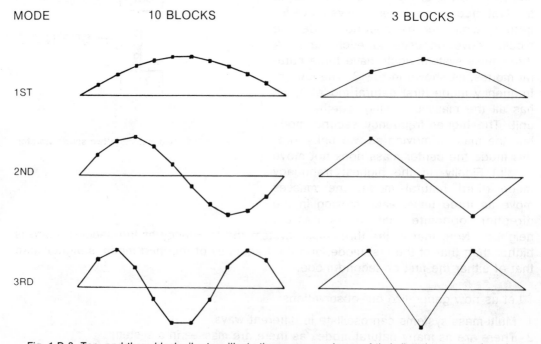

Fig. 1.D-3. Ten- and three-block vibrators illustrating correspondence of their first three natural modes.

30

You may have noticed that for several of the modes of vibration one block may remain at rest and not vibrate at all (e.g., the center block in the second mode of the three-block system). When a block (or a point of a continuous system like a string) always remains at rest, it is termed a *node*. A point of maximum vibration is termed an *antinode*. Note that for any of the vibrating systems shown the number of antinodes present is always equal to the mode number.

Compound Vibration

Obviously there are many other ways in which a set of blocks on a string could vibrate besides the natural modes we have discussed. Careful analysis of a set of blocks vibrating in some complicated manner leads us to the discovery that no matter how elaborate the oscillations may appear to be they can be represented as a composite of its natural modes, each mode having a different amplitude and phase. As an example consider the three-block system shown in Fig. 1.D-4. Suppose each of the three possible natural modes were independently set into oscillation with the amplitude shown in (a) through (c). Now suppose that the system were vibrating in a complex manner, such as detailed in (d). This complex motion is nothing more than the second mode (having an amplitude of 0.5 cm) added to the first mode (having an amplitude of 1.0 cm). To obtain this complex vibration we merely impose the second mode upon the first mode and the combination gives the initial displacement shown. When this system vibrates it vibrates with two different frequencies, the frequency of the first mode having an amplitude of one cm and the frequency of the second mode having an amplitude of one-half cm. In a similar manner we can see that the motion detailed in (e) is merely the sum of the first and third natural modes. Again, the system vibrates with two frequencies, the lower-frequency first mode having a one-cm amplitude and the second mode having a 0.25-cm amplitude.

The above information on the frequencies and amplitudes present in the vibrating system can be conveniently presented by a "vibration recipe." The vibration recipe, as a culinary recipe, gives the various ingredients present and the amount of each ingredient. For vibrating systems this information is usually presented on a bar graph of amplitude versus frequency, called the *spectrum*. Note that the horizontal axis tells us which ingredients (frequencies) are present, while the vertical axis tells us the amount of each ingedient. (See Fig. 1.D-5.)

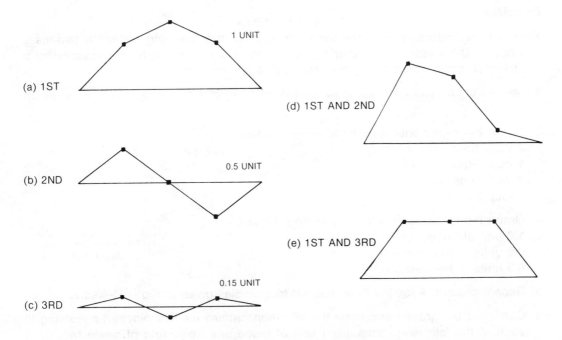

Fig. 1.D-4. Three blocks on a string system: (a)-(c) natural modes; (d)-(e) compound modes.

31

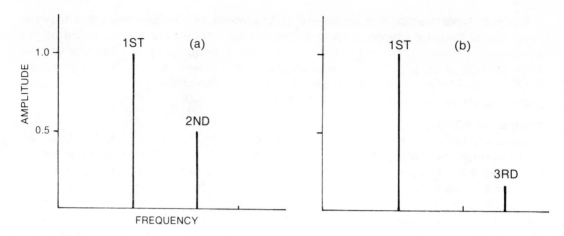

Fig. 1.D-5. Spectrum (vibration recipe) for compound modes shown in Figure 1.D-4: (a) spectrum of 1.D-4d, (b) spectrum of 1.D-4e

Resonance of Compound vibrators

Forced vibration has the same meaning for compound vibrators as for simple vibrators. However, in the two-mass system there will be two different frequencies to which the compound vibrator responds strongly, as compared to one in the simple vibrator. When a compound vibrator is forced to vibrate at one of the frequencies in which it would vibrate if it were free, a large displacement of the vibrator results. If we drive a compound vibrator with a variable-frequency driving force, we get a response curve similar to Fig. 1.C-4. However, the response curve for the compound vibrator will exhibit multiple peaks corresponding to the resonant frequencies associated with each of its natural modes. Resistance plays the same role for compound vibrators as for simple vibrators. The resistance may be different for different natural modes, but generally the resistance is larger for higher frequency modes because of the greater number of back-and-forth motions in these modes. The larger the resistance asociated with any mode, the more rapidly it decays when in free vibration, or the broader its resonance when undergoing forced vibration.

Exercises

1. How many natural modes are there for each of the following cases? A two-mass vibrator; three masses on a string; four masses on a string; twenty masses on a string; N masses on a string; two connected air volumes.

2. Make a chart similar to Fig. 1.D-2 for (a) four blocks on a string; (b) five blocks on a string.

3. Graph the ingredients for the following recipe.
   2 cups flour
   1 cup sugar
   ½ cup eggs
   ½ cup butter

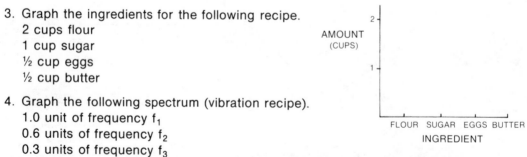

4. Graph the following spectrum (vibration recipe).
   1.0 unit of frequency f_1
   0.6 units of frequency f_2
   0.3 units of frequency f_3

5. Repeat problem 4 for the three natural frequencies given in Fig. 1.D-4a,b,c.

6. Construct bar graphs and draw the vibration pattern for two blocks on a string in each of the following cases: (a) 1 unit of mode one; (b) 1 unit of mode two; (c) 1 unit of mode one and 1 unit of mode two.

32

7. Show the vibration pattern and construct bar graphs for each combination of the modes of Fig. 1.D-4 given: (a) Add modes b and c. (b) Add modes a, b, and c.

8. Graph the ''vibration recipes'' (in bar graph form) for the following vibrating systems.

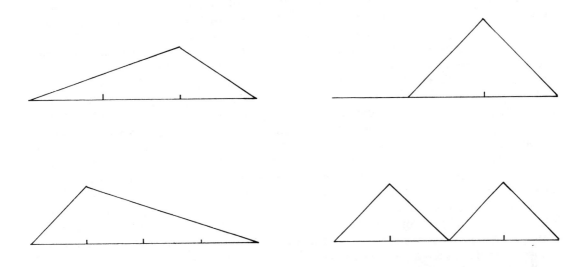

*9. Compound vibrators and forced vibration

    a. Suspend a mass from a spring. Then attach another spring and mass to the first mass to form a compound vibrator. Set the system into oscillation in one natural mode. What is the period of this mode? What is the frequency of this mode?

    b. Set the system into oscillation in the other natural mode. What are the period and frequency of this mode?

    c. Attach a driving apparatus (such as your finger) to one of the masses or springs. Adjust the driving frequency until a resonance, or maximum amplitude of oscillation, is obtained. Measure the period and frequency for this condition. Are these approximately the same period and frequency as those obtained for one of the natural modes? Why?

    d. Repeat c, but adjust the driving frequency until a different resonance is found.

    e. Give some other examples of compound vibrators.

Further Reading
Backus, chapters 3, 4.
Benade, chapters 6, 7.
Crawford, chapters 2, 3.
French, chapter 5.

Audiovisual
1. *Coupled Oscillators;* Part III, ''Normal Modes'' (Loop #80-2694, EFL)

Demonstrations
1. Two-mass and spring vibrator
2. Two masses on a string. Use fluorescent clay masses mounted on a string which is attached to a loudspeaker driver.
3. Two connected air volumes vibrator.

1.E. MEASUREMENT OF VIBRATING SYSTEMS

Transducers

Many of the phenomena of vibrating systems cannot be studied and measured directly because they happen too fast and/or they are not visible to the observer. For instance, the strings of a violin move much too rapidly for us to make any successful measurements of their displacements as they vary with time. Also, the sounds that move through the air are invisible to our eyes and we need some means for turning the invisible waves into some visible form so that we can make measurements on them. In this section we will be concerned with some of the instruments which aid us in our study and measurement of sounds.

Any device which transforms energy from one form to another is known as a *transducer*. The transducers most useful for our purposes are those which involve acoustical energy, in either their input or their output. An *acoustical generator* is a transducer that produces acoustical disturbances in matter, while an *acoustical receiver* detects them. The human voice is an acoustical generator, since it converts an input of "nerve energy" into acoustical energy (speech), while the ear serves as an acoustical receiver, converting acoustic signals (sound) into nerve impulses. Two categories of transducers that are of particular significance in their application to sound are microphones and loudspeakers, which are discussed in the following section. Other transducers and instruments that are of considerable interest in the present context are discussed in later sections.

Not all transducers are sensitive to the same frequencies. The human ear responds to a frequency range of approximately 16 to 16,000 Hz, while a dog's ear responds roughly to the range from 15 to 50,000 Hz. Since the dog can hear much higher frequencies than a human, the dog will respond to high-frequency dog whistles which we cannot hear. A good recording microphone may respond to the frequency range from 10 to 20,000 Hz, while the transducers in the telephone respond only to the more limited frequency range of about 300 to 3000 Hz.

Microphones and Loudspeakers

A *microphone* is a transducer which is actuated by acoustical energy and outputs electrical energy in the form of a time-varying voltage (voltage that fluctuates in a manner so as to reproduce the vibrations of the air). The basic operation of a microphone can be represented as shown in Fig. 1.E-1. The sound impinges on a diaphragm, which causes the diaphragm to vibrate. The mechanical motion of the diaphragm is coupled to the transducer, which produces the varying voltage. The original acoustic vibrations are thus preserved in the time-varying voltage. All microphones in use today may be classified as being one of two types, or combinations of these two types. In this section we will consider the dynamic mike, which is one example of the more common type.

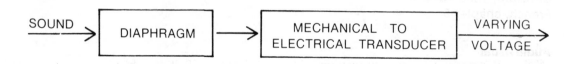

Fig. 1.E-1. Basic processes in a pressure microphone.

A dynamic microphone consists of a diaphragm attached to a coil of wire which is free to move between the poles of a magnet, as shown in Fig. 1.E-2. A dynamic mike employs the coil moving in a magnetic field as its transducing element. The electrons in the coil move with the coil and experience a force (as described with the magnetic force law) which produces a varying electric current in the coil. The varying electric current results in a varying voltage at the ends of the coil.

A loudspeaker is a transducer actuated by electrical energy that produces acoustical vibrations proportional to the time-varying electrical input. One common type of loudspeaker, the direct radiator dynamic loudspeaker, is the reverse of the dynamic mike. The varying input voltage is connected to a coil of wire situated in the magnetic field of a permanent magnet. The varying voltage produces moving charges which experience a force due to the presence of the magnetic field. This force causes the coil to move, and since the coil is attached to a large diaphragm, the motion of the coil drives the diaphragm, which causes the air to vibrate. Because of the simplicity of construction, the small space requirements, and the fairly uniform frequency response, this type of speaker is used almost universally for radios, small phonographs, and intercom systems.

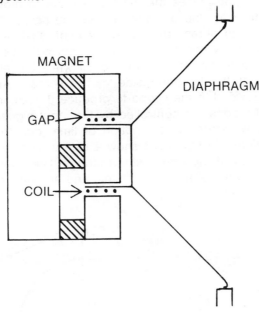

Fig. 1.E-2. Elements of a dynamic microphone.

Other Transducers

A transducer which has proven to be useful in acoustics research is the *accelerometer*—a device useful for measuring the acceleration (and indirectly the velocity and displacement) of an object. Sufficiently small accelerometers are useful for measuring the acceleration of the tubes of wind instruments, the bodies of stringed instruments, the acceleration of the human nose (for determining relative amounts of nasal energy), and so on. In general, they need to be much smaller than the object whose acceleration they are measuring so that they do not interfere with the motion of the vibrating object. The output from a typical accelerometer is in the form of a varying voltage that is proportional to the acceleration of the device.

A *magnetic pickup* is another transducing device useful for measuring the velocity of strings (which must be capable of carrying an electric current). From the information on the velocity of the string, other parameters of the motion, such as displacement and acceleration, can be inferred. It is possible, then, to use a magnetic pickup to generate a voltage that is proportional to the velocity of a vibrating string and thereby to obtain a signal that can be used to describe the string motion.

Piezoelectric materials produce voltages when compressed or stretched. A *piezoelectric pickup* produces an output voltage that is proportional to the applied force. It can be used as a pickup device in certain stringed musical instruments to measure the force that a string is exerting on the bridge. From that force the string motion can be inferred. The output voltage from the pickup can be amplified and played through a loudspeaker to produce a much more intense sound than is otherwise possible. Certain electric guitars and electric pianos use piezoelectric pickups in this fashion.

Other Instruments

The *oscilloscope* is one of the most useful laboratory instruments available. With an oscilloscope it is possible to take a time-changing voltage and display it so that it can be seen. You can imagine the workings of the oscilloscope in the following terms. The scope contains an "electron gun" that shoots out a continuous stream of electrons (much as a garden hose shoots out a stream of water). The electron stream is fired at a screen (like a TV screen) that has been treated with special material that phosphoresces and gives off light when electrons strike it. The scope has a deflecting mechanism that causes the electron stream to move from left to right on the screen and to draw a "line of light" across the screen. The rate at which the electron stream is moved across the screen is controlled by various knobs on the front panel of the scope. When the electron stream reaches the right-hand side of the screen, the stream is momentarily turned off while the electron gun is abruptly moved back to aim at the left-hand side of the screen. An additional deflecting mechanism is also provided in the scope to produce deflections in the vertical direction. The vertical deflections are controlled by the input signal, and thus the vertical deflections draw the waveform as a function of time (due to the horizontal deflections). The three sketches in Fig. 1.E-3 show a side view of the scope and the electron stream, a front view of the scope with a straight line drawn because there is no input voltage to control the vertical deflections, and another front view of the screen with a waveform displayed.

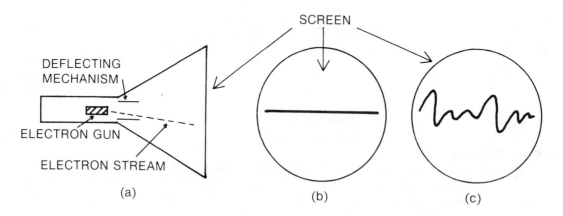

Fig. 1.E-3. Elements of an oscilloscope: (a) side view, (b)-(c) front views of screen.

A *spectrum analyzer* is useful for converting time-varying voltages into their corresponding spectra (vibration recipes) so that the amount of energy present at each of many different frequencies is determined. Fig. 1.E-4 illustrates the process of spectral analysis. Note that the input signal is in the form of voltage versus time and that the spectrum is in the form of voltage versus frequency. More will be said in later chapters about the application of spectrum analyzers for determining the frequencies present in the sounds produced by the human voice or musical instruments.

A *function generator* is a device that provides a variable frequency oscillating voltage corresponding to one or more of the following wave types: sine, sawtooth,

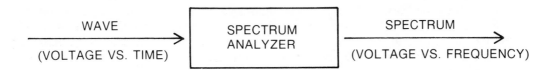

Fig. 1.E-4. Function of a spectrum analyzer.

square, triangular. These wave types are shown in the left column of Fig. 1.E-5. We have seen that any complex vibration of a compound oscillator can be constructed from various amounts of its natural modes. Likewise, the wave types present in the function generator can be built up from "natural vibrations" in the form of sinusoids. In Fig. 1.E-5 the spectra (vibration recipes) for each of the above waveforms are presented in two forms. The middle column shows how the spectrum of each wave appears after analysis by a spectrum analyzer, while the right column gives an idealized version of the same spectrum. Note that the "real world" spectra are not narrow lines as in the idealized spectra. There also is a dark background of low amplitude underlying the peaks. The dark background results mostly from noise in the system, while the width of the peaks is due to a more subtle consideration. The spectrum analysis was performed on a limited portion of the wave (about eight cycles). When the temporal extent of a wave is decreased the widths of the frequency components of its spectrum are increased, which results in wider lines than in the ideal spectra. (This is one form of a well-known law, the uncertainty principle.) To obtain narrower frequency components we would need to include a greater number of cycles in the analysis. The peculiar structure of the background, most evident in Fig. 1.E-5a, is also a result of time limiting the extent of the wave.

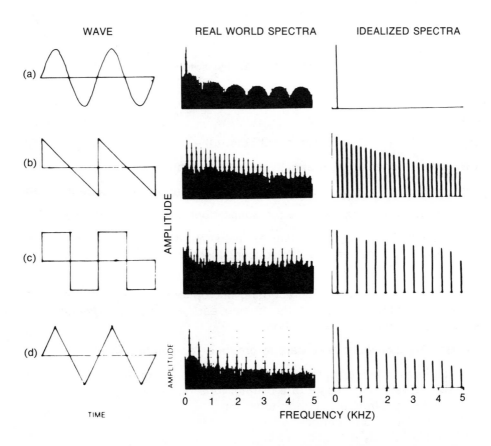

Fig. 1.E-5. Waves and their spectra: (a) sine, (b) sawtooth, (c) square, (d) triangular.

Recent acoustical research has made extensive use of the digital computer for such things as analysis and synthesis of musical instrument tones and speech and for modeling acoustical systems. A digital computer, however, functions by manipulating numbers, while acoustical signals, even after being transduced, are in the form of varying voltages. If we want to input speech or musical sounds to a computer we can attach a microphone to the computer and input the acoustical signal to the microphone. The microphone produces a time-varying voltage which represents the sound. However, some sort of device is needed to convert the voltages into numbers which the computer can accept. Such devices have been built—they are known as *analog-to-digital* (A/D) *converters* (voltage-to-number converters). Similarly, after a computer has been used to synthesize numbers representing a sound, a *digital-to-analog* (D/A) *converter* (numbers-to-voltage) must be used to change the numbers from the computer into a varying voltage, which can be then transduced to sound waves by means of a loudspeaker. The diagrams in Fig. 1.E-6 illustrate the basic behavior of these converters.

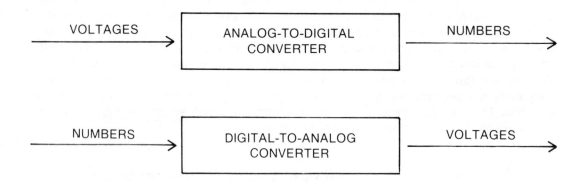

Fig. 1.E-6. Functions of analog-to-digital and digital-to-analog converters.

Exercises

1. Why were hi-fi and stereo impractical before the 1940s?

2. In what sense does an auto engine act as a transducer?

3. Name some items useful in everyday life that depend on transducers.

4. What role do transducers play in the application of the "scientific method"?

5. An X-ray machine is a useful tool for medical and dental diagnostics. In what way can it be considered a transducer? Identify the input and output energy forms.

6. Why is it necessary for a hi-fi recording microphone to respond over essentially the same frequency range as the human ear, while a more limited frequency range is acceptable for the telephone?

7. Why is a D/A converter necessary to produce "computer music"?

8. Name and describe the transducers appropriate to do the following.
   a. Show the varying pressure of a clarinet tone.
   b. Show the spectrum of an oboe tone.
   c. Show the spectra of different speech sounds.
   d. Show the velocity of a guitar string.
   e. Measure the acceleration of the wall of a sounding trumpet.

9. Complete the following table.

"Transducer"	"Energy" in	"Energy" out
movie camera	light	photo
human ear	acoustical	nerve
human voice	nerve	acoustical
oscilloscope		
D/A converter		
A/D converter		
TV camera		
TV receiver		
microphone		
loudspeaker		
accelerometer		
magnetic pickup		
piezoelectric pickup		
spectrum analyzer		

10. Which of the "transducers" in problem 9 are acoustical generators? Which are acoustical receivers?

*11. Transducers and measuring instruments
    a. Connect a microphone to an oscilloscope. Draw a diagram of your setup showing what forms of energy exist at different places.
    b. Speak into the mike and observe the patterns on the scope. What do the peaks on the scope display represent? What do the valleys represent?
    c. Speak a steady vowel into the mike. What features do you observe in the oscilloscope display?
    d. Speak a "hissy" speech sound into the mike. What features do you observe?
    e. Speak a nasal sound into the mike. What features do you observe?
    f. Whistle into the mike. What features are present?
    g. Snap your fingers or clap your hands by the mike. What features do you observe in these waveforms?
    h. Now connect a function generator to a loudspeaker and to an oscilloscope. Draw a diagram of this setup and label it as before.
    i. Turn the function generator control to produce sine waves. Describe the appearance and sound of the sine waves. Repeat for triangular waves and for square waves.
    j. Connect the function generator to a spectrum analyzer and observe the spectra of different waves. (If you don't have a spectrum analyzer available, use the spectra of Fig. 1.E-5.) Describe the pertinent features of each spectrum you observe. Why are the "real world" spectra different from the idealized versions?

Further Reading
Backus, chapter 15
Culver, chapter 16
Gerber, chapter 3
Olson, chapters 9, 10
White, chapter 7
Kamperman, G. W. 1977. "Sound and Vibration Measuring Equipment." Sound & Vibration *11*, no. 1, 8–9.

Audiovisual
1. *The Oscilloscope* (Loop #89-3966, EFL)

Demonstrations
1. Function generator connected to oscilloscope and loudspeaker
2. Microphone connected to oscilloscope
3. Function generator connected to spectrum analyzer
4. Microphone connected to spectrum analyzer

Chapter 2

Characteristics of Sound Waves

2.A. WAVES IN MATTER

The previous chapter was concerned with simple and complex vibrations that stayed at one place and didn't travel. Many vibrations, including those we hear, propagate from a vibrating source to other places. In this section we will consider how vibrations travel through solids, liquids, and gases.

Models of Matter and Wave Propagation

A *pulse* (or impulse) is a burst of energy or a disturbance of short time duration. It is often useful to pulse a system and observe how the system responds to the pulse. In this manner some of the interesting features of the system can be observed. Suppose, for example, that we bang a piece of iron once with a hammer—that is, we pulse the iron. If it were possible to observe the system consisting of the individual atoms we would notice that the pulse which we imposed on the system does not remain stationary but travels from the point of impact outward. To learn how the pulse travels we must consider the atomic structure of the solid. Fig. 2.A-1 is a two-dimensional representation of the atoms or molecules in a solid. (In an actual solid the structure would, of course, be three-dimensional.) The atomic structure is characterized by an orderly arrangement of molecules which are held in place by electrical forces, represented by springs. The electrical force law governs the forces that bind the atoms or molecules in a solid together. Each atom is composed of a positively charged nucleus surrounded by negatively charged electrons. The positively charged nuclei of different atoms repel each other, but they are attracted by the negatively charged electrons located between them and thus bound to each other. The electrical binding forces can be thought of as analogous to springs, as shown in the diagram.

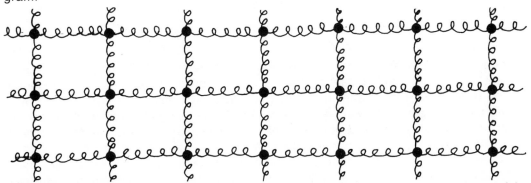

Fig. 2.A-1. Two-dimensional diagram of molecules in a solid. The binding forces are represented by springs.

When one of the molecules is displaced from its normal position the forces represented by the springs return it to that position. Suppose, for example, that one molecule is displaced to the right. Then the "spring" to the right of the molecule is compressed while the "spring" to the left is expanded. There are, then, two forces acting to return the molecule to its normal position: the force due to the compressed "spring," which tends to push the molecule back into place, and the force due to the expanded "spring," which tends to pull the molecule back into place. Each molecule has an equilibrium position and may undergo back and forth excursions, or oscillations, in a manner quite similar to that of a mass on a spring. When the system is disturbed by a pulse from a hammer blow, several molecules are displaced and several "springs" are correspondingly compressed and expanded. As the forces move the molecules back toward their equilibrium positions, the molecules oscillate for a short time. Since a molecule is attached to each end of a "spring," and each molecule has several "springs" attached, displacement of one molecule will soon cause more molecules, which are attached to the other ends of the "springs," to be displaced. This process will continue from each molecule to its neighbors and the pulse, rather than causing oscillation at only one point in the solid, will cause a moving disturbance—that is, a disturbance that propagates through the solid.

Disturbances may also propagate through liquids and gases. The propagation mechanism for liquids and gases is somewhat different from the mechanism in solids. Fig. 2.A-2 is a representation of the molecules in a gas. Again, there are electrical forces, represented by the springs, but the forces are not strong enough to keep the molecules in a nice, orderly arrangement. The forces act only when the molecules are very close, as during a collision. For this reason the "springs" are shown attached to only one molecule. When two molecules get sufficiently close, the "springs" cause the molecules to repel; otherwise they do not affect each other. Suppose now that we pulse a gas by bumping some of the molecules on the left and thus displace them toward the right. Since the molecules are not attached, there is no restoring force, but the molecules bumped toward the right encounter other molecules and bump them toward the right. The process will continue through the entire region where the gas is enclosed; so a pulse can be propagated in a gas as well as in a solid or a liquid.

Fig. 2.A-2. Two-dimensional diagram of molecules in a gas

The structural models of solids and gases given here are greatly simplified to make evident some of the more salient properties of wave propagation and its dependence on the properties of the matter through which the wave passes. It is perhaps worth noting that we view solids as composed of closely packed molecules that are bound in place, liquids as composed of closely packed molecules that are free to slide past each other, and gasses as composed of widely separated molecules that are in a constant state of random motion and collision with one another. The strong intermolecular forces ("springs") of solids cause a solid to maintain its shape, whereas liquids and gases typically take on the shape of their containers because of the weaker intermolecular binding forces. Increasing the temperature of a gas causes an increase of the molecular speeds, which results in a higher propagation velocity for disturbances in a hot gas than in a cold gas.

Wave Types

Often it is not a single pulse which disturbs a material medium, but a continuous succession of pulses called a *wave train* or, more simply, *wave.* The medium is then disturbed in a continuously-recurring manner and the disturbance is propagated as a wave. The particles which compose the medium vibrate or "wave" back and forth as the disturbance passes.

When a jump rope is pulled taut and flipped, a disturbance is caused which travels along the rope. The individual parts of the rope move up and down as the disturbance (a wave) passes. A wave of this type, in which the particles of the medium move transverse (perpendicular) to the direction in which the wave travels, is called a *transverse wave.* If the molecules of the solid represented in Fig. 2.A-1 are jiggled up and down at the left end (as shown in Fig. 2.A-3a), this disturbance travels through the solid toward the right because the molecules are bound to each other. If one molecule of a gas (as in Fig. 2.A-2) were jiggled up and down at the left, the disturbance would not travel to the right because the molecules are not bound together. Transverse waves can propagate in solids but not in gases and liquids because gases and liquids have no restoring force which would act to return a displaced molecule to its former position.

If the left wall of a gas container is jiggled (as in Fig. 2.A-3b), however, this disturbance does travel through the gas. The wall bumps the nearest molecules, thus exerting a force which is propagated to other molecules. The forces between molecules occur only during collisions, at which time the "springs" are momentarily compressed. The disturbance, which consists of molecules shaking to and fro along the direction the disturbance propagates, spreads because the jiggling molecules bump into their neighbors. A wave in which the particles of the medium move parallel to the direction the wave travels is called a *longitudinal wave.* If a solid is jiggled in and out on its left side this back and forth disturbance can also travel along the solid to the right. Longitudinal waves can therefore propagate in solids, liquids, and gases.

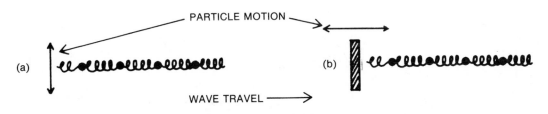

Fig. 2.A-3. Wave types: (a) transverse, (b) longitudinal.

Wave Properties

For a wave to exist we must first have something to transport the wave. Without a substance (such as a gas or solid) to "wave" as the disturbance passes, there would be no wave propagation. The substance which transports a wave is known as the *medium*: the medium for sound waves is air, while the medium for transverse waves on a taut string would be the string itself.

The speed that a wave maintains as it propagates through a medium is determined by the physical properties of that medium. For instance, the wave speed in a string is given by the relation $v = \mathrm{SQRT}\,(F/\delta)$, where v is the wave speed, F is the stretching force (or tension) applied to the string, and δ is the string's density (given as mass per unit length). Note that increasing the density has the opposite effect. The speed of sound in air is given by the relation $v = \mathrm{SQRT}\,(1.4P/\delta)$ where P is the atmospheric air pressure and δ is the density of the gas (given as mass per unit volume). Note that this expression is quite similar to the equation for the string, with pressure analogous to force. A change in the temperature of the air, however, will change both the atmospheric pressure and the density of the air slightly. The changes are of such a nature that as temperature increases the speed of sound increases in direct proportion. The relation $v = 331.7$ m/sec $+ 0.6$ (T°C) may be used to determine the speed of sound in air (v) at any temperature (T). At room temperature the speed of sound is approximately 334 m/sec.

A periodic wave, in which the wave motion repeats, has a frequency associated with it that is determined by the vibrating source producing the wave. A periodic wave also has a wavelength associated with it.

The relationships among the three quantities—speed, frequency, and wavelength—for a periodic wave are now illustrated. Consider a vibrating string with a disturbance caused by an oscillator (which moves up and down sinusoidally) attached to its left end. The wave motion propagates to the right, as shown in Figure 2.A-4. The frequency of the oscillator is represented by f (the number of vibration cycles per second). The period of vibration (the time for one cycle of vibration) is related to the frequency by the expression $T = 1/f$. The *wavelength* is represented by λ, as shown in Figure 2.A-4, and is defined as the distance any part of the wave travels during a time equal to one period. (A related definition of wavelength is the distance between like parts of a wave.) Since the definition of speed is $v = d/t$, we can write that the wave speed is the distance (d) the wave travels during a time t. Using the definition of wavelength, it can be seen that a time equal to one period (T) is required for a point on a wave to travel a distance equal to one wavelength. Putting this information in the definition of speed, we obtain $v = \lambda/T$. Since $T = 1/f$, the expression for speed can be written as $v = \lambda f$. Hence the speed of propagation of a wave is equal to the product of the frequency of vibration and the wavelength. Since the speed of a wave is a constant determined by the physical properties of the medium, if we change the frequency of the oscillator the wavelength changes in such a manner that the product of frequency and wavelength is constant. For instance, if the frequency is doubled, the wavelength will be halved, and vice versa.

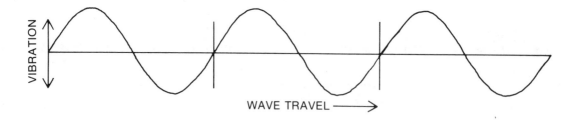

Fig. 2.A-4. Wave motion in a vibrating string

Exercises

1. Describe how a pulse travels through a solid and how the "internal" forces of a solid help to transmit the pulse. What keeps the solid from coming apart?

2. Answer question 1 for a gas.

3. Name several disturbances that travel: (a) in a gas; (b) in a liquid; (c) in a solid.

4. What materials (solid, liquid, gas) will transmit longitudinal waves? Why does a gas not transmit both kinds of waves?

5. If a transverse wave cannot be propagated through liquids, how do you explain water waves? What is the medium for these waves?

44

6. Complete the following table.

Instrument	Longitudinal or transverse?	Free or forced vibration?	Does wave travel in solid or gas?
clarinet reed			
vocal cords			
vocal tract			
piano string			
violin string			
oboe tube			
drum head			
chime			

7. Assuming that a wave train has a frequency of 500 Hz and a wavelength of 0.01 m, what is the speed of the wave?

8. Given a frequency of 100 Hz and a speed of 1.0 m/sec, compute the wavelength of a certain wave.

9. Given a speed of 10.0 m/sec and a wavelength of 0.10 m, find the frequency of the wave.

10. If the tension in a string is 0.1 N, and the string has a mass of 10^{-5} kg and a length of 1.0 m, what is the speed of waves in the string?

11. If the ambient pressure in air is 10^5 N/m^2 and the density of air is 1.3 kg/m^3, calculate the speed of sound.

12. Take the speed of sound in air to be 340 m/sec. What is the wavelength in air if f = 340 Hz? What is the frequency when the wavelength is 0.10 m? What is the speed in helium if f = 1,000 Hz and the wavelength is 0.97 m?

13. The velocity of sound in air is about 34,000 cm/sec. If the space between the earth and the moon were filled with air, how long would it take sound to travel from the moon to the earth (a distance of 4×10^{10} cm)?

14. What is the speed of sound at the following temperatures? (a) 70°C; (b) 32°C; (c) 12°C; (d) 0°C; (e) 20°C; (f) –10°C.

*15. Find the frequencies of a few tuning forks by comparing them with the output of a sine wave generator; that is, adjust the frequency of the generator until it matches that of the tuning fork you are using. How do the labeled frequencies on the tuning forks compare with the sine wave generator setting? How do you account for discrepancies?

*16. Measurement of wavelength and sound velocity.
   a. Set up the apparatus as shown below.

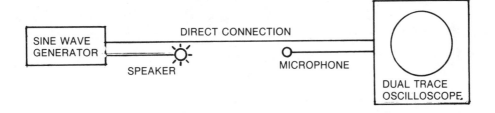

   b. Two sine waves will be seen on the screen of the oscilloscope, one from the direct connection and one coming via the microphone. Try changing the distance between the speaker and the microphone; you will notice that one of the sine waves moves relative to the other.

c. Set the microphone close to the speaker and note the relative phases of the two sine waves on the scope. Slowly move the microphone away from the speaker until the traces again have the same relative phase. As you did this the microphone has moved by a wavelength. What is the wavelength in meters?

d. What is the frequency reading on the sine wave generator?

e. Compute the sound speed in air from the relationship $v = f\lambda$. How well does your result agree with the accepted value?

Further Reading
Backus, chapter 3
Culver, chapters 3, 14
Denes and Pinson, chapter 3
Krauskopf and Beiser, chapter 9
Sears, Zemansky, and Young, chapter 21

Audiovisual
1. *Discovering Where Sounds Travel* (11 min, color, 1965, PAROX).
2. *Sounds and How They Travel* (11 min, color, 1965, PAROX).
3. *Sound Waves and Their Sources,* 2nd ed.(10 min, B&W, 1950, EBE).
4. *Waves and Energy* (11 min, color, 1961, EBE).

Demonstrations
1. Longitudinal and transverse waves in a slinky
2. Longitudinal and transverse waves in a one-dimensional array of spring-coupled masses
3. Longitudinal and (failure of transverse) waves in a one-dimensional array of uncoupled masses
4. "Sound in Vacuum," AAPT, pages 162, 243
5. "Velocity of Sound," AAPT, pages 57–58, 123
6. "Velocity of Sound," Meiners, section 19-2
7. "Wave Machine," Meiners, section 18-2
8. "Wave Models," Meiners, section 18-8
9. "Wave Speed," Meiners, section 18-3

2.B. DESCRIPTION OF WAVE PHENOMENA

In this section we will confine our attention to sine waves traveling in solids or gases. We will consider two different vibrating systems: (1) transverse waves in a string, and (2) longitudinal waves in an air-filled tube. Recall that the speed of a wave is related to frequency and wavelength by the expression: $v = f\lambda$.

Representation of Waves

Waves on a string are transverse displacement waves; that is, the waves travel from left to right (as shown in Fig. 2.B-1a) but the particles of the string execute up-and-down excursions as the waves pass. Sound waves in air are longitudinal waves; the air molecules are displaced back and forth in the direction of wave travel as a sound wave passes. For sound waves in air, however, it is more convenient to speak

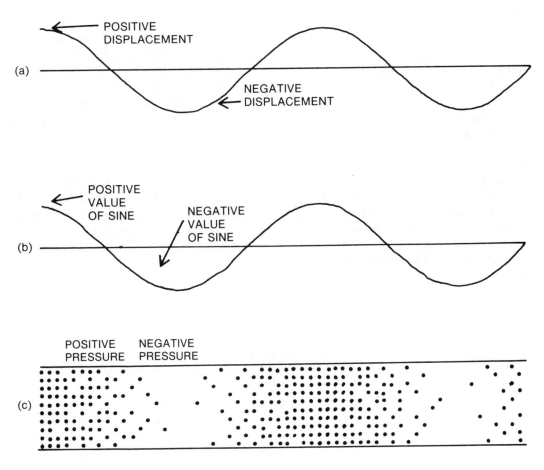

Fig. 2.B-1. Sine wave used to represent displacement waves on a string or pressure waves in a tube.

of the pressure wave produced rather than the displacement of the air molecules. When the air molecules are forced closer together than normal by the impinging disturbance, thus creating a slight excess of air molecules, there is a slight increase of pressure (above normal atmospheric pressure), termed a *condensation*. When the air molecules are pulled slightly farther apart than normal there is a slight decrease in pressure, termed a *rarefaction*. We will refer to pressures greater than atmospheric pressure as positive (+) pressures and pressures less than atmospheric as negative (–) pressures.

Transverse waves on a string can be represented by sinusoids if we adopt the convention that positive (or negative) parts of the sinusoid represent positive (or negative) displacement of the string. Likewise, the longitudinal pressure waves in a tube of air can be represented by sinusoids if we adopt the convention that the positive part of the sinusoid now represents pressure above atmospheric (+) and the negative part represents pressure below atmospheric (–). As an example, consider a vibrating piston, such as a loudspeaker diaphragm, mounted on the left end of a long tube filled wih air, as shown in Fig. 2.B-1c. As the piston moves in and out at a certain frequency it alternately produces condensations and rarefactions. The disturbance created by the vibrating piston then moves from left to right in the tube with a certain speed. The wavelength of the wave train thus produced is the distance from one condensation to the next. The series of condensations and rarefactions in the tube at any instant of time can be represented by a sinusoid shown in Fig. 2.B-1b.

Now consider a long coiled spring through which a longitudinal wavetrain is passing. When the spring is compressed we have an increased "pressure" and when the spring is expanded we have a decreased "pressure." The spring with a set of condensations and rarefactions is shown in Fig. 2.B-2. Note that the "condensations" and "rarefactions" of the spring do *not* correspond to the displacements of the coil. For instance, at each "condensation" and "rarefaction" (positive and negative "pressure") the displacement is zero. That is, at condensation points the spring is pulled in from either side, while at rarefaction points the spring is expanded from either side. In either case there is no displacement at the center of the condensation or rarefaction. Where, then, is the displacement of the spring a maximum? The displacement of the spring in the positive direction (toward the right) is maximum at a point halfway between a rarefaction and a condensation (such as point A in Fig. 2.B-2). The displacement of the spring in the negative direction (toward the left) is maximum at a point halfway between a condensation and a rarefaction (such as point B in Fig. 2.B-2). Note that the pressure wave is a sinusoid, but it is displaced from the corresponding displacement wave by one-quarter of a cycle. This is a general result which holds for pressure waves—the pressure wave leads the displacement wave by one-quarter cycle.

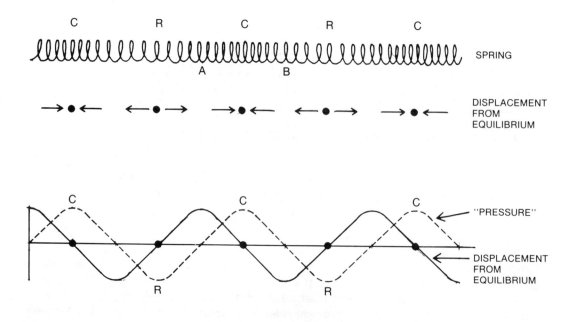

Fig. 2.B-2. "Pressure" and "displacement" waves in a spring illustrating the relative phases between the two.

48

Wave Phenomena

All waves have in common several properties of wave motion: reflection, refraction, diffraction, and interference. Although these properties are common to all wave motions, we will be concerned mainly with their application to sound waves.

One property exhibited by waves is that of *reflection.* When a wave encounters a surface (or any discontinuity in the medium) it is either reflected or absorbed (or some of each). Reflections may be one of two types, regular or diffuse. A *regular* reflection occurs when a wave encounters a hard, smooth surface. All of the wave then bounces off the surface in the same direction and the wave merely changes direction. For example, a mirror reflects light waves in a regular manner, as shown in Fig. 2.B-3a. A *diffuse* reflection occurs when the waves encounter a rough surface, as shown in Fig. 2.B-3b. After the reflection the reflected waves no longer travel in one direction, but in many different directions. This type of reflection occurs when light encounters a wall—the light reflects, but not in a regular manner as from a mirror. When sound waves are reflected in a relatively regular manner we perceive an echo if the reflecting surface is sufficiently far away. When the sound is diffused by walls we may perceive a continuation of the sound called *reverberation.*

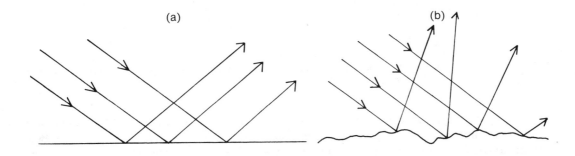

(a)   (b)

Fig. 2.B-3. Regular and diffuse reflections from smooth and rough surfaces respectively.

Buildings should be properly designed to distribute reflected sound energy more or less uniformly over the entire listening area and to avoid focusing of sound. The Mormon Tabernacle in Salt Lake City is an example of a building where the sound is focused because of the approximately elliptical shape of the building. When a sound, such as a pin being dropped, is produced at one focal point of the ellipse (near the front of the building) the sound energy is focused at the other focal point (near the rear), so that even a soft sound is easily heard. The path the sound rays take is shown in Fig. 2.B-4a. This focusing effect of sound, while interesting, is highly undesirable for an auditorium because the sound is concentrated in one small region of the room instead of being distributed uniformly over the entire audience. A similar phenomenon is that of the so-called "whispering gallery." The whispering gallery is a circular shaped enclosure. When a person stands by one wall and whispers, the sound is clearly heard on the opposite side of the enclosure, as well as at many other places along the wall. The sound is being reflected along the wall, so that it "creeps" from side to side, as shown in Fig. 2.B-4b.

A second property of waves is *refraction,* the bending of waves when they pass from one medium into a medium having a different wave speed. Light, for example, bends when it passes from air to water, since light travels more slowly in water than in air. Refraction of sound is most commonly observed when the density of air changes by being either heated or cooled. When the ground is covered with snow the air near the ground is cooler than higher up, and consequently the air near the ground is more dense. In such circumstances a sound wave will be bent downward, because sound travels more slowly closer to the ground, where the air is cooler, and the wave is bent toward the direction of lower speed. Under such conditions the sound will travel great distances along the ground, an effect you may have noticed during a nocturnal walk through the snow on a quiet night. In the summer the air is generally cooler higher above the surface, and sound is consequently bent upward.

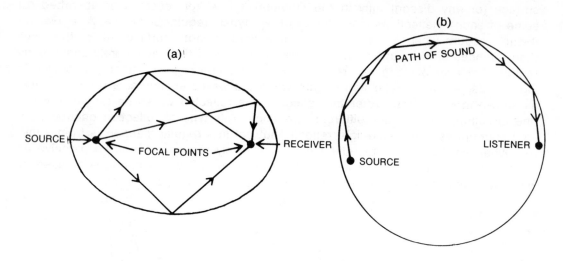

Fig. 2.B-4. Effects of regular reflections of sound: (a) focusing of sound in elliptical enclosure, (b) creeping of sound around a circular enclosure.

A third property of waves is *diffraction*, the bending of waves (in the same medium) around ostacles or through openings. Consider a plane wave train incident upon an opening from the left, as shown in Fig. 2.B-5, the crest of each wave being represented by a solid line. If the opening is quite large compared to the wavelength, as portrayed in Fig. 2.B-5a, the wave passes through the opening and continues much as before. When the opening is smaller, however, as in Fig. 2.B-5b, some bending of the wave is observed at the edges. When the opening is smaller than a wavelength, the diffraction of the waves is so substantial that the opening becomes essentially a new source of waves, as represented in Fig. 2.B-5c. Note that diffraction always occurs, but to be noticeable the opening must be about the same size as, or smaller than, a wavelength. Furthermore, the smaller the opening, the greater the diffraction. Fig. 2.B-5d shows that for a longer wavelength and a larger opening the same amount of diffraction is observed; i.e., the diffraction depends on the ratio between the size of the opening and the wavelength. Since sound waves can be conveniently measured in meters, diffraction of sound waves is an important effect which is easily observed. Diffraction of light waves is much more difficult to observe because light waves are only abut 0.0000005 meters in length. This explains why when you go to a concert and get stuck in a seat behind a wide column you can still hear the concert even though you cannot see it.

A fourth property of waves, that of *interference*, has to do with the way waves can add together. This phenomenon is so important for sound that an entire section (2.C) is devoted to it.

Doppler effect

Almost everyone has experienced the Doppler effect (at least for sound) even if one didn't understand it. Recall the last time you were hitchhiking home. As the cars on the road zoomed past, you may have noticed a change in the pitch of the sound as the cars approached you and then receded from you. This was an example of the Doppler effect. The *Doppler effect* is the change in the apparent frequency of a sound due to a relative motion between the sound source and the listener. Since we

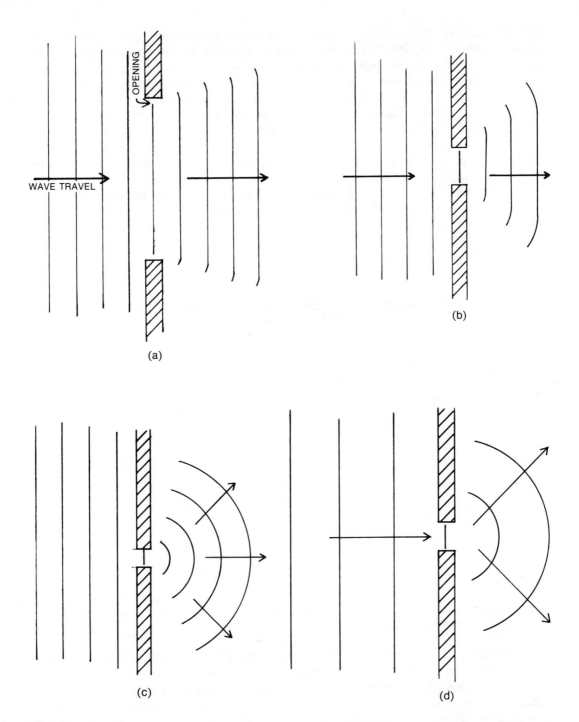

WAVE TRAVEL

OPENING

(a)

(b)

(c)

(d)

Fig. 2.B-5. Diffraction effects of waves: (a) wavelength small compared to opening, (b) wavelength and opening about same size, (c) wavelength large compared to opening, (d) longer wavelength and larger opening.

perceive frequency primarily as pitch, we hear a pitch change as the source of sound passes us. For example, when the horn on an approaching car is sounded, the pitch is higher than when the car is at rest. After the car passes us, the pitch of the horn is lower than normal.

The cause of such a pitch (or frequency) change can be illustrated by a still pool of water into which you throw your pet cockroach, Bugsy. If Bugsy remained stationary in the water he would create a concentric set of waves every time he splashed with his legs, as shown in Fig. 2.B-6a. As he swims toward the right he creates new waves farther to the right; so as the waves move outward they no longer form concentric circles but rather look like those shown in Fig. 2.B-6b. Note that the waves are compressed (are closer together) in the direction in which Bugsy is moving and are farther apart in the direction away from which he is moving. Since the wavelength is the distance between waves, we can see that Bugsy's motion through the water shortens the wavelength of the waves in the direction toward which he moves and lengthens those in the direction away from which he moves, as shown in Fig. 2.B-6b. If you are sitting in the water at location A you experience a longer wavelength than when Bugsy was at rest in the water, while if you are located at position B you experience a shorter wavelength. Recall, however, that wavelength and frequency are inversely related—the longer the wavelength, the lower the frequency and vice versa. Hence, at position A (a source moving away from you) you hear a lower frequency, while at position B (a source approaching) you hear a higher frequency. The same situation holds for sound waves; thus the Doppler effect.

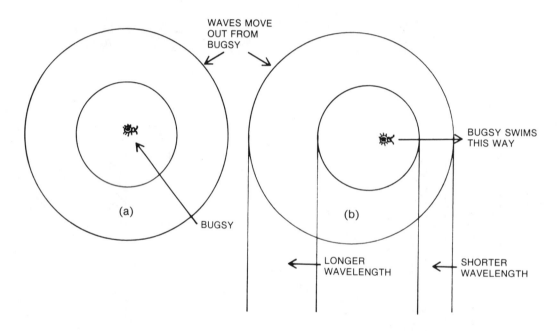

Fig. 2.B-6. (a) Concentric wavefronts for a vibrator at rest in the medium, (b) Displaced circular waveforms for a vibrator moving in the medium.

As a further analogy to the Doppler effect, suppose that three monks decide to open a fish and chips stand. The stand is to be completely automated; one monk cooks the chips (he is the chip monk) and places them on a conveyor belt for the second monk, who prepares the fish (he is the fish fryer) and serves the plates over the counter. (The third monk, the sole brother, only serves filets.) One day the chip monk, who always serves exactly one plate per second, begins walking toward the fish fryer while dishing out the chips (still at a rate of one plate/sec). The fish fryer, who is accustomed to receiving exactly one plate/sec is now confused because the frequency of plates has increased to two plates/sec. Hence, motion of the source of plates (chip monk) toward the receiver resulted in an increased frequency of plates, even though the source is outputting the plates at the same frequency (one plate per second). When the chip monk stops his excursions the frequency received by the fish fryer returns to one plate/second. The fish fryer, however, decides to investigate the cause of the disturbance. He begins to walk toward the chip monk while still picking up one plate/second. Soon he notices that he has to pick up more plates per second because he is moving toward the now stationary source. Hence, an increase of frequency also results from a stationary source and a moving receiver. Similarly, relative motion between the source and the receiver so that they are moving apart results in a decrease in frequency, regardless of whether the source or the receiver moves.

Exercises

1. What are the meanings of *displacement, velocity,* and *acceleration* for: (a) waves in a string; (b) sound waves in air?

2. What happens to the wavelength when a wave travels from a less dense to a more dense medium?

3. Explain the difference between refraction and diffraction.

4. Why can we hear around corners but not see around corners?

5. Is there any practical significance to the fact that sound waves can be refracted?

6. If you can clap your hands and hear an echo 0.2 sec later, how far away is the reflecting surface?

7. Fill in the following table for the frequencies heard by a listener for various possibilities of relative motion between the source of sound and the listener. Use the appropriate equation from Appendix 4. Take the speed of sound to be 340 m/sec and assume the speeds in the table are in m/sec. Negative speeds in the table mean that the listener (or source) moves away from the source (or listener).

f(source)	v(listener)	v(source)	f(listener)
1.0	34.	0.0	
10.0	–68.	0.0	
5.0	0.0	17.0	
3.0	0.0	–34.0	
2.0	34.	–34.0	

*8. Fill a flat glass cake pan with lightly colored water and light it from below. Drop a small object, such as a marble, into it at different places relative to the sides of the pan and observe the reflected wave patterns.

*9. Use the setup of exercise 8, but form barriers of thin aluminum in various shapes and observe the results.

*10. Doppler effect
   a. Take a small tape recorder and record the sounding of an auto horn. Next, station the recorder beside a relatively straight stretch of road. Have the auto get up to a speed of 10-20 m/sec (30-50 mph) and sound its horn as it approaches and then passes the recorder. Also, note the speedometer reading.
   b. Play the recorder into a spectrum analyzer and measure the most significant frequencies for the auto at rest, for the approaching auto, and for the receding auto.
   c. Using the frequency formula for the Doppler effect (see Appendix 4), calculate the approach speed of the auto from the measured frequencies of the horn for the auto at rest and approaching.
   d. Compute the receding velocity of the author in a manner similar to that of part (c).
   e. How closely do the calculated speeds compare with each other and with the speedometer reading?

Further Reading
Backus, chapter 3
Culver, chapter 3

Denes and Pinson, chapter 3
Krauskopf and Beiser, chapter 9
Olson, chapter 1
Sears, Zemansky, and Young, chapter 23
Stevens and Marshofsky, chapter 1
White, chapter 2

Audiovisual
1. *Diffraction and Scattering of Waves around Obstacles* (Modern Learning Aids RT-16)
2. *Doppler Effect in a Ripple Tank* (Loop #80-2371, EFL)
3. *Reflection of Circular Waves from Various Barriers* (Modern Learning Aids RT-2)
4. *Reflection of Waves from Concave Barriers* (Loop #80-233, EFL)
5. *Refraction of Waves* (Loop #80-234, EFL)

Demonstrations
1. Diffraction of sound, Meiners, section 19-7
2. Doppler effect, Meiners, section 19-6
3. Refraction of sound, AAPT, pages 139–40
4. Refraction of sound, Meiners, section 19-8
5. Ripple tank wave phenomena, Meiners, section 18-6
6. Small loudspeaker connected to oscillator, with speaker on strong cable so it can be whirled about

2.C. INTERFERENCE AND STANDING WAVES

Interference of Waves

In this section we will consider a long string and a long air-filled tube as two vibrating systems of interest. The *superposition principle* states that when two or more waves simultaneously occupy the same medium (string or air in a tube) the resulting wave is given by adding the individual waves. When two sets of wave trains simultaneously traverse the same medium, however, interference effects may be observed.

Constructive interference occurs when the waves add together in phase and create a larger wave, as shown in Fig. 2.C-1. If the waves add out of phase a smaller wave is created, as shown in Fig. 2.C-2, resulting in *destructive interference.*

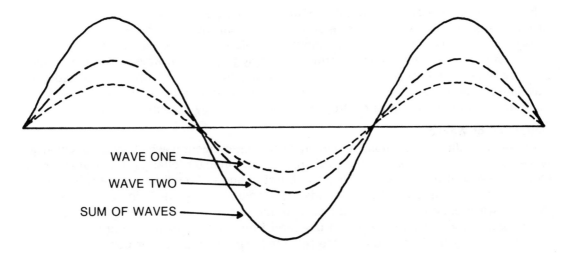

Fig. 2.C-1. Constructive interference illustrated by two waves adding in phase.

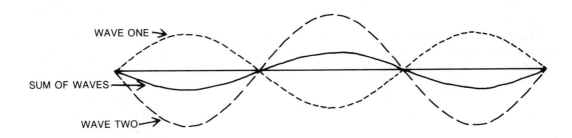

Fig. 2.C-2. Destructive interference illustrated by two waves adding out of phase.

Consider, for example, two identical positive pulses moving toward each other on a string. As the pulses pass through each other a larger positive pulse results, as shown in Fig. 2.C-3. On the other hand, if one of the pulses is negative and the pulses pass through each other, a momentary cancelation occurs, as shown in Fig. 2.C-3. The first case is an example of constructive interference because the waves add together and produce a larger wave. The second case illustrates destructive interference because the waves momentarily add to zero.

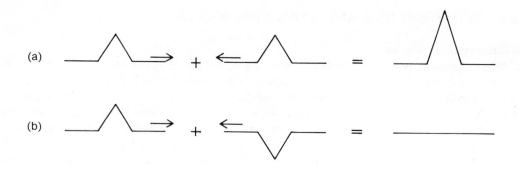

Fig. 2.C-3. Constructive and destructive interference of pulses on a string.

Sound waves, however, are not crests and troughs (as pictured above), but consist of regions of condensation and rarefaction, which can be represented as a series of crests and troughs. On this basis it should be evident that the same results regarding constructive and destructive interference apply to pressure waves in a gas. When two pressure waves of equal frequency and amplitude but of opposite phase are traveling through the same medium, destructive interference (or silence) results. This does not mean that energy is destroyed; rather the energy is redistributed.

Reflection of Waves

As seen in the last section, reflection is the phenomenon which occurs when a wave comes to a point where the properties of the medium change. For waves on a string, the end of the string provides an abrupt change of medium for the wave wise, the end of a tube provides a change of medium for a wave in a tube of air. When a wave encounters a change of medium, part of the wave (in some cases all of the wave) is reflected back in the direction from which it came. If the incident (incoming) wave is positive (a crest), the reflected wave may be either positive or negative (a trough), depending on how the medium changes. Consider a wave pulse (half of a triangular wave) on a string. Assume that the wave travels to the end of the string, which is fastened in place so it cannot move. As the wave encounters the fixed end of the string, the wave attempts to pull the string upward. The string, however, cannot move. A reflected wave is then created in such a way that the reflected wave and the incident wave add together at the fixed end of the string so that there is no displacement of the string.

The reflected wave must be reversed from the incident wave so that when the two are added together at the fixed end they produce zero displacement by destructive interference, as shown in Fig. 2.C-4.

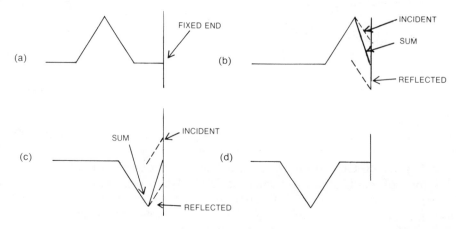

Fig. 2.C-4. Wave reflection from the fixed end of a string: (a) incident wave, (b) wave encounters fixed end and reflection begins, (c) reflection almost complete, (d) reflected wave.

56

If a wave encounters a free end of a string, a different kind of reflection occurs. With the end of the string free to move, an upward motion of the string occurs as the wave reaches the end of the string. Since there is no force constraining the string, a reflected wave is produced which is in phase with the incident wave; that is, the reflected wave adds to the upward displacement of the string. The reflected wave then appears identical to the incident wave at the free end of the string, as shown in Fig. 2.C-5. Note in Fig. 2.C-5 that because of the constructive interference of the incident and reflected waves the displacement of the end of the string is greater than it would have been had the wave just passed through. This additional displacement of the end of the string is what leads to the familiar effect of snapping a whip. When the wave pulse is reflected at the free end of the whip the large displacement causes the tip of the lash to move faster than the speed of sound in air, thus causing the familiar "crack" sound.

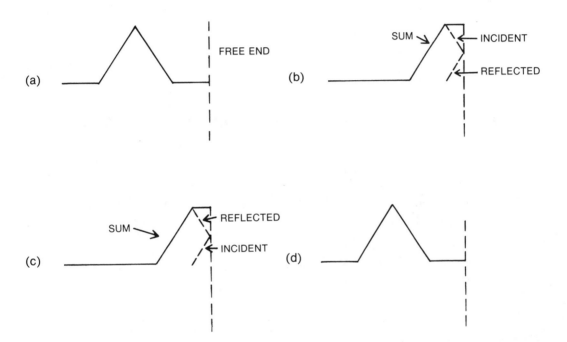

Fig. 2.C-5. Wave reflection from the free end of a string. (Same labels apply as in Fig. 2.C-4.)

Standing Waves

Standing waves result from both constructive and destructive interference between incident and reflected waves. As the waves pass through each other, moving in opposite directions, the wave pattern which results from alternate constructive and destructive interference does not move along the string or tube and is termed a *standing wave.*

Suppose we generate a triangular wave at the left end of a string, as shown in Fig. 2.C-6a. The right end of the string is fixed so that it cannot move. As the wave encounters the boundary, it is reflected out of phase. The positive pulse of the wave, which encounters the boundary first, is reflected as a negative pulse, as shown in Fig. 2.C-6b. The negative pulse which follows the positive pulse in the incident wave is reflected as a positive pulse, which now follows the negative pulse in the reflected wave, as shown in Fig. 2.C-6c. The result is that the reflected wave has the same appearance as the incident wave (Fig. 2.C-6d) because two reversals took place: one due to reflection from a fixed end, and the other due to the fact that the first pulse to encounter the fixed end is the first pulse reflected.

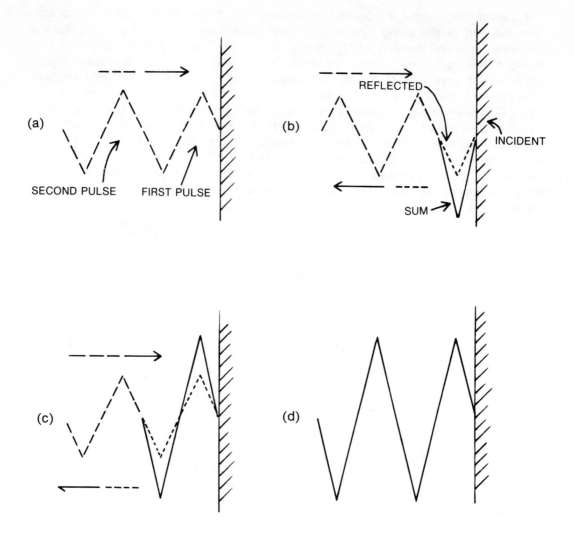

Fig. 2.C-6. Interference of incident and reflected waves to form a standing wave: (a) incident wave, (b) positive pulse reflects as negative pulse, (c) negative pulse reflects as positive pulse, (d) standing wave after reflection of two wavelengths.

When the reflection from only one end of a string (or a tube) is considered, as we have done here, standing waves are produced for all frequencies of vibration. However, when we consider reflections from both ends of a string (or tube), which are typical of musical instruments, we will find that standing waves are produced only for certain selected frequencies. This matter will be discussed in the next section.

Beats

When two sinusoids of slightly different frequencies are sounded together the resulting wave fluctuates between large and small amplitudes because the sine waves alternately interfere constructively and destructively. The pulsations in amplitude are termed *beats,* and the number of beats per second is equal to the difference in frequency of the two sinusoids. For sound waves, the phenomenon of beats results in a fluctuation in the loudness of the sound; the sound is loud, then soft, then loud, then soft, etc., the number of fluctuations per second being equal to the number of beats. Fig. 2.C-7 shows how two sine waves alternately interfere constructively and destructively to produce beats.

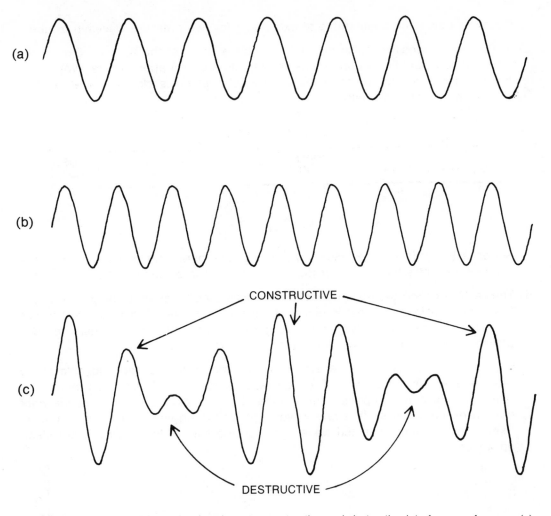

(a)

(b)

CONSTRUCTIVE

(c)

DESTRUCTIVE

Fig. 2.C--7. Production of beats by the alternate constructive and destructive interference of waves: (a) wave one, (b) wave two, (c) sum of waves one and two.

The phenomenon of beats is often used for the tuning of musical instruments. A piano tuner tunes one string of three to a tuning fork by eliminating the beats. He then tunes the other two strings to the first string by the same process.

Exercises

1. Consider two waves having the same speed and wavelength to be approaching each other from opposite ends of a string, as shown. Sketch the resulting waves for the following conditions: Points A & A' coincide; B & B' coincide; C & C' coincide; D & D' coincide; and E & E' coincide.

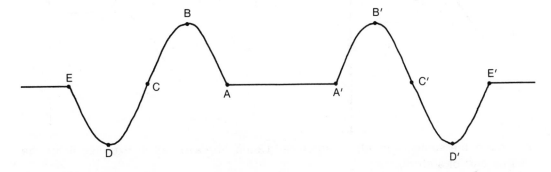

2. Describe a situation comparable to exercise 1 for two pressure waves in a tube.

3. a. Consider a positive pulse in a tube which is open at the far end. Sketch the reflected pulse. For an open tube the results are only approximate. Why?
   b. Consider a positive pressure pulse in a tube which is closed at the far end. Sketch the reflected pulse.

4. Suppose the pulses in exercise 3 were negative instead of positive. Sketch the incident and reflected waves for the open and for the closed tubes in this case.

5. The results for the pressure waves in exercises 3 and 4 seem to be just opposite from what we might expect intuitively from the examples of displacement waves in a string in the text. Why? Would we get different results if we considered displacement waves in the tube rather than pressure waves?

6. Suppose a triangular wave generator is attached to the far left end of a string (the generator is not shown). The right end of the string is fixed, as shown, so that it cannot move. Suppose the sketches below show just enough of the right portion of the string so that one wavelength of the original wave train is visible. Sketch the resulting reflected wave and then the sum of the original and reflected waves.

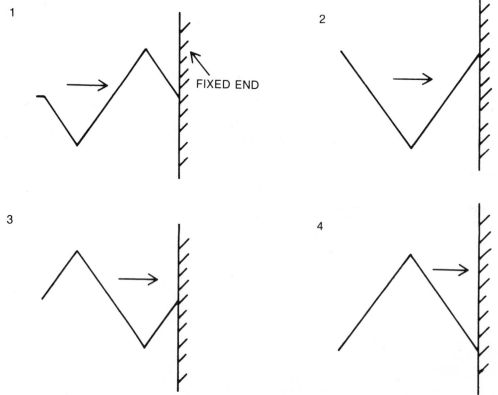

7. Sketch the same incident, reflected, and sum of waves as in exercise 6 for the free end of a string.

60

8. Repeat problem 6 for pressure waves in an open-ended tube.

9. Repeat problem 7 for pressure waves in a closed tube.

10. Compute the beat frequencies for the following tone pairs.

Tone pair	f_1(Hz)	f_2(Hz)	f_{beat}(Hz)
1.	100	101	
2.	300	306	
3.	400	392	
4.	1,000	998	

11. A tuning fork has a frequency of 440 Hz. When sounded with another tuning fork, 5 beats per second are heard. What is the frequency of the second fork? How could you tell whether the second fork has a frequency greater or less than the first fork?

*12. Interference of sine waves—beats
   a. Connect a sine wave generator and a function generator to a mixer. Connect the mixer to an oscilloscope and to a loudspeaker. (This will enable you to both see and hear the results.)
   b. Set the function generator to produce sine waves and adjust its frequency to about 200 Hz. Set the amplitudes of the two generators to be about equal. Adjust the frequency of the sine wave generator so that you get about one beat per second. Describe the waveform that you see on the scope.
   c. Adjust the sine wave generator so that you get two or three beats per second. How does the result differ from the result in 12b?

Further Reading
Backus, chapter 3
Culver, chapter 4
Olson, chapter 1
Sears, Zemansky, and Young, chapter 22

Audiovisual
1. *Standing Waves in a Gas* (Loop 80-3874, EFL)
2. *Standing Waves on a String* (Loop 80-3866, EFL)
3. *Superposition of Pulses* (Loop 80-2397, EFL)
4. *Vibrations of a Drum* (Loop 80-3924, EFL)
5. *Vibrations of a Metal Plate* (Loop 80-3932, EFL)
6. *Vibration of a Rubber Hose* (Loop 80-3890, EFL)

Demonstrations
1. Acoustical interference, AAPT, p. 65
2. Beating sine waves
3. Interference of waves, Meiners, sections 18-4
4. Wave machine, Meiners, section 18-2

2.D. WAVES ON STRINGS, IN TUBES, AND IN OTHER STRUCTURES

Standing Waves on Strings

As noted in the previous section, standing waves occur for any frequency on a very long string because only the waves incident and reflected at one end of the string need to be considered. However, for short strings the situation is complicated by the fact that incident and reflected waves must be considered at both ends of the string. Standing waves can then exist only for certain frequencies (or wavelengths) at which the incident and reflected waves interfere in such a way as to satisfy the end-point conditions of the string. The resulting natural modes (or allowed manners of vibration of the string) occur at frequencies for which the string resonates and maximum vibration amplitude results. Strings are usually fixed at both ends; so the allowed standing wave must have a node (point of zero amplitude) at each end.

As an example of standing wave patterns, consider a vibrating string which is fixed at each end. If we shake the string, producing a wave train, the wave is reflected from both ends of the string, producing interference effects. If we shake the string at just the right frequency we can produce a standing wave, where parts of the string, called the *nodes,* are stationary. The simplest pattern which can be produced is that shown in Fig. 2.D-1a, consisting of a node at each end of the string and an *antinode* (point of maximum vibration) in the center. The length of the string in this case is one-half wavelength, and the frequency corresponding to this wavelength is called the *fundamental frequency* of the vibrating string, so called because it is the lowest frequency associated with the first natural mode which can form a standing wave pattern on the string. If the string is shaken more rapidly (thus increasing the frequency) another standing wave pattern eventually results, as shown in Fig. 2.D-1b. This pattern consists of a node at each end and one in the center, and the length of the string is equal to one wavelength. Since the wavelength is half as long as in the preceding case, the frequency is twice as great. The next standing wave pattern which can be produced is shown in Fig. 2.D-1c; the frequency for this case is three

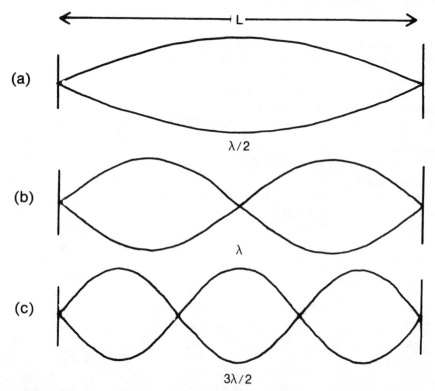

Fig. 2.D-1. Natural odes of a string fixed at both ends: (a) first mode, $\lambda = 2L$, $f = f_1$; (b) second mode, $\lambda = L$, $f = 2f_1$; (c) third mode, $\lambda = 2L/3$, $f = 3f_1$.

times the fundamental frequency. This set of frequencies is the first three natural mode frequencies of the string.

The natural frequencies can be calculated from the relation $v = f\lambda$, where the wave speed in the string is determined from the string's density and tension. The wavelength can be expressed in terms of the length of the string. For the fundamental mode $\lambda_1/2 = l$ (length of string), as can be seen in Fig. 2.D-1a, which gives $\lambda_1 = 2l$ and results in $f_1 = v/\lambda_1 = v/2l$ for the fundamental frequency of a string fixed at both ends. Similarly, for the second mode, and so on.

Note that for a string fixed at both ends the frequency of each mode is that mode number multiplied by the fundamental frequency. If the fundamental frequency is 100 Hz, the second mode has a frequency of 200 Hz, the third 300 Hz, etc. We can summarize the above situation by saying that for a string fixed at each end the frequency of any mode is n times the fundamental frequency, where n is the mode number. The natural frequencies of a fixed-fixed string, then, are given by $f_n = n \times f_1$ (where $n = 1, 2, 3 \ldots$). In this case the frequencies are *harmonic* to the fundamental because they are integer multiples of it.

Consider now the physically unrealistic, but instructive, situation of a string which is fixed at one end and free at the other. The left side of the string must always be a node, since it is fastened in place. Similarly, the right side of the string will always be an antinode, since there is no force to hold the string in place. Fig. 2.D-2a shows the

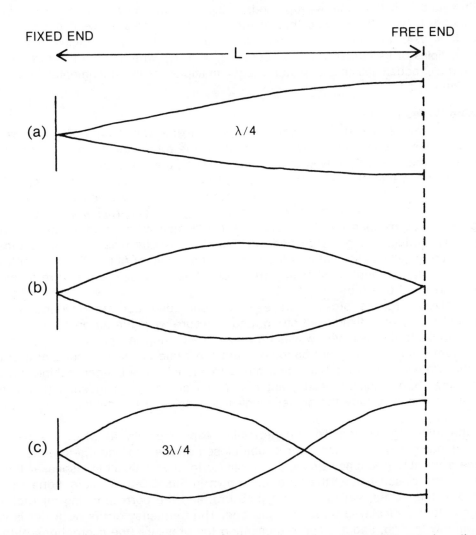

Fig. 2.D-2. Natural modes of a string fixed at one end and free at the other: (a) first mode, $\lambda = 4L$, $f = f_1$; (b) unallowed mode; (c) second mode, $\lambda = 4L/3$, $f = 3f_1$.

smallest part of a wave which can "fit" on the string and satisfy the endpoint conditions. Clearly this is one-quarter of a wavelength, so the fundamental wavelength is four times the length of the string. Now try to fit a standing wave on the string that is equal to one-half of a wavelength. Such a wave would require a node at each end of the string, as shown in Fig. 2.D-2b, but this is not possible since the free end is an antinode. Hence there is no mode for the fixed-free string where $f = 2 f_1$. The next smallest fraction of a wavelength which "fits" on the string is shown in Fig. 2.D-2c. For this second mode three-fourths of a wavelength fits on the string, and the second mode has a wavelength equal to one-third that of the fundamental and a frequency three times that of the fundamental, as shown in Fig. 2.D-2c.

The modal frequencies of a fixed-free string can be calculated in a manner similar to that used for a fixed-fixed string. Referring to Fig. 2.D-2a, we see that the wavelength of the lowest mode is related to the length of the string as $\lambda/4 = l$ or $\lambda = 4l$ resulting in $f_1 = v/\lambda = v/4l$ as the frequency of fundamental mode. For the second allowed mode (Fig. 2.D-2c) $3\lambda/4 = l$ or $\lambda = 4l/3$, resulting in $f_2 = v/\lambda = 3v/4l$ for the frequency of the second mode, which we note is three times the frequency of the first mode. By studying higher modes we discover that the higher mode frequencies are odd integer multiples of the fundamental frequency.

Two interesting features become apparent when we compare a fixed-fixed string with a fixed-free string. First of all, for fixed-fixed and fixed-free strings of equal length, l, and with the same wave speed, v, the fundamental frequency of the fixed-fixed string, $f_1 = v/2l$, is twice that of the fundamental frequency of the fixed-free string, $f_1 = v/4l$. Secondly, the fixed-fixed string has natural frequencies that are integer multiples of its fundamental frequency ($f_n = n f_1$), whereas the fixed-free string has natural frequencies that are odd integer multiples of its fundamental frequency ($f_n = (2n - 1) f_1$).

Standing Waves in Tubes

For tubes whose diameters are small compared to the wavelength of waves traveling in the tube, the pressure waves in the air in the tube exhibit many of the same features as the standing waves in a string, except that pressure is approximately zero at the open end of a tube; that is, a pressure node exists at an open end. This is because at the open end of a tube a pressure wave is reversed upon being reflected, much as a displacement wave on a string reflects from a fixed end of a string. Hence at an open end the incident and reflected wave add together destructively, thus producing a pressure node at this point. At the closed end of a tube a pressure pulse reflects so that it is the same as the incident pulse. The incident and reflected pulse then add together constructively, thus producing a pressure antinode at the closed end of a tube.

For analysis purposes, pressure waves in an open-open tube are analogous to displacement waves on a fixed-fixed string, and pressure waves in an open-closed tube are analogous to displacement waves on a fixed-free string. A general rule applicable here is that when the endpoint conditions are the same at the two ends of a string or at the two ends of a tube the higher mode frequencies are integer multiples of the fundamental; and when the end point conditions are different at each of the two ends of a string or a tube the higher mode frequencies are odd-integer multiples of the fundamental.

By the following method we can determine experimentally at which frequencies strong standing waves will occur for a tube closed at one end and open at the other. We use a variable frequency sinusoidal oscillator to drive a small loudspeaker that is placed in the closed end of the tube, as shown in Fig. 2.D-3. A microphone also is inserted in the closed end to measure pressure, and the corresponding microphone voltage output is displayed on an oscilloscope. The frequency of the oscillator is varied from low to high, and a graph representing the pressure (the microphone output) at the closed end of the tube is made as shown in Fig. 5.D-4 for a tube 17 cm long.

64

A means (not shown in Fig. 5.D-3) is provided to cause the loudspeaker to move constant amounts of air at all frequencies so that the variations in the microphone output as frequency changes is due to the tube modifications made on the signal. The pressure response measured in this way for constant amounts of air movement gives us what is called the *input impedance* of the tube. The input impedance at a particular frequency tells us how the tube responds at that frequency. High values of input impedance occur for frequencies at which strong standing waves are produced; low values occur for weak standing waves. Note in Fig. 2.D-4 that the frequencies at which the input impedance is largest are just the resonance frequencies of a closed-open tube; the higher resonance frequencies are odd integer multiples of the lowest. The frequencies of the first three peaks are 500, 1500, and 2500 Hz.

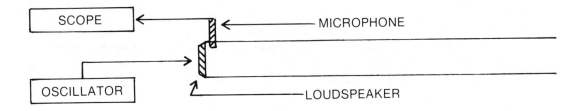

Fig. 2.D-3. Apparatus for measuring the input impedance of a tube.

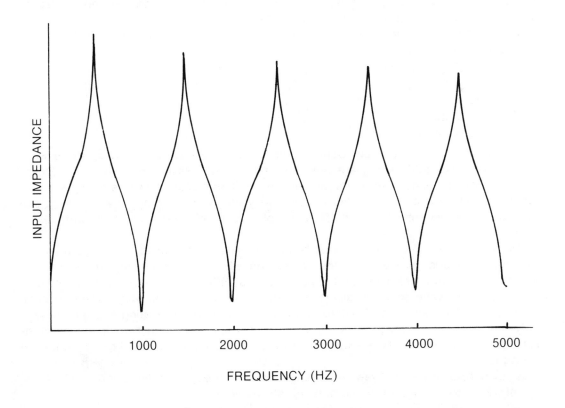

Fig. 2.D-4. Input impedance of cylindrical tube.

Standing waves on membranes and in rooms

Waves moving from one end of a string to the other end and back again are one-dimensional waves, i.e., they are confined to move along the string.

Waves moving in a drumhead are two-dimensional because the membrane has extent in two dimensions, thus permitting the waves greater freedom of motion.

We first consider two-dimensional displacement waves in a rectangular membrane which is fixed along all of its edges. Its lowest mode, as we might expect, is one in which all parts of the membrane move together; displacement nodes exist at all points along the edges. The vibration frequency of the lowest mode is given by $f_{1,1}$ = $(v/2)$ SQRT $((1/l1)^2 + (1/l2)^2)$, where $f_{1,1}$ is the frequency of the lowest mode, v is the wave speed in the medium, and $l1$ and $l2$ are the lengths of the membrane in each of its two dimensions. The wave speed v depends on the mass per unit area of the membrane and on the force per unit length along the edge of the membrane. For comparison recall that the lowest mode frequency of a string fixed at both ends is f_1 = $v/2l$, which is analogous to the above expression. Note that the natural mode frequency for the membrane is double subscripted. Each subscript refers to the conditions on the membrane in one of its two dimensions. The fact that each subscript is a "1" indicates that the membrane vibrates as one unit in that direction. The vibration frequencies of other modes of the membrane are given by $f_{n1,n2}$ = $v/2$ SQRT$((n1/l1)^2 + (n2/l2)^2)$, where $f_{n1,n2}$ is the vibration frequency of the n1,n2 mode, v is the wave speed in the medium, $l1$ and $l2$ are the lengths of the membrane in each of its two dimensions, and n1 and n2 are integers indicating the mode. When n1 = 1 and n2 = 1 we have the first mode discussed above. With n1 = 2 and n2 = 1 we have a mode in which the membrane vibrates in two parts along the first dimension and in one part along the second dimension. The frequencies of the higher modes may or may not be integer multiples of the lowest mode frequency, depending on the particular proportions of the rectangular membrane.

As examples of the first few mode frequencies, consider a membrane with v = 40 m/sec, $l1$ = 0.1 m, and $l2$ = 0.2 m.

$$f_{1,1} = 40/2 \text{ SQRT}((1/0.1)^2 + (1/0.2)^2) = 223.6 \text{ Hz}$$
$$f_{1,2} = 283 \text{ Hz}$$
$$f_{1,3} = 360 \text{ Hz}$$
$$f_{1,4} = 447 \text{ Hz}$$
$$f_{2,2} = 447 \text{ Hz} = 2f_{1,1}$$

Sound waves in rectangular rooms moving back and forth between opposite walls or between floor and ceiling are also one-dimensional waves. However, two-dimensional waves (striking four surfaces) or three-dimensional waves (striking six surfaces) can also exist in rooms. As a result, very complex standing wave patterns can exist in rooms because of the interference of waves reflected from room surfaces or from objects in the room.

Consider three-dimensional pressure waves in a hard-walled rectangular room. Pressure antinodes will exist at the walls. The vibration frequencies of the various modes of the room are given by $f_{n1,n2,n3}$ = $v/2$ SQRT$((n1/l1)^2 + (n2/l2)^2 + (n3/l)^2)$, where v is the wave speed in air, $l1$, $l2$, and $l3$ are the lengths of the room in each of its three dimensions, and n1, n2, and n3 are integers indicating the mode. With only one of the n's nonzero, one-dimensional modes exist; with two of the n's nonzero, two-dimensional waves exist; and with all three n's nonzero, three-dimensional waves exist.

As examples of the first few three-dimensional modes, consider a room with $l1$ = 4.0 m, $l2$ = 5.0 m, and $l3$ = 6.0 m (v = 340 m/sec).

$$f_{1,1,1} = 340/2 \text{ SQRT}((1/4)^2 + (1/5)^2 + (1/6)^2) = 61 \text{ Hz}$$
$$f_{2,1,1} = 96 \text{ Hz}$$
$$f_{1,2,1} = 85 \text{ Hz}$$
$$f_{1,1,2} = 79 \text{ Hz}$$
$$f_{2,2,2} = 122 \text{ Hz}$$

Normal modes exist on membranes and in rooms that are other than rectangular in shape. It is not possible, however, to write simple expressions for the natural frequencies in these cases. Hard-walled rooms having parallel sides give rise to very strong standing waves, which means that the pressures measured at different points in the room may be quite different, depending on whether the measurement is made at a node or an antinode. The so called "dead spots" occur at nodal positions in rooms and are especially prevalent when strong standing waves exist. Anything that can be done to reduce the standing wave and to make the sound pressures in the room more uniform will help to alleviate the problem of dead spots. Standing waves can be reduced by placing absorbing materials in the room and by making the room more irregular.

Exercises

1. Sketch the third, fourth, and fifth modes for a fixed-fixed string. Express the frequency of each mode in terms of the fundamental frequency.

2. Sketch the third, fourth, and fifth modes for a fixed-free string. Express the frequency of each mode (which is possible) in terms of the fundamental frequency, f_1.

3. Draw the first four modes for pressure waves in an open-open tube.

4. Draw the first five modes that might be considered pressure waves in an open-closed tube. Place an X through any diagram where the modes are not possible.

5. Fill in the table given below.

System	f_1	f_n	features of system
fixed-fixed string			
fixed-free string			
open-open tube			
open-closed tube			

6. What gives rise to "dead spots" in rooms?

7. Describe how treating walls with absorbing materials, making a room of irregular shape, and placing objects in a room help to alleviate "dead spots" in the room.

8. Calculate the frequencies for (1,1), (1,2), (2,1), and (2,2) modes of a square membrane of a side length 0.5 m if the wave speed is 30 m/sec. What special relationship exists between the (1,2) and (2,1) modes that does not hold for rectangular membranes in general?

*9. Standing waves in strings

   The apparatus shown below was set up to produce standing waves in a wire. A 60-Hz driver was attached to the left end. The mass of the wire was δ 0.000642 kg/m. The length of the wire was 1.0 m. The tension applied to the wire could be controlled by adding mass.

   a. When a mass of 0.3 kg was suspended from the wire, the fourth mode was produced as shown. What was the wave speed for this mode? What was the wavelength?

   b. What mass must be suspended to produce the second mode? The first mode?

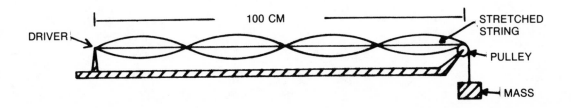

*10. A loudspeaker was set in the corner of a hard-walled, rectangular room (dimensions of 3.0 × 4.0 × 5.0 m). A sine wave generator was used to drive it. A sound level meter was placed in a diagonally opposite corner to measure sound pressure levels.

   a. A maximum sound pressure level was observed to occur at a frequency of 78 Hz. What room mode was being excited? What was the approximate sound speed in the room?

   b. For what mode and at what next higher frequency would another SPL maximum occur?

   c. Repeat 10b. for the next mode.

Further Reading

AAPT, pp. 83–85
Backus, chapters 4, 9
Benade, chapters 7, 9
Olson, chapters 1, 8
Roederer, chapter 4
Sears, Zemansky, and Young, chapter 22
Stevens and Warshofsky, chapter 4

Audiovisual

1. *Standing Waves in a Gas* (Loop 80-3874, EFL)
2. *Standing Waves on a String* (Loop 80-3866, EFL)
3. *Vibration of a Rubber Hose* (Loop 80-3890, EFL)
4. *Vibrations of a Drum* (Loop 80-3924, EFL)
5. *Vibrations of a Metal Plate* (Loop 80-3932, EFL)

2.E. COMPLEX WAVES

A *complex wave* is any wave other than a sinusoid. But, as we shall soon discover, any complex waveform can be constructed from sinusoidal components. The study of complex waves can be approached from two opposite directions: by *analysis,* which is the breaking down of a complex wave into its components, or by *synthesis,* which is the building up of complex wave from its components. We will start with analysis.

Analysis of Complex Waves

The manner in which complex waves are related to sinusoids is rather interesting. Recall that in section 1.D we discovered that any complicated motion of a compound vibrator could be formed by adding various amounts of its natural modes. The superposition principle (discussed in section 2.C) tells us that when two or more waves occupy the same region of a medium the resultant wave is just the point-by-point sum of each individual wave. We already know that the natural modes of waves on strings or in tubes are the sinusoids with frequencies which are multiples of the fundamental frequency. Can we generalize the results for a compound vibrator and, by applying the superposition principle, conclude that complex waves are constructed by addition of the natural mode sinusoids? In fact, this result was obtained (by mathematical means) over 150 years ago by a French mathematician, Joseph Fourrier. His conclusions can be stated as *Fourier's Theorem:* Any repetitive wave pattern, no matter how complex, may be broken down into constituent sinusoids of different amplitudes, frequencies, and phases. If the complex wave under consideration exists on a string or in a tube, it can be shown that the sinusoidal components are in fact the natural modes of the system. Each sinusoid helping to make up a complex wave is called a *partial.* The first partial (the lowest-frequency sinusoid) is known as the *fundamental.* The higher partials (second, third, etc.) of the complex tone can be either one of two different types: harmonic or inharmonic. Higher partials whose frequencies are integer (whole number) multiples of the fundamental frequency are called *harmonics* (or harmonic partials). When the frequencies of the higher partials are not integer multiples of the fundamental frequency they are termed inharmonic partials. The term *overtone* is sometimes used to refer to those harmonics having frequencies higher than the fundamental. With this system the second harmonic is known as the first overtone, the third harmonic being the second overtone, etc. Because discussion of the overtone system is not necessary here, and because it often leads to confusion, we will not consider it further.

As noted above, each ingredient of a complex wave is a sinusoid having a different frequency. The spectrum (vibration recipe) for a particular wave tells us the amount of each ingredient present. The analysis of a complex wave, then, means determining the spectrum of the wave. One way to obtain the spectrum of a complex wave is by means of electronic filters. A *filter* is a device which allows certain frequencies to pass through unchanged, while other frequencies are eliminated. You may think of a set of filters as having a function analogous to a set of screens, each with a different mesh size. The screens are used to sort a conglomerate of gravel into coarse, medium, and fine components. Likewise, a set of three filters (each having different characteristics) could be used to separate a complex wave into low, medium, and high frequency components. Fig. 2.E-1 shows how a complex wave could be resolved into its components by passing through a set of filters.

The spectrum analyzer described in section 1.E can be thought of as composed of such a set of filters. However, there are about 200-400 filters in a typical spectrum analyzer, as compared to the three filters illustrated. The greater the number of filters, the more finely the frequency components can be resolved during a spectrum analysis.

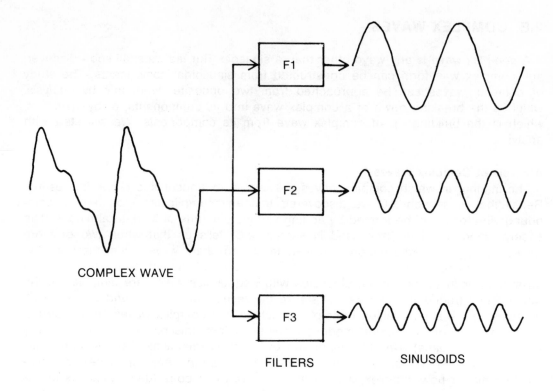

COMPLEX WAVE

FILTERS   SINUSOIDS

Fig. 2.E-1. The analysis of a complex wave, with a set of filters tuned to different frequencies.

As an example of Fourier analysis of a complex wave, consider the three wave-forms shown in the left column of Fig. 2.E-2. The waveforms shown are the pictures obtained on an oscilloscope when three different "instruments" are sounded: (a) a tuning fork, (b) a clarinet, and (c) a trumpet. Because of the complex nature of these waveforms, very little information can be gleaned merely by observing them. If how-ever, they are analyzed to obtain their spectra, as shown in the right column of Fig. 2.E-2, the pertinent information is displayed at a glance. We see, for instance, that the tuning fork waveform consists of only one component, the fundamental, while the clarinet displays predominantly odd harmonics. The trumpet is seen to have all har monics, but the third harmonic has the greatest amplitude.

Synthesis of Complex Waves

A sine wave, representing pressure, can be written

$$p = A \sin (ft + \phi),$$

where p is the quantity you would plot on a graph of pressure versus time, A is the pressure amplitude of the sinusoid, f is the frequency, t is time, and ϕ is the phase (in fractions of a cycle). For example, Table 2.E-1 was constructed given that A = 1.0, f = 10 Hz, and ϕ = 0. From this table of data the graph in Fig. 2.E-3a was plotted. Note that the time intervals (from 0.0 sec to 0.1 sec) were chosen to cover one cycle of the vibration, and the sinusoid plotted below starts with a phase of zero at zero time.

WAVEFORM   SPECTRUM

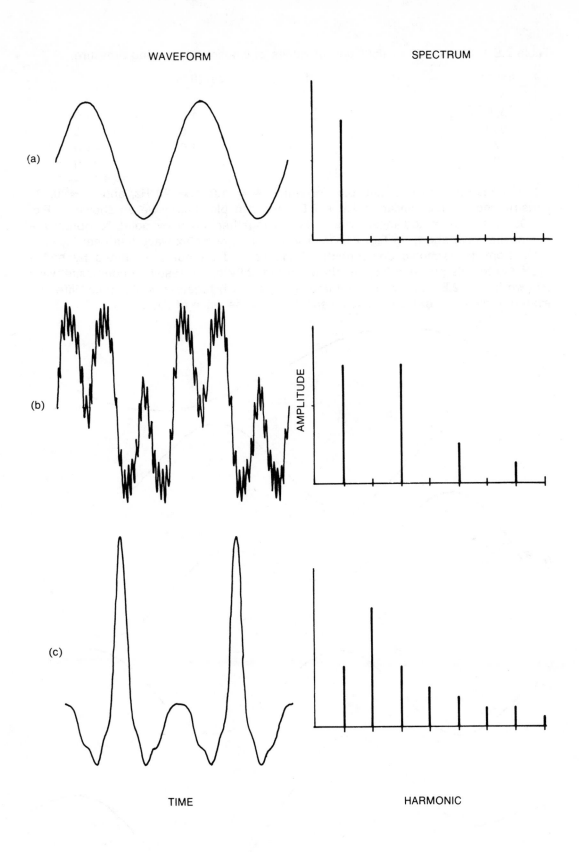

TIME   HARMONIC

Fig. 2.E-2. Idealized waveforms and spectra of several "musical" instrument tones: (a) tuning fork, (b) clarinet, (c) trumpet.

Table 2.E-1. Steps in the calculation of values of sinusoidally varying pressure.

t(sec)	ft	ft + ϕ	sin (ft + ϕ)	p
0	0	0	0	0
.025	.25	.25	1.0	1.0
.050	.50	.50	0	0
.075	.75	.75	−1.0	−1.0
.100	1.00	1.00	0	0

Now consider a second sinusoid, for which A = 1.0, f = 20 Hz, and ϕ = 0. By constructing a table similar to Table 2.E-1 we can plot this wave as shown in Fig. 2.E-3b. Let us now add these two sinusoids together, point by point, to obtain the composite wave shown in Fig. 2.E-3c. The resulting complex wave has been synthesized from its harmonic components. If the second harmonic is shifted by half a cycle (so that its phase is 0.5), as shown in Fig. 2.E-3d, the resulting composite wave (shown in Fig. 2.E-3a), obtained by adding (a) and (d) together, looks quite different, even though it has exactly the same spectrum as the wave of Fig. 2.E-3c.

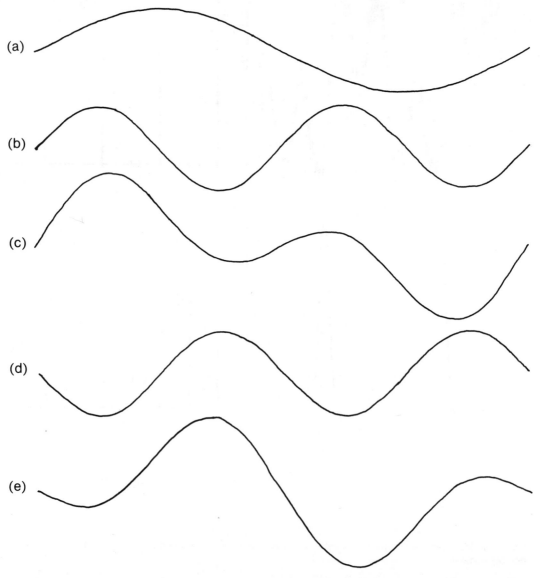

Fig. 2.E-3. Modes and compound modes: (a) first mode with phase of 0.0, (b) second mode with phase of 0.0, (c) sum of (a) and (b), (d) second mode with phase of 0.5, (e) sum of (a) and (d).

72

Now consider the synthesis of a pulse-like wave formed by adding the following three sinusoids: $p_1 = 1.0 \sin (10t + 0.75)$; $p_2 = 1.0 \sin (20t + 0.25)$; and $p_3 = 1.0 \sin (30t + 0.75)$. The values of p_1, p_2, and p_3 at various times t, as well as the sum $p = p_1 + p_3$, are given in Table 2.E-2. The waveforms for p_1, p_2, and p_3 are graphed in Fig. 2.E-4a-c. Part (d) of the figure gives their sum. Note that this complex wave could be synthesized by adding the numerical values of Table 2.E-2 directly or by adding the waves of Fig. 2.E-4a,b,c graphically.

Table 2.E-2. Tabulation of P_1, P_2, P_3, and their sum.

t	P_1	P_2	P_3	$P = P_1 + P_2 + P_3$
0.000	−1.0	1.0	−1.0	1.0
0.005	−.95	.81	−.59	−0.73
0.010	−.81	.31	0.31	−0.19
0.015	−.59	−.31	0.95	0.05
0.020	−.31	−.81	0.81	−0.31
0.025	0.0	−1.0	0.0	−1.0
0.030	.31	−.81	−0.81	−1.31
0.035	.59	−.31	−0.95	−0.67
0.040	.81	.31	−0.31	0.81
0.045	.95	.81	−0.31	2.35
0.050	1.0	−1.0	1.0	3.0

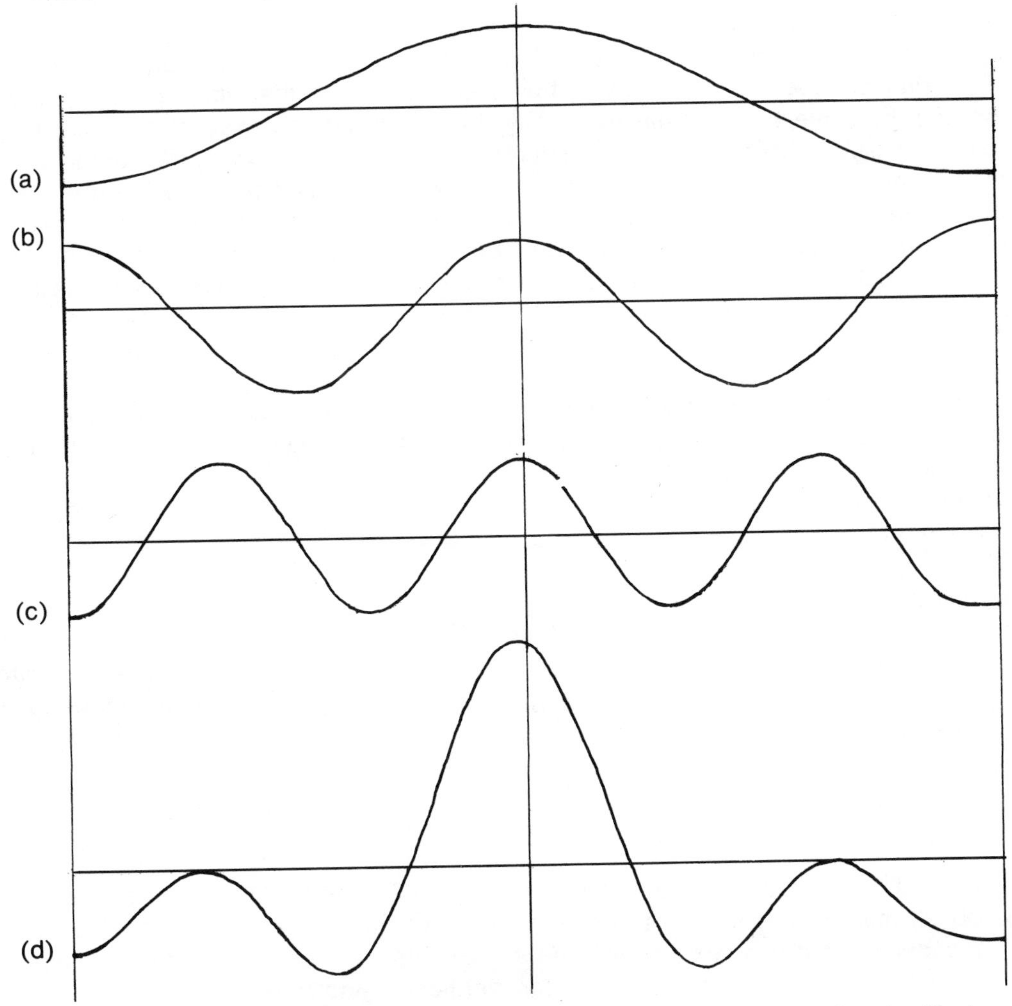

Fig. 2.E-4. Natural and compound modes, (a) first mode with phase of 0.75, (b) second mode with phase of 0.25, (c) third mode with phase of 0.75, (d) sum of (a),(b), and (c).

Any musical sound, no matter how complicated, can be represented by a complex wave. But, since the complex wave is no more than a sum of simple sinusoids, it stands to reason that the very complex waveform produced by an entire orchestra is composed of hundreds and hundreds of simple waves. The fact that all of this information can be represented by one complex wave which changes with time is what makes phonographs and tape recorders possible. The grooves on a phonograph record are nothing more than a very complex waveform which contains the information (the sum of all the individual waves) recorded at the studio. When the record is played, the complex vibration is reproduced by the needle and eventually causes the loudspeaker diaphragm to vibrate in the same manner. Our ears receive the complex wave and we interpret it as an orchestra performing.

Timbre and Spectrum

In chapter 1 we mentioned that the pitch of a sound is related to its frequency while the loudness is related to the wave amplitude. Obviously, more than just pitch and amplitude information is needed to describe different sounds. We can imagine a situation in which a flute and violin each play the same note (same frequency) at the same loudness level. We are able to distinguish between these two sounds; in fact, we say they each have a different tonal quality, or timbre. The timbre of a complex sound is related to the spectrum of that sound: different spectra produce sounds of differing tonal qualities. The subject of timbre and tone quality will be considered in greater detail in section 3.B, while in section 6.A we will discuss other characteristics of musical tones which are important for distinguishing various instruments. For now, let it suffice to say that different qualities of tone are due to different spectra. By considering the spectra of Fig. 2.E-2, for example, we see that although each waveform has the same fundamental frequency, the three different waveforms each have a different timbre because of their different spectra. The tuning fork has a rather simple spectrum (only one component), and the sound is considered "pure." The clarinet, with a spectrum consisting predominantly of odd harmonics, has a tone which can be described as "hollow" (presumably because of the missing even harmonics). Finally, the trumpet has a spectrum consisting of many harmonics, the third harmonic having the greatest amplitude. The trumpet tone is often referred to as "bright" and "full," and in light of the spectrum shown this seems to be an apt description.

In summary, the differences in timbre of complex sounds depends primarily (but not entirely) upon the presence, the number, and the relative amplitudes of the partials. The relative phases of the partials have little effect upon the resultant quality (except in extreme cases), even though phase differences may alter the shape of a waveform drastically.

Exercises

1. Is a free or a forced system more likely to produce harmonic partials? What system is more likely to produce inharmonic partials? Why? (Hint: Forced systems give rise to periodic waves. Free systems most often give rise to nonperiodic waves.)

2. Consider two sinusoids given by sin (ft) and sin (ft + ϕ), where ϕ is the phase. Plot these two sine waves for f = Hz and ϕ = 0.25 over an interval of 1 second.

3. What is the amplitude of each wave in exercise 2? The frequency? The phase?

4. Do all musical notes that have the same fundamental frequency and the same number of higher partials sound the same? What besides the number of higher partials would have to be the same for identical sound to result?

5. What determines the quality of a musical sound? Are inharmonic partials ever desirable in musical sounds?

6. Outline a method by which a musical tone may be artificially created.

7. Explain why identical notes plucked on a guitar and a banjo have distinctly different sounds.

8. The conventional spectra below represent six waves, some simple and some complex. (These waves are just made up and may or may not represent "real world" waves. The number preceding each "sin" is the amplitude or the amount of that particular component present.)
   a. What is the fundamental frequency of each wave?
   b. What are the frequencies and amplitudes of the higher partials for each wave?
   c. Are the higher partials harmonic or inharmonic relative to the fundamental?

Wave 1: $1.0 \sin (f_1 t)$	$f_1 = 100$ Hz
Wave 2: $1.0 \sin (f_1 t) + 1.0 \sin (f_2 t)$	$f_1 = 100$ Hz $f_2 = 200$ Hz
Wave 3: $1.0 \sin (f_1 t) + 1.0 \sin (f_2 t + 0.5)$	$f_1 = 100$ Hz $f_2 = 200$ Hz
Wave 4: same as wave 2 except $f_2 = 205$ Hz	
Wave 5: $1.0 \sin (f_1 t) + 1.0 \sin (f_1 t + 0.5)$	$f_1 = 100$ Hz
Wave 6: $1.0 \sin (f_1 t) + 0.11 \sin (f_3 t + 0.5)$ $+ 0.4 \sin (f_5 t)$	$f_1 = 100$ Hz $f_3 = 300$ Hz $f_5 = 500$ Hz

9. Wave 6 in exercise 8 is approximately a triangular wave. Plot one cycle of it on graph paper by plotting each component separately and then graphically adding the three components to form the complex wave.

10. Wave 5 in exercise 8 is two sinusoids differing in phase by 0.5 cycle. What is the net result?

11. Waves 2 and 3 have the same spectra. However, they differ in phase. If both of them are plotted and the wave shapes compared, they look different; however, they will sound the same. Plot the two waves; then plot their spectra as a bar graph.

12. Plot the spectrum for wave 6 as a bar graph.

13. Complex waves: analysis. A function generator was connected to a band-pass filter and its output observed on an oscilloscope.
   a. When the function generator was set to produce sine waves at a frequency of 200 Hz the voltages in the "sine" row of the table were measured on the oscilloscope at the frequencies shown.
   b. Similar measurements were made when the function generator produced square waves, triangular waves, and sawtooth waves, as shown in the table.

Function	200 Hz	400 Hz	600 Hz
Sine	10	.01	.02
Square	10	.03	3.3
Triangular	10	.02	1.1
Sawtooth	10	.05	3.3

c. What can you say about the spectrum for each of the four wave types?

d. Why were voltages only slightly different from zero measured?

14. Complex waves: synthesis. The relative amplitudes for the first few partials of a sawtooth wave and a square wave are given in the table below as A1, A2, and so on. The relative phases for all partials are zero.

Wave Type	A1	A2	A3	A4	A5	A6
Sawtooth	1	1/2	1/3	1/4	1/5	1/6
Square	1	0	1/3	0	1/5	0

a. Plot one cycle of the sine wave representing the first partial of the sawtooth wave on a sheet of graph paper. Next plot two cycles of the second partial on the graph paper with the proper relative amplitude. Repeat this for the remaining partials. Now graphically add all of these sine waves together to produce a fairly good approximation of a sawtooth wave.

b. Repeat 14a for the square wave.

*15. Interference of sinusoids with a complex wave. Connect a sine wave generator and a function generator to a mixer. Connect the mixer to an oscilloscope and to a loudspeaker.

a. Set the function generator to produce square waves at a frequency of 200 Hz. Set the sine wave generator to a frequency of about 200 Hz. Do you hear strong beats? Why?

b. Repeat 15a with the sine frequency set to 400 Hz. Do the spectra shown in problem 13 or in section 1.E provide any clues?

c. Repeat 15 with a sine frequency of 600 Hz.

d. Repeat 15a-c with a triangular wave.

e. Repeat 15a-c with a sawtooth wave.

Further Reading

Backus, chapter 4
Benade, chapter 3
Culver, chapter 1
Denes and Pinson, chapter 3
Olson, chapter 1
Roederer, chapter 4

Audiovisual

1. *Superposition* (Loop 80-3858, EFL)

Demonstrations

1. Spectrum analysis of various waves
2. Fourier synthesis of various waves
3. Variable frequency sinusoid beating against complex wave

2.F. SOUND RADIATION AND THE DB SCALE

The existence of a complex wave implies a source of vibration and a medium to transmit the energy. If we consider vibrating solids or air columns as a source and air as the medium, a sound wave is, under certain limitations, the vibration of the air. When a source sets the molecules of the air into vibration, a varying pressure having the same frequency as the source is produced. This varying pressure wave can be considered a sound if it is within the audible frequency range of the human ear, whether or not there is actually an ear present to hear it. A sound wave, then, is any vibration of the air between 16 Hz and 16,000 Hz. Air vibrations below 16 Hz are called "infrasound," while those above 16,000 Hz are termed "ultrasound."

Almost any sort of rapid motion through the air generates sound waves: e.g., the buzzing wings of a fly or the vibrations of cymbals struck together. Part of the energy of motion becomes sound waves which travel out from the source in all directions, eventually striking walls or objects in a room. Yet most sounds do not cause the walls or objects in a room to vibrate noticeably, even though we might perceive those sounds as being quite loud. Furthermore, even the feeble power of a buzzing fly can seem quite loud on a quiet night when we are trying to sleep. We are able to hear very weak sounds indeed, but even loud sounds are not very powerful compared to other energy sources, such as heat and light.

Sound Intensity

The energy of a source of sound, such as a musical instrument, can be increased by doing work on the vibrating system, which is the same as adding energy to the system. If energy is added rapidly a large power (energy per unit time) is required. For instance, the wattage rating of electric light bulbs is a power rating—a 200-watt bulb requires twice as much power, or twice as much energy, per second as a 100-watt bulb. The power we encounter for sounds is very much smaller than typical electrical powers (which testifies to the remarkable sensitivity of the ear), but the same principles apply and can be used to rate the sound output (in watts) of various sources. Consider, for example, a 100-watt light bulb. While perfectly adequate to provide light for an average-sized room, it is not a very powerful source of light. Yet those familiar with high fidelity equipment will recognize that a 100-watt amplifier would rock the entire house! As a matter of fact, a 10-watt amplifier is capable of producing more sound than the average housewife will tolerate.

Visualize, now, 100 watts of sound power spreading out in all directions in an ever enlarging sphere. Although the total sound power remains constant on the surface of the sphere, it is spread over a larger area as the size of the sphere increases. The *intensity* of a spherical sound wave is the power per unit area; that is, the total sound power divided by the surface area of the sphere. Fig. 2.F-1 is a representation of sound power being spread over larger and larger areas as it travels away from the source. (It is comparable to spreading the same amount of butter over larger and larger pieces of toast.) As we move away from the source, then, the sound must become less intense (the same power spread over a larger area) and the sound gets softer. In Fig. 2.F-1 we can see that if a certain amount of sound power falls on a 1 m^2 surface located at a distance of 10 m from the source, the same sound power would be spread over $2 \times 2 = 4$ m^2 at a distance of 20 m and $3 \times 3 = 9$ m^2 at a distance of 30 m. The intensity, then, decreases as the inverse square of the distance from the source; when the distance from the source is doubled, the sound intensity falls to one-fourth its previous value. (This assumes of course that there are no reflecting walls and that the sound spreads out uniformly in all directions.)

In addition to the loss of sound intensity due to "spreading out," sound energy and intensity are decreased because of (1) the absorption of sound energy when a sound wave contacts an object, (2) the absorption of sound energy in a medium such as air as a sound wave moves through, or (3) the radiation of sound energy

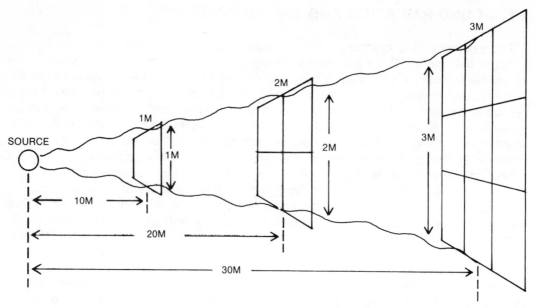

Fig. 2.F-1. Diagram illustrating the decrease of sound intensity with distance because the sound power is spread over larger areas.

from strings, membranes, or holes in a tube. The first and second causes of energy loss are important for rooms and will be discussed in chapter 4. The absorption of sound energy at the walls of wind instruments (due to friction) can be very significant, often consuming in excess of 90 percent of the energy supplied by a player.

Radiation of Sound

Radiation is the emitting or giving off of energy. A loudspeaker provides a familiar example and illustrates two important features of acoustic radiators: (1) a large radiating surface is required to radiate long wavelengths (low frequencies) efficiently; and (2) radiators do not radiate high-frequency sound uniformly in all directions. A large surface area is required to radiate low-frequency sound efficiently because a large mass of air must be set into vibration. The low frequencies (long wavelengths) are radiated approximately uniformly around a loudspeaker because of diffraction. Fig. 2.B-5 could represent a loudspeaker rather than a hole in a wall. The long wavelength sound waves are greatly diffracted by "passing through" the loudspeaker, which is of much smaller dimensions than the sound wave. High-frequency sounds are radiated more strongly in the direction in which the speaker is "aimed" and less strongly to the sides because they have short wavelengths compared to the size of the speaker. Fig. 2.F-2 shows approximate distributions of sound energy emerging from a loudspeaker (mounted in a wall) for low, medium, and high frequencies. The low-frequency waves can be regarded as having wavelengths several times the size of the speaker. Wavelengths of the high frequency waves are only a fraction the size of the speaker. The medium-frequency waves have wavelengths of about the same size as the speaker.

Radiators of the voice are the mouth and nose. Radiators of wind instruments are the tone holes and/or the bell openings. Radiators of the string instruments are their bodies. Whether an instrument radiates energy uniformly in all directions or beams most of it in a particular direction depends on its radiator size relative to the wavelength of the sound being radiated. Many radiators, such as the strings on a piano, do not efficiently transfer their vibrations to the air without some assistance. By coupling the piano strings to a sound board, its vibrations are transferred to the air much more efficiently. Many musical instruments provide examples of this type of coupled vibration in that they consist of two vibrating systems: the first determines the frequency of vibrations and forces the second to vibrate at this same frequency; the second system radiates the sound to the air.

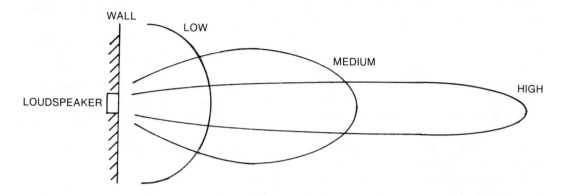

Fig. 2.F-2. Radiation patterns from a loudspeaker for low, medium, and high frequencies. (Low frequencies have long wavelengths compared to speaker size and radiate uniformly in all directions. High frequencies have short wavelengths compared to speaker size and radiate mostly to the front of the speaker.)

Decibel Scale

The human ear can respond to sound pressures that vary by a factor of 1 to 10,000,000 between the lowest pressure to which the ear responds and a pressure that produces pain. This tremendous range of sensitivity is rather hard to imagine. If you had a scale sensitive enough to weigh a single human hair and the scale responded to the same range as the human ear it could also be used to weigh a 30-story apartment building! It has been known for quite some time that our ears have a particular way of dealing with this fantastic range of sound powers. When the intensity of a sound is increased about ten times, the sensation of loudness is approximately doubled, regardless of whether the sound was soft or loud. In general, we judge the relative loudness of sound in terms of ratios of sound pressures, not in terms of differences of sound pressures. If we listen to a given sound (such as a buzzing fly) and the sound pressure is suddenly doubled (a larger buzzing fly) our ears perceive a certain increase in loudness. If we listen to a loud buzz saw and then hear an even louder buzz saw we may find the sound pressure again doubled. Although there is a much greater pressure difference between the sounds of the two buzz saws than between the two flies, the increase in loudness in each case seems to be the same. In other words, the increase in loudness is very nearly independent of the sound pressure of the original sound. Thus a particular difference in sound pressure that is negligible in comparing two loud sounds can be tremendous in comparing two faint sounds.

The above discussion suggests that it may be convenient to use a measuring scale for sound based on sound pressure ratios rather than absolute sound pressures. The *decibel* (dB) scale is just such a scale. While the dB scale did not have its origin in the above considerations, it is possible to say that the dB scale does approximate the way our ears perceive sound much better than a linear pressure scale would. The decibel scale expresses a ratio between two sound pressures in dimensionless units known as decibels (dB). Mathematically, the *sound pressure level* (expressed in decibels) is defined as 20 times the logarithm of the ratio of the pressure to some standard pressure:

$$dB = 20 \log (P/P_s),$$

where P is the pressure and P_s is a standard reference pressure. By convention, the reference pressure is taken to be $P_s = 0.00002$ N/m². This is approximately the smallest pressure that a human can detect under ideal conditions and in the frequency region where the ear is most sensitive. When a sound wave has a pressure of 0.00002 N/m², the sound pressure level will be 0 dB. Note that 0 dB does not mean there is an absence of sound; rather a 0 dB sound pressure level corresponds to a sound pressure equal to the standard reference pressure.

Although the dB scale may seem like an artibrary contrivance designed to confuse students, there is some justification for it. A change in pressure level of one dB is about the smallest change that a human can perceive, and a change of 10 dB corresponds roughly (at least at a frequency of 1,000 Hz) to a doubling of perceived loudness. Since the human ear can detect pressures over the entire range from "just perceptible" to "pain," the dB scale is typically used between about 0 dB (threshold of hearing) and 140 dB (threshold of pain). Furthermore, since the ear perceives changes of pressure ratios rather than absolute pressure changes, the dB scale mimics the subjective processes of the brain.

A change in decibel level should not be confused with a change in loudness, even though the two phenomena are related. Loudness, as we will see in section 3.B, is a subjective sensation, while the decibel level measures the energy of a sound wave. In comparing the energies of several sound sources, we proceed as follows. If one sound has a pressure amplitude 10 times that of another sound we can compute that the more intense sound has a level 20 dB greater than the first sound. Suppose that the sound pressure level of a trombone is recorded at 80 dB. When two identical trombones play the same note exactly in phase the sound pressure will double, but the sound pressure level goes up to only 86 dB. If 76 trombones could all play the same note in phase, the pressure level would increase by a factor of 76, and the sound pressure level would go up only to 118 dB. To reach the threshold of pain at 140 dB would require no fewer than one thousand trombones all playing in unison (although your threshold of discomfort or dismay might be reached with a much smaller number of trombones). Whenever the pressure increases by a factor of 10 the sound pressure level increases by 20 dB.

Table 2.F-1 shows some sound pressure ratios and their decibel equivalents. Note that, because of the huge range of pressures to which the ear is sensitive, the decibel scale is also somewhat more convenient to use than a linear pressure scale.

Table 2.F-1. Sound pressure ratios and their decibel equivalents.

Sound pressure ratio	Decibel equivalents
1 : 1	0
10 : 1	20
100 : 1	40
1000 : 1	60
10,000 : 1	80
100,000 : 1	100
1,000,000 : 1	120

A *sound level meter* is a device (usually portable) used for measuring sound pressure levels. It consists of a microphone, an amplifier, various filters, and a meter for displaying the measurement results. The A-weighting filter modifies the incoming sound so that the meter will respond in a manner similar to the way our ears respond. As we will see in section 3.B, the ear attenuates the low frequencies of weak sounds. It therefore requires more low-frequency sound energy than high frequency energy to cause a given deflection on the meter. Presumably, then, a decibel reading on the A-scale, written dB(A), is a fairly accurate gauge of how humans will perceive the sound. The C-filter treats all frequencies the same; none are attenuated. The B-filter attenuates low frequencies, but to a lesser extent than the A-filter.

It is necessary to exercise considerable care when using the sound level meter for accurate measurements. An appreciable error in the reading may occur because of the operator himself. It is desirable that the operator move from place to place while performing the measurement in order to insure a truly representative sample of the overall sound field. When the measurements are performed inside a building the observer must be especially careful because of the many reflected sounds, all of which can influence the reading.

Table 2.F-2 gives some sound levels produced by typical sound sources. Note that quiet background noise corresponds to a dB(A) level of about 40 dB, while 120 dB is the threshold of feeling, in which case the sound is so loud that we feel a buzzing sensation in our ears.

Table 2.F-2. Typical A-weighted sound levels measured with a sound-level meter. (Distances from source in meters shown in parentheses.)

Decibels re 0.00002 N/m²	
140	Threshold of pain
130	Jet takeoff (60 m)
120	Threshold of feeling
	Riveting machine (operator's position)
110	
	Buzz saw (operator's position)
100	
	Bellowing moose (2m)
90	Printing press plant
	Pneumatic drill (15 m)
80	Inside sports car (50 mph)
	Vacuum cleaner (3m)
70	Near freeway (auto traffic)
60	Normal conversation (1m)
50	Light traffic (30m)
	Night noises in city
40	Buzzing fly (2m)
	Soft whisper
30	
	Studio (speech)
20	Studio for sound pictures
10	Threshold of hearing
	1,000-4,000 Hz Range
0	

Exercises

1. If a tree falls in a forest and there is no one to hear it, is a sound produced? Explain.

2. What is the intensity of sound at a point two meters from a source if the intensity is 10^{-6} watt/m² at a distance of one m? (Assume uniform spherical radiation: the surface area of a sphere is $4\pi r^2$, where r is the radius.) What is the intensity at a distance of 5 m?

3. Draw a simplified clarinet diagram and label points of energy input and energy loss.

4. Several of the transducers that we termed acoustic generators in chapter 1 radiate sound energy. What are some of these?

5. Name several musical instruments and give an estimate of the size of their radiators.

6. Which of the instruments named in exercise 5 radiate low frequencies the most efficiently?

7. Estimate the size of the two most significant radiators associated with the vocal system. Which is generally the most efficient of these vocal radiators?

8. What is done in hi-fi systems to radiate low frequencies efficiently?

9. Do we expect brass instrument tones to be more "brilliant" sounding when we sit in front of the instrument or to the side of the instrument? Why?

10. Plot the spectrum of tone 6 in exercise 8 of section 2.E in terms of sound pressure level.

11. What is the difference in dB between the pressure level of two sounds having pressure levels of 0.001 and 0.2 N/m²?

12. An intensity level can be defined in a manner analogous to that of pressure level. The relation is dB (intensity level) = $10 \log (I/I_s)$, where I is the intensity of interest and I_s is a standard intensity equal to 10^{-12} watts /m². Calculate the intensity level for each of the three cases in exercise 2.

13. The C-network of a sound level meter weights all frequencies about equally. For a complex sound composed primarily of low frequencies, will the A-scale reading be larger than, smaller than, or about equal to the C-scale reading? Answer the same question for a complex sound composed mainly of high frequencies.

*14. Intensity level variation with distance and direction

A single diaphragm, enclosed-cabinet, direct-radiator loudspeaker was placed out of doors away from all reflecting objects. It was driven with sinusoids of 100 Hz and 1,000 Hz, as shown in the table. Sound pressure levels were measured with a sound level meter using the C-scale (which gives approximately equal weightings at all frequencies). Measurements were made in front and to the side of the loudspeaker as shown in the "distance" and "position" columns, and the values of sound level were recorded in the table.

a. What approximate rule can you state in regard to intensity level variation with distance from source?

b. What can be said about the directional effects of the loudspeaker?

Frequency (Hz)	Distance (w)	Position	Sound level (dB)
100	2	front	81
100	4	front	74
100	8	front	69
100	8	side	65
1,000	8	front	71
1,000	8	side	65

Further Reading
Backus, chapters 3, 4, 5
Culver, chapters 3, 5, 8
Denes and Pinson, chapter 3
Gerber, chapter 2
Kinsler and Frey, chapters 7, 10
Olson, chapters 4, 6

Audiovisual
1. *The Science of Sound* (Album #FX6007, 1959, FRS)
2. *Sound, Energy, and Hearing* (30 min, color, 1957, EBE)

Demonstrations
1. Speaker array to demonstrate directivity of sound
2. Sound level meter

Chapter 3

The Ear and Hearing

3.A. ANATOMY AND FUNCTIONS OF THE EAR

The process of hearing may be considered as taking place in three stages, the outer-ear stage, the middle-ear stage, and the inner-ear stage. In the outer-ear stage, the incoming sound is converted to mechanical motion of the eardrum. In the second stage, the mechanical motion is transmitted to the inner ear, where it is subsequently, in the third stage, converted to a series of nerve impulses to the brain. The range of sound pressures over which the ear can respond is truly phenomenal. It cannot only withstand extremely intense sounds, but it is capable of responding to pressures which are so small that displacement of the eardrum is less than the diameter of the air molecules striking it. Also, the ear can respond to frequencies from about 16 Hz to 16,000 Hz. But the ear is much more than a wide-range microphone; it also acts as a sophisticated time and frequency analyzer. Undoubtedly the hearing mechanism is one of the most intricate and delicate mechanical structures found in our bodies.

Anatomy and Mechanisms

The ear consists of three main parts: the outer, middle, and inner ears, as shown in Fig. 3.A-1. The outer ear consists of the visible portion of the ear, the ear canal, and the eardrum. The middle ear consists of a small chamber containing the three tiny bones which transmit the vibrations to the inner ear. The inner ear is a com-

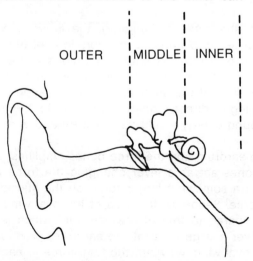

OUTER MIDDLE INNER

Fig. 3.A-1. Simplified anatomy of the Human Ear.

plexly shaped cavity filled with a liquid and the necessary structures to convert mechanical vibrations into nerve pulses.

The outer ear serves to protect the sensitive middle and inner ear mechanisms against harsh external environment and to maintain internal temperature and humidity somewhat independent of external conditions. Also, since the ear canal is an acoustic resonator open at one end and closed at the other, it enhances sounds whose frequencies are close to its resonance frequency of about 3000 Hz.

Consider now the schematic representation of the ear shown in Fig. 3.A-2. The middle ear cavity is connected to the throat by means of the Eustachian tube. Under normal conditions the Eustachian tube is closed, thus sealing the cavity, but the tube may open during swallowing or yawning. The opening of the tube serves as a mechanism to equalize the pressure in the middle ear chamber and the external air pressure. Most people who have flown in airplanes or who have driven a car up a mountain have noticed the discomfort which often accompanies rapid changes in altitude. The discomfort of the ear is caused by an increase or decrease in the external air pressure relative to the pressure in the middle ear. Swallowing or yawning opens the Eustachian tube, allowing the middle ear pressure to become equal to that of the surrounding atmosphere, and thus relieves the discomfort.

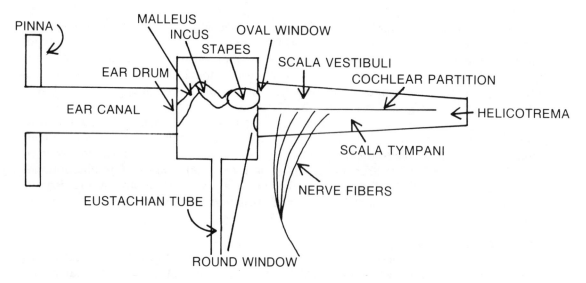

Fig. 3.A-2. Schematic representation of the human ear.

The three small bones—the *malleus* (hammer), the *incus* (anvil), and the *stapes* (stirrup)—of the middle ear serve to transmit the vibrations of the eardrum to the fluid of the inner ear. They not only transmit sound vibrations to the inner ear, but also amplify the pressure from the outer ear to the inner ear. Such a mechanical linkage is necessary to cause the fluid of the inner ear to vibrate appreciably. If the incident vibrations of the air impinged directly on the opening to the fluid-filled inner ear (*cochlea*), most of the sound energy would be reflected and very little would actually enter the inner ear.

The combination of the eardrum and the three bones amplifies the sound pressures in two ways. First, the bones act as a lever changing the force exerted by the eardrum on the hammer into a somewhat larger force on the stirrup. This accounts for a small increase in the force. Secondly, the area of the oval window (the entrance to the cochlea) is small compared to that of the eardrum, which is about 25 times as large. A small pressure over a large area (at the eardrum) produces a large pressure over a small area (at the oval window) when the total force in each case is the same. These two effects result in a pressure at the oval window about 30-40 times greater than at the eardrum. This means that we can detect sounds about 1,000 times weaker (in an energy sense) than we could without the combination of the eardrum and middle ear bones.

The inner ear (cochlea) is a transducer shaped something like a snail's shell, which transforms mechanical vibrations into nerve impulses; as in Fig. 3.A-2, the cochlea is shown uncoiled. The cochlea can be viewed as a fluid-filled, rigid-walled tube divided along its length into an upper compartment (the scala vestibuli) and a lower compartment (the scala tympani) by the slightly flexible cochlear partition. The cochlear partition extends along almost the entire length of the cochlea, ending only near its far end. This opening, which provides a path for fluid flow between the upper and lower compartments, is known as the *helicotrema.* At the junction between the scala vestibuli and the middle ear is a flexible membrane, called the oval window, to which the stapes is attached. Where the scala tympani joins the middle ear there is another flxible membrane called the round window. When the oval window is pushed inward by the stirrup, the round window is pushed outward by the fluid of the inner ear (which is incompressible). Similarly, an outward motion of the oval window gives rise to an inward motion of the round window. Through the complicated mechanical linkage of the middle ear, then, air vibrations are transformed into vibrations of the fluid in the inner ear.

Energy Conversion and Encoding

The cochlear partition is actually a fluid-filled duct, as shown in Fig. 3.A-3. We have seen that a sound wave impinging on the ear will produce vibrations in the cochlear fluid. Since the cochlear partition is not an inflexible divider, it will respond to the vibrations of the fluid by acquiring induced mechanical vibrations of its own.

The conversion of the mechanical vibrations of the cochlear partition into electrical impulses traveling to the brain is accomplished by the organ of corti which lies on the basilar membrane in the lower part of the cochlear partition. The organ of Corti consists of about 30,000 small cells, called hair cells, which are attached to the nerve transmission lines going to the brain. The other ends (the "hairy" ends) of the

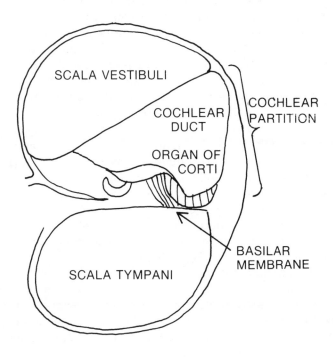

Fig. 3.A-3. Simplified cross sectional view of cochlea.

hair cells are embedded in the tectorial membrane, as shown in Fig. 3.A-4. When the basilar membrane vibrates, the "hairs" of the hair cells are bent, which produces electrochemical pulses, thus causing a set of nerve impulses which travel to the brain and which we interpret as sound.

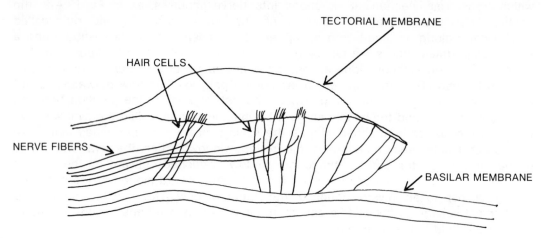

Fig. 3.A-4. Simplified details of organ of Corti.

Frequency information is encoded in part by the region along the cochlear partition that undergoes the largest motion when vibrations occur in the cochlear fluid. Incident waves of different frequencies cause different regions of the cochlear partition to respond. If a high-frequency sinusoid strikes the ear, the area of the cochlear partition adjacent to the oval window responds most strongly. If a low-frequency sinusoid is incident, the far end (near the helicotrema) of the partition responds most vigorously. Georg von Bekesy won a Nobel prize for obtaining experimental evidence from actual mammalian cochleas that demonstrated this behavior. He was able to observe actual traveling waves on the cochlear partition: the place of maximum amplitude of the wave was dependent on the frequency of the excitation. Fig. 3.A-5 portrays simplified computer simulations of the response of the cochlear partition to sinusoids of different frequencies. The diagrams show "time exposures" of several cycles of vibration; so we do not see the wave at each instant, but only an overall wave envelope which shows the point of maximum amplitude.

The encoding of intensity of a sound wave is determined by the number of nerve impulses produced each second. Any hair cell must be stimulated by some minimum vibration amplitude or it does not produce a nerve impulse (firing). When stimulated by large-amplitude vibrations (resulting from intense sounds) a hair cell fires more often than with smaller-amplitude vibrations. (On the average, each hair cell is limited to a maximum of about 1000 firings each second. Also, more intense sounds cause a greater extent of the cochlear partition to vibrate, which results in the firing of still more hair cells. Simplified representations of the encoding of frequency and intensity information by the ear is shown in Fig. 3.A-6a,b.

Some frequency information is also encoded in the repetition rate at which packets of hair cells fire. When a complex sound (consisting of a variety of frequencies) strikes the ear, many different regions of the basilar membrane are excited. Fig. 3.A-6c is a simplified representation for the encoding of frequency, amplitude, and spectrum of a complex tone. The many frequencies of a complex tone not only stimulate different regions of the basilar membrane, but for each region of the membrane stimulated, the same overall nerve pulse repetition rate results. That is, each region of the basilar membrane which is stimulated sends out nerve impulses in a sort of

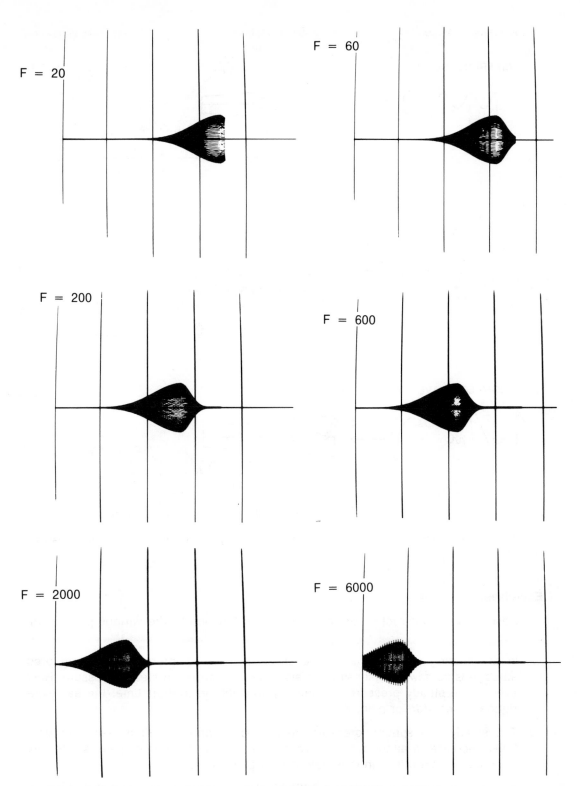

Fig. 3.A-5. Simplified computer simulations of the response of the cochler partition to sinusoids of 20, 60, 200, 600, 2,000, and 6,000 Hz. The grid is marked off in 1 cm increments; the whole "uncoiled" cochlea is about 3.5 cm long. The oval window is on the left in each diagram.

morse code, with different codes being used for each partial, but with the codes having the same repetition rate. The brain then receives the collection of codes illustrated in Fig. 3.A-5c and interprets it in some fashion as a complex tone.

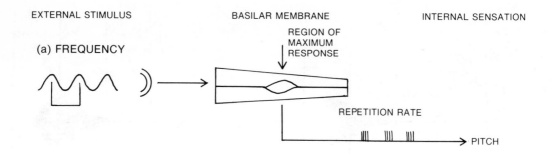

(a) FREQUENCY

EXTERNAL STIMULUS   BASILAR MEMBRANE   INTERNAL SENSATION

REGION OF MAXIMUM RESPONSE

REPETITION RATE

PITCH

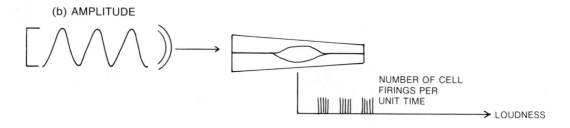

(b) AMPLITUDE

NUMBER OF CELL FIRINGS PER UNIT TIME

LOUDNESS

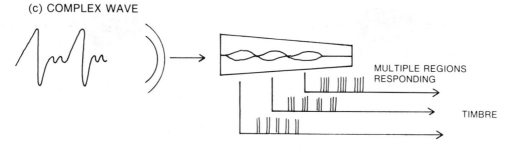

(c) COMPLEX WAVE

MULTIPLE REGIONS RESPONDING

TIMBRE

Fig. 3.A-6. Simplified representation of the encoding of frequency, amplitude, and complexity of waves by the ear.

Exercises

1. Describe the construction of and discuss the function of the various parts of the ear.

2. In the hearing process the ear senses and helps to interpret oscillating pressures. It is somewhat insensitive to slow, small changes in steady pressure. However, if the steady pressure changes appreciably in a short time, the ear experiences discomfort or pain. Why?

3. The steady atmospheric pressure decreases by about 1/30 in value for each 300m increase in altitude. Does this explain why discomfort or pain is often experienced in aircraft or in mountain climbing? Explain.

4. Assume the outer ear canal is a cylindrical tube, open at one end and closed at the other, with a length of 3 cm. Calculate the resonant frequency of this tube.

5. If you consider the ear to be a mechanical system, what is probably the nature of its response to a transient acoustic pulse?

6. Discuss ways in which the ear and a microphone are analogous in terms of their respective inputs and outputs.

7. How are nerve impulses transmitted to the brain?

8. When sine waves strike the ear, what parts of the cochlear partition respond most strongly to high, medium, and low frequencies? Show this on a sketch of an "uncoiled" cochlea.

9. How do the encoding of frequency and intensity information differ? Discuss the mechanisms.

10. Does the maximum rate of hair cell firings place any limits on the sending of frequency information or of intensity information to the brain? Discuss.

11. Prior to the "place model" of hearing discussed in this section, there was an accepted "resonance model" of hearing. According to this model the cochlear partition acted like a group of many small piano strings, each string having a different resonance frequency. An incoming sinusoid would cause the corresponding string to resonate. Complex waves would excite several strings; thus the frequency components of the wave would be resolved. Criticize this model. (Hint: More information on this model is contained in the book by Denes and Pinson.)

12. Do the "resonance" and "place" models of hearing (see exercise 11) deal pri-
Are the positions of maximum displacement distributed linearly or logarithmically relative to frequency?

13. Qualitative mechanical behavior of the inner ear.
Do the following by referring to Fig. 3.A-5.
    a. Determine the distance from the oval window at which maximum displacement occurs for each frequency.
    b. How well do the values you have determined agree with those in the published literature?
Are the positions of maximum displacement distributed linearly or logarithmically relative to frequency?

Further Reading
Gerber, chapter 1
Denes and Pinson, chapter 5
Flanagan, chapter 4
Stevens and Warshofsky, chapters 2, 3
Fletcher, chapters 7, 14
Roederer, chapter 2

Audiovisual
1. *Ears and Hearing,* 2nd edition (22 min, color, 1969, EBE)
2. *Simulated Basilar Membrane Motion* (11 min, bw, 1966, GSF)

Demonstrations
1. Mechanical model of the ear

3.B SUBJECTIVE SENSATIONS AND THEIR PHYSICAL BASES

In the previous section we mentioned the mechanism of hearing. We now attempt to relate our subjective interpretations of sound to actual physical attributes of the sound wave. It has often been noted that the eye is a remarkably precise instrument, and it is. In many ways, however, our auditory processes are even more remarkable and more sensitive than the eye. Consider first that the ear can detect frequencies ranging between 16 and 16,000 Hz, a frequency range of approximately 10 octaves. The eye, on the other hand, responds to light having a frequency range of barely one octave. Furthermore, as we will see at the end of this section, even sound frequencies beyond our direct auditory perception can exert influences on us. The range of intensities to which the ear can respond is equally phenomenal, and once again the eye must be relegated to second place. Finally, the actual pressure variations to which the ear can respond are almost indescribably minute. For instance, if you stand one meter away from a workman hammering on a steel plate, you will be subjected to a very intense sound field—almost at the threshold of feeling. If this energy were converted to electrical power, however, it would be barely enough to run a one-watt light bulb. When the ear responds to the very softest sound, sound on the threshold of hearing, the eardrum executes an excursion which is only a fraction of the diameter of the air molecules striking it, and the displacement of the basilar membrane is only about one-tenth of this. As a matter of fact, if our ears were any more sensitive than they are—that is, if they responded to even softer sound—we would hear a permanent soft buzzing sound because of the random thermal motions of the air molecules.

In chapter 2 we considered objective physical characteristics of sound waves, such as frequency, sound level, and spectrum. Let us now consider the corresponding subjective quantities which are perceived by a listener when sound waves enter his ear. The three subjective quantities which correspond most closely with the three physical characteristics given above are pitch, loudness, and timbre. We do not mean to imply that pitch, for instance, depends only on frequency, or that loudness depends only on sound level—because they do not. Nor do we mean to imply that there is a linear relationship between the physical and subjective quantities—i.e., that doubling the intensity doubles the loudness—for that most assuredly is not the case. The actual relationships, as we will discover, are actually much more complicated and subtle.

Loudness and Intensity

Let us begin by considering the relation between loudness and intensity (or sound level). As was pointed out in chapter 2, the sound level of a sound wave is conveniently expressed in decibels, 0 dB being the sound level of a pressure of 0.00002 N/m^2. The 0-dB sound level corresponds to just audible sound (under quiet conditions) for an "average person" when listening to a frequency of 1000 Hz. By presenting tones having different frequencies and sound levels to listeners, the threshold of hearing can be determined at different frequencies. Results are averaged for many listeners having "normal" hearing to obtain the curves shown in Fig. 3.B-1. The threshold curve (lowest curve of Fig. 3.B-1) is, in fact, the sound level which is just barely perceived at each frequency. For just audible sound at a frequency of 100 Hz the level must be about 40 dB and about 60 dB at 40 Hz.

In order to differentiate loudness level from sound level a new unit, the *phon* (rhymes with *John*) is used to specify loudness level. Sounds having the same phon level would sound equally loud regardless of their frequencies. "Zero phons" is defined as the threshold of hearing for an "average" person. A loudness level of 10 phone is defined as the loudness level produced by a 1000-Hz tone at a sound level of 10 dB. Likewise, the loudness level (in phons) of any 1000-Hz tone is defined as being numerically equal to its sound level (in dB).

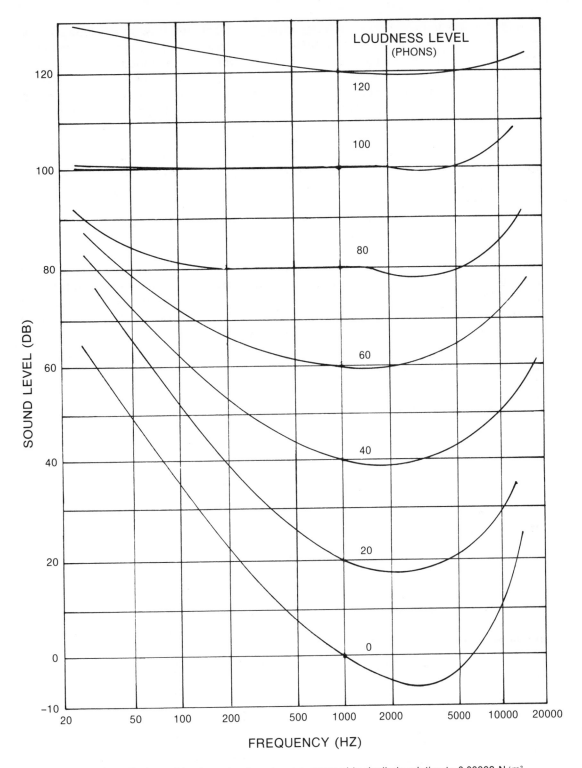

Fig. 3.B-1. Simplified equal loudness level contours expressed in decibels relative to 0.00002 N/m².

By further psychoacoustic experimentation, tones of various frequencies are matched in loudness level. The results are the equal loudness level contours shown in Fig. 3.B-1. The sound level is plotted along the vertical axis and frequency along the horizontal axis. The contours shown are the lines of equal loudness level (given in phons). Thus from this graph sounds of various frequencies and levels may be compared and their relative loudnesses determined (i.e., we can predict which of two sounds will be louder, but we cannot from this graph tell how much louder). You may notice from the graph that as higher phon levels are approached the contours seem to flatten out to some degree. The curve which corresponds to 120 phons is called the threshold of feeling, because these sound levels produce a tickling sensation in the ear. At still higher levels (135 dB) the sensation is one of pain. Note that the equal loudness level contours are not straight lines; this is because loudness level depends on frequency as well as intensity, and not all equally intense sound are equally loud. The dependence of loudness on duration is discussed later. The shape of the threshold (0 phon) curve may be the result of a number of conditions. The resonance of the ear canal around 3000 Hz makes the ear more sensitive at frequencies between about 2000 and 4000 Hz. A sparser density of hair cells in the low-frequency part of the cochlear partition makes the ear less sensitive at low frequencies. A filtering effect of the combined middle ear and cochlea makes the ear less sensitive at high frequencies. Loudness level tells us which of two sounds is louder, namely the one having a greater loudness level.

However, loudness level does not specify how much louder one sound is when compared to another. In order to describe how many times louder one sound is than another, a new quantity, called the *sone*, was developed. The sone is defined so that when the perceived loudness doubles, the loudness (expressed in sones) also doubles. The relationship between phons and sones is given in Table 3.B-1.

Table 3.B-1. Relationship between loudness levels and corresponding loudness. (Loudness level is in phons and loudness in sones with a loudness level of 40 phons arbitrarily assigned a loudness of 1 sone.)

Loudness level (phons)	Loudness (sones)
20	.08
30	.3
40	1
50	2.8
60	6
70	12.5
80	25
90	45
100	80
110	150
120	250

Table 3.B-2. Relationship between some frequencies and the corresponding pitches for sinusoids. (A frequency of 1000 Hz at a sound pressure level of 40 dB is arbitrarily assigned a pitch of 1000 mels.)

Frequency (Hz)	Pitch (mels)
10	0
20	20
50	75
100	150
200	300
500	600
1000	1000
2000	1500
5000	2300
10000	3000

Pitch and Frequency

Frequency is measured in Hz. Our perception of frequency is pitch, and since there is not a one-to-one relationship between frequency and pitch, a new unit, the *mel,* is defined for pitch. The mel then plays the same role for pitch that the sone plays for loudness—that is, the mel is defined so that doubling the number of mels represents a doubling of the perceived pitch, regardless of the change in frequency. The relationship between perceived pitch and frequency for sinusoids is shown in Table 3.B-2. The mel is defined so that a pitch of 1,000 mels is produced by a 1,000-Hz tone at a level of 40 dB. Note that when the frequency doubles to 2,000 Hz the pitch has not quite doubled, but has increased to 1,500 mels. The pitch of pure tones depends to some extent on their loudness, with louder tones having somewhat lower pitches (as much as 10% lower for a 150-Hz tone of 4 sones). The dependence of pitch on duration is discussed later. It should be pointed out that for complex tones having many frequency components there is more nearly a one-to-one correspondence between pitch and frequency than for sinusoids.

Tone Quality, Timbre, and Spectrum

When considering tones from different musical instruments each of which has the same pitch and the same loudness as the other, we are able to distinguish among them on the basis of tone quality or timbre. The foregoing terms are often used interchangeably, but we will be somewhat more restrictive in our definitions, following the lead of Seashore.

We define *timbre* to be that attribute of a steady state tone which permits it to be distinguished from other tones having the same loudness and pitch. Timbre depends on the spectrum of a tone, which can be characterized in different ways. (Timbre may also depend to some extent on the relative phases of the partials, although the ear is insensitive to phase except in extreme cases.) Harmonic spectra in which the partial frequencies are integer multiples of the fundamental frequency exist for tones that are periodic. Harmonic spectra can be characterized by specifying the frequency and sound level of each partial. This would be an especially convenient characterization if the harmonic structure were approximately the same for all notes of an instrument (i.e., if the harmonic partials had the same intensities relative to each other). However, in examining spectra of the voice and of musical instruments at many different fundamental frequencies, a spectral description of fixed relative intensities among the partials does not hold.

A "formant model" is an alternative to a "harmonic model" for describing musical tone spectra. A *formant* is a pronounced peak (either narrow or broad) at an approximately fixed frequency position in a spectrum. The formant spectral model claims that the spectrum for a particular instrument is characterized by one or more formants (or regions of energy) at appropriate frequency positions for that instrument. In this model the partial levels do not maintain the same relative values for different notes; the partials that take on the largest values are those that lie within a formant. For instance, if an oboe has a formant at 1000 Hz its fourth partial would be most intense when the oboe is sounded at a frequency of 250 Hz, but when sounding at 500 Hz the second partial would have the highest level. The formant model of spectra appears to be more nearly adequate than the harmonic spectrum model. It provides a very good description of speech spectra and has found considerable application in the description of musical instrument spectra. (See, for example, the references in sections 5.C and 6.A.)

Tone quality encompasses aspects of successive changes and fusions of timbre, pitch, and loudness of a tone. Timbre, pitch, and loudness may be thought of as characterizations of an instantaneous snapshot of a tone, while tone quality results from the progression and changes of these instantaneous pictures and could be considered analogous to the effect of stringing a series of still photographs together to get a movie. Transients in musical tones also play a significant role in determining tone quality. Attack transients, the beginning of a musical tone, play a particularly important role in the identification of nonpercussive musical instrument tones (see section 6.A). For percussion tones the decay transient, the conclusion of the tone, is

93

more important. Uncertainty in musical tones due to vibrato, tremolo, and choral effects are also important parts of tone quality. Other aspects of quality might well include tone duration, rate, and rhythm. If you are searching for a clear and concise statement about tone quality, however, we have none to offer: the area of quality is still defined only in rather nebulous terms. All we can do at present is to speak vaguely of a sound being oboe-like, or "ee"-like, etc. An attempt to quantify a particular tone quality by specifying the appropriate transients and all the time-varying spectra of this musical tone, while a very complex operation in its own right, would still be doomed to failure by reason of its being overly simplistic. Some additional aspects of this complicated subject will be discussed in section 6.A.

Ultrasound and Infrasound

The preceding discussion has been concerned with audible sound (from 16 to 16,000 Hz). But ultrasound, having frequencies above 16,000 Hz, and infrasound, with frequencies below 16 Hz, can influence living organisms, even though we can't hear these sounds. Ultrasonic sound waves can, however, be heard by many animals. Some food processing plants are even using an intense ultrasonic generating device, audible only to rodents, as a "rat repellent." The applications of ultrasonics are wide ranging and many areas of science and technology benefit from this tool. High-frequency sound is important for two reasons: (1) these high-frequency vibrations can be focused, and (2) very high sound intensities can be generated easily. Industrial uses of ultrasonic radiations range from cleaning and drilling to welding. Ultrasonics can also be used to locate flaws in metals, or as a sheet metal thickness gauge. These waves have also been used as burglar alarms and as a substitute for chemical pesticides. Some of the applications in biology and medicine include localized internal hearing, destruction of bacteria, and cancer therapy. In spite of all the useful applications, however, we know far too little about the possibly harmful effects of ultrasound on humans.

For frequencies below 16 Hz we enter the still somewhat mysterious region of infrasound. Although we cannot hear these sounds we can still sense them by feeling. Infrasonic energy comes to use from many sources, such as thunder, sonic booms, or vibrations from heavy vehicles. The applications for infrasound seem to be mostly destructive and sinister rather than constructive or useful. Recent studies suggest that because of the sheer power of these waves they could provide an alarming new weapon of war, literally shaking people and property to pieces.

Military weapons research has indicated that very low frequencies of sound having high intensity can produce profound physical disturbances inside the human body. Vibrations of 8 Hz at a level of about 130 dB can cause a person's chest walls to vibrate by resonance. This vibration produces internal pain from the intense friction caused by the stomach, heart, and lungs rubbing against each other. Other experiments which dealt with sound pressure levels between 95 and 115 dB in the frequency range from 2 to 16 Hz showed that exposure times between 5 and 15 minutes were sufficient to produce a sense of ill-feeling, including dizziness and loss of balance. At a frequency of 7 Hz violent and sudden nausea was generally produced at somewhat lower pressure levels. The above symptoms indicate a disturbance of the organs of balance, located in the inner ear. In fact, sounds in the range from 2 to 20 Hz are within the natural resonance frequency range of the semicircular canals, thus causing large oscillations in the fluid.

The hazard of infrasound is that it is produced, often at a considerable intensity, in a wide range of man-made environments. Reports show, for example, that a car traveling at 80 km/hr will generate fairly intense infrasonic waves at frequencies below 16 Hz. The infrasound, which the driver may not notice, could produce an effect similar to drunkenness and may be the explanation for many otherwise inexplicable car accidents.

Perhaps even more ominous are the possible effects on people of naturally produced infrasound in the atmosphere. Thunderstorms are known to produce significant levels of infrasound; perhaps this accounts for the bad tempers that thunderstorms arouse in some people. It is also known that before an earthquake some people and many animals experience feelings of unease. Can these psychological disturbances be attributed to the infrasonic waves generated as a precursor to the earthquake? Although we cannot answer such questions with any degree of certainty, the limited experiments performed to date indicate that infrasound, in addition to the physical effects mentioned previously, also has a very profound psychological effect. One study, for instance, did find a statistically significant relationship between the presence of strong infrasonic waves, generated by natural phenomena, and increases in absenteeism and car accidents. Perhaps, then, many behavioral problems and unpleasant feelings of malaise previously blamed on the climate may be due to infrasound. Since our mechanized society is continuously producing infrasound, and because people are being exposed to more of it, these preliminary findings warrant more detailed study.

Exercises

1. What intensity level is required to produce just audible sound for frequencies of 50, 100, 500, 1000, 5000, and 10,000 Hz?

2. Why is a lower intensity level required to produce just audible sound at 3,000 Hz than at other frequencies? What does the mechanism of the ear, discussed in section 3.A, have to do with this effect?

3. What is the meaning of "threshold of feeling" and "threshold of pain"?

4. What is the lowest audible frequency for an "average" person? What is the highest audible frequency for an "average" person?

5. Will an 80-phon loudness level be twice as loud as a 40-phon loudness level? Will an 80-sone loudness be twice as loud as a 40-sone loudness?

6. What pitch (measured in mels) must a simple tone have to have a pitch double that of a 100-mel tone? What frequency must a tone have to have double the pitch of a 100-mel tone?

7. How much greater sound level (in dB) must a 100-Hz tone have to sound as loud as a 1,000-Hz tone if the 1,000-Hz tone is at a level of 0, 40, or 80 dB?

8. What is the range of sound levels between threshold of hearing and threshold of feeling to which the ear responds at frequencies of 40 Hz; 100 Hz; 1,000 Hz; 10,000 Hz?

9. If the sound level rises from 40 dB to 80 dB at a frequency of 200 Hz, what is the change in loudness (expressed in sones)?

10. Fill in the following table.

Frequency	dB	Phons	Sones
100		20	
100		40	
100		60	
1000	40		
10,000	60		
10,000		60	
10,000			60

11. Suppose that sounds from a band at the following frequencies produce the sound levels shown when the band is 4 m from a listener. When the band is 128

m from the listener, what intensity levels are produced at the listener's position? (Assume that each doubling of the distance decreases the level by 6 dB.) Which sounds are audible if we assume that they become inaudible at the 20-phon level? (This shows how "band color" can change as a function of distance from the listener.)

Frequency (Hz)	Level (in dB at 4 m)	Level (in dB at 128 m)	Audible?
100	80		
200	70		
500	70		
1000	60		
5000	60		
10,000	50		

12. The text discusses loudness as being primarily dependent on intensity. However, loudness depends also on frequency and spectrum. Discuss the dependence of loudness on frequency using the equal loudness contours of Fig. 3.B-1. How might loudness depend on spectrum?

13. In section 2.F sound levels as a function of the distance from the source were discussed. An approximate rule of thumb is that each doubling of the distance decreases the intensity level by 6 dB, or halving the distance increases the level by 6 dB. Suppose at a distance of 10 meters you measure a level of 60 dB for a 100-Hz tone. What is its loudness in sones? If you now move to a distance of 200 meters, what level might you measure? What is the loudness at the distance of 200 meters?

14. Why does a dog react to the blowing of a dog whistle, while humans hear nothing?

15. Explain how the focusing of ultrasonic waves can be used to produce localized heating within a person's body.

16. How can infrasound make you get sick?

17. How does infrasound warn animals of an impending earthquake?

18. Could infrasound kill you? In particular, consider how intense infrasonic waves could cause internal bleeding.

19. If a tube open at each end were to be used for producing infrasonic waves in air at a frequency of 8 Hz, how long would the tube have to be?

*20. Comparison of subjective and objective measures for hearing
   a. 1,000-Hz sinusoids were produced through a loudspeaker at a 60-dB level. What was the loudness level? What was the loudness of the tone? The sound level was increased to 90 dB. What was the loudness level? What was the loudness? Was the second sound half again as loud as the first?
   b. The amplifier gain was adjusted until a tone was produced that sounded twice as loud as the first one. What was its loudness? What was its loudness level? What was its sound level?
   c. Function generators were used to produce a sawtooth wave and a square wave of 300 Hz. Their intensity levels were adjusted so that they sounded equally loud to a listener who also observed that they had the same pitch. However, the observer said that the two waves did not sound the same. Why might this be so? Use your knowledge of the spectra of the two waves in giving your explanation. (Refer to the spectra in unit 1.E.)

Further Reading

Backus, chapters 5, 6, 7

Benade, chapters 5, 13, 14

Chedd, chapters 5, 6

Denes and Pinson, chapters 5, 6

Fletcher, chapters 8, 10, 11, 12, 14

Gerber, chapters 5, 6, 7, 9

Roederer, chapters 2, 3, 4

Seashore, chapters 2, 5, 6, 7, 8, 9

Stevens and Warshofsky, chapters 4, 5

White, chapters 5, 6

Winckel, chapter 2

Brown, R., J. Harlen, and J. Gable. 1973. "New Worries about Unheard Sound," New Scientist *60* 414–16.

Anastassiades, et al. 1973. "Infrasonic Resonances Observed in Small Passenger Cars Travelling on Motorways," J. Sound and Vibrations *296*, 257–59.

Evans, M. and W. Tempest. 1972. "Some Effects of Infrasonic Noise in Transportation," J. Sound and Vibration *22* (1), 19–24.

Audiovisual

1. *The Science of Sound* (Album #FX6007, 1959, FRS)
2. *Descriptive Acoustics Demonstrations* (Tape section 3.b, 1977, GRP)
3. *Sound Energy and Hearing* (30 min, 1957, EBEC)

Demonstrations

1. Function generator connected to loudspeaker
2. Waveforms of various musical instruments displayed on oscilloscope.

3.C. ADDITIONAL ASPECTS OF SOUND PERCEPTION

As we have seen, our ears (in conjunction with the associated analysis system of the brain) can detect and interpret a considerable variety of sounds varying over substantial ranges of sound level and frequency. Although physicists are generally interested in describing the physical characteristics of sounds, the resulting phys- iological effects and psychological sensations are particularly important to the under- standing of sound. In the previous section we explained the correlation between the three main physical characteristics of sound waves (frequency, amplitude, and spec- trum) and their reception by the human. In the present section we will consider sev- eral additional features of sound perception which, although subtle, play an extreme- ly important role in understanding and interpreting the otfen complex maze of sonic information which bombards our ears.

Binaural Effects

Because normal hearing is binaural (two ear), we have an ability to locate a sound in space which is similar to our visual depth perception which results from having two eyes. This *localization* of sound means that our two ears are able to tell the di- rection (in a horizontal plane) from which a sound originates. In the vertical direc- tion, however, our ability to localize sound is less accurate. In normal hearing situa- tions the sound reaching one ear differs slightly from the sound impinging on the other ear because the ears are separated by a hard sphere (the head) with a diame- ter of about 20 cm. Consequently, when we listen to pure tones there may be slight differences of sound intensity and of the time at which each ear receives a corre- sponding portion of the wave. Fig. 3.C-1 shows how sound localization for pure tones depends on frequency. The y-axis represents the average error in degrees. No- tice that man can locate high and low frequencies rather well, but there is a curious hump (between 2,000 and 4,000 Hz) where larger errors are made. The theory to ex- plain these results follows.

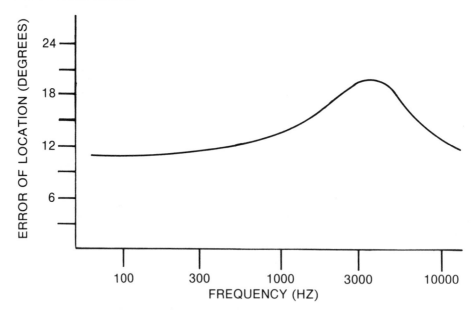

Fig. 3.C-1. Location of pure tones by man. (After Steven and Newman 1934.)

Low-frequency tones, having a long wavelength in comparison to the size of a head, diffract around the head, while wavelengths substantially smaller than the head do not diffract appreciably. When the sound wave does not diffract there is an ap- preciable ''sound shadow'' due to the head, which results in the sound being slightly less intense to the ear within the shadow. Since the magnitude of this intensity difference between the two ears varies with the direction of sound propagation, we are able to localize the source of sound. The shorter the wavelength (relative to the head), the more distinct the sound shadow created. For low-frequency tones there is no appreciable sound shadow, and consequently differences of intensity at each ear are negligible. For long wavelengths, however, the slight time delay between the reception of a particular portion of a wave by one ear and its reception by the other ear becomes important. Because of the spatial separation of the two ears, a long wave will have a slightly different phase at one ear than at the other. For each frequency, the magnitude of this phase difference varies according to the position of the sound source, and thus we can localize it. When high-frequency sounds appear equally intense to each ear, or when low-frequency sounds arrive at each ear simultaneously, the apparent location of the sound is directly in front of or behind the listener. The above considerations show that since intensity comparisons help localize high frequencies and phase comparisons help localize low frequencies, there will be a frequency range in which neither method will be very effective; thus the hump in Fig. 3.C-1.

The localization of sound is of considerable importance when a signal is detected in the presence of other interfering sounds. Almost everyone has had the experience of listening to one conversation at a noisy party while excluding the irrelevant background noises. That this effect (sometimes termed the "cocktail party effect") is related to our ability to localize sound can be demonstrated by a simple experiment. If you record a monaural tape of a friend talking to you at a loud party, on the playback you most likely won't even be able to hear him, much less understand his words. Our ability to localize enables us to "focus" our hearing on a particular conversation coming from a specific direction and to discriminate against other sounds. In order to achieve this effect, however, the listener must decide which sound he wishes to hear and then concentrate upon it, thus introducing a new psychological concept, "attention," as a hearing parameter. In listening to music we use our ability to localize sound in order to listen selectively to certain instruments. We may decide to concentrate on the sound of a solo violin, even though it is much quieter than the concurrent sound of the brass section. Stereophonic sound systems, by reproducing a binaural effect, restore the listener's sense of "presence" at a musical performance by enabling him to listen selectively to the music.

Masking and Critical Bandwidth

Although we can use our ability to localize sounds as an aid to hearing a conversation in spite of irrelevant background noises, there are some situations in which background noise cannot be effectively screened because of the phenomenon of *masking*. For example, we have all probably experienced the difficulty of trying to hear someone speak when we are surrounded by noisy machinery. The reason for the decrease in ability to "hear" in a noisy environment is that the sound of the noise drowns out, or *masks*, other sounds. Experiments conducted on the masking effect of noise on pure tones show an approximately linear relationship between masking and noise level (the greater the noise, the greater the masking). Fig. 3.C-2 shows the amount of masking (the increase of sound level needed for detection over that required when there is no noise) vs. the noise level in dB. The masking occurs over the entire range of audible frequencies, but since noise can be considered as many unrelated sinusoidal components sounding together, we can consider the masking to be due to many simultaneous pure tones. As might be expected, the most effective maskers are the noise components closest in frequency to the masked tones.

Masking experiments with sinusoids help us understand sound perception and the capabilities of the inner ear. Three of the most important results of these masking experiments are: (1) tones close together in frequency mask each other more than tones widely separated in frequency; (2) low-frequency tones mask high-frequency

99

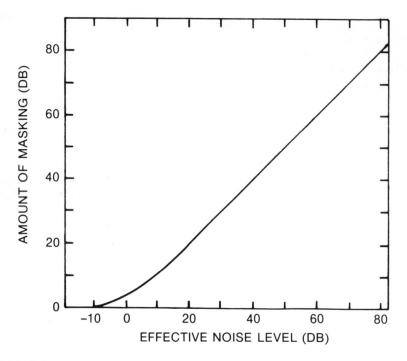

Fig. 3.C-2. Relation between noise level and amount of masking. (After Hawkins and Stevens, 1950.)

tones more than vice versa; and (3) the greater the intensity of the masking tone, the broader the band of frequencies for which masking is evident. These masking results can be explained by considering the manner in which the basilar membrane responds to different frequencies. In Fig. 3.C-3 we consider the qualitative response of the membrane to different frequencies. Looking at the upper part of the diagram, note that tones widely separated in frequency do not stimulate the same part of the basilar membrane and thus do not interact and mask each other appreciably. In the lower diagram it can be seen that tones close together in frequency stimulate the same part of the basilar membrane, thus interacting and masking each other. Note in the lower diagram that a low frequency stimulates the same part of the membrane that responds maximally to high frequency, whereas the high-frequency pattern does not extend into the region of largest response of the low frequency. This is because the stimulation pattern of the membrane decreases more steeply on the low-frequency side than on the high-frequency side. Because the low-frequency stimulation region extends into the high-frequency region, a low-frequency tone masks nearby high-frequency tones more effectively than vice versa. Intense masking sounds cause a larger region of the basilar membrane to be stimulated than less intense sounds do. This stimulation of a larger region on the basilar membrane results in the masking of a broader range of frequencies by more intense sounds.

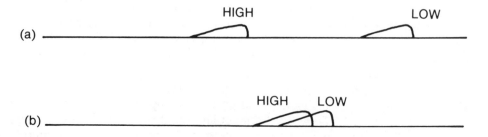

Fig. 3.C-3. Qualitative response of basilar membrane when (a) tones are widely separated in frequency and (b) a low frequency is masking a high frequency.

100

The *critical bandwidth* is an idea associated with the masking of one sound by another. In essence, this critical bandwidth is the frequency difference within which tones interact and affect each other appreciably; if they are farther apart in frequency than this critical bandwidth, they have little effect on each other. Fig. 3.C-3 shows how the existence of critical bandwidths can be explained in terms of the behavior of the basilar membrane. Tones that are relatively close together in frequency will cause responses on the basilar membrane which overlap to some extent. When the tones are separated enough in frequency so that there is no appreciable interaction on the basilar membrane, they are separated by more than a critical bandwidth. Table 3.C-1 gives some critical bandwidths for various frequencies.

Table 3.C-1. Critical bandwidths for different frequencies.

Frequency (Hz)	Critical bandwidth (Hz)
100	90
200	90
400	100
1000	150
2000	280
4000	650
10,000	1200

The phenomenon of masking is often observed in music performance. For example, one instrument can mask a second instrument playing the same note if the first instrument is played more loudly. On the other hand, when the fundamental frequencies are substantially different (e.g., a piccolo and a tuba playing simultaneously) we can easily distinguish each sound.

Nonlinear Systems and Combination Tones

Until now all the vibrating systems we have considered have been linear, meaning that the response of the system is directly proportional to the input (or stimulus). For example, if we were to graph the amount of displacement (response) versus the force applied (input) to a mass and spring system we would get a straight line. This would show us that the relation between displacement and force is linear (or one of direct proportionality): doubling the force doubles the displacement of the mass. Many vibrating systems, particularly in music and speech, are slightly nonlinear, and in some instances very nonlinear. Imagine, for example, that we find our mass and spring system getting progressively "stiffer" when displaced by large amounts—that is, more force is required to produce an additional unit of displacement when the displacement is already large than when it is small. The displacement is no longer directly proportional to the force; this mass and spring system is now a nonlinear system.

When linear systems are driven by applied sinusoidal forces, only those frequencies associated with the applied sinusoids appear in the motion of the system. An interesting property of non-linear systems is that frequencies other than the applied frequencies may appear in the motion of the system when it is driven. Typically, the greater the driving force applied to a nonlinear system, the more apparent the additional frequencies become. This has some interesting implications for the ear. If the ear were a perfectly linear system we would hear only those frequencies supplied to it. Because the ear is somewhat nonlinear, however, we perceive frequencies in addition to those supplied. When two sinusoids of different frequency are presented to the ear, a new tone, called a *combination tone,* may be heard in conjunction with the original tones. The combination tone is due to the nonlinear action of the ear. For example, when two frequencies, f_1 and f_2 are supplied to the ear, combination tones such as $f_2 - f_1$, $2f_1 - f_2$, etc., known as difference tones, and $f_1 + f_2$, $2f_1 + f_2$, etc., known as summation tones, may be perceived. What nonlinerity causes the

101

combination tones? Formerly it was believed that the middle ear was nonlinear, but it has been shown that no appreciable nonlinearity exists in this portion of the ear, so that any nonlinearity must occur in the inner ear. A study of direct neural pulse measurements has established that there are regions on the basilar membrane which are activated at positions corresponding to the frequencies of combination tones. The combination tones are thought to be caused by a distortion of the wave form in the cochlea itself. Although summation tones have been assumed to exist for many years, very few researchers have been able to detect them. It may be that they get masked more severely than do the difference tones. Difference tones like $2f_1 - f_2$ are not only heard, but heard at levels in excess of those expected based on the non-linear explanation.

Difference tones have been used to explain the phenomenon of perceiving a pitch associated with a missing fundamental frequency. For example, suppose we have a moderately intense complex tone having components of 400, 600, and 800 Hz. If we consider only difference tones produced by nearest neighbor pairs we get a difference tone of 200 Hz from two pairs: $600 - 400 = 200$ Hz and $800 - 600 = 200$ Hz. But 200 Hz is just the fundamental frequency to which the 400-, 600-, and 800-Hz components are higher partials. In fact, the perceived pitch of this complex tone corresponds to a frequency of 200 Hz, even though there is no 200-Hz component in the original tone.

More recent research suggests that the missing fundamental is perceived even when the higher partials are presented at a sound level sufficiently low that no difference tones are produced. A possible explanation for the perception of the missing fundamental is that in a complex tone we perceive the *periodicity* and not the lowest frequency as the fundamental. Periodicity is a measure of how often a wave form repeats each second. (See Fig. 3.A-6 for a simplified view of the periodicity of the nerve impulses coming from the cochlea.) For example, if a complex tone has frequencies of 400, 600, and 800 Hz, the tone will repeat itself 200 times per second, even though there is no frequency of 200 Hz present. Hence, if the ear perceives pitch on the basis of periodicity it will perceive a pitch associated with 200 Hz as the fundamental, even though a 200-Hz component is not present. Another interesting and common example of the perception of pitch associated with a missing fundamental frequency is found in telephone speech. Suppose a male voice has a fundamental frequency of about 125 Hz with the associated higher partials of 250 Hz, 375 Hz, 500 Hz, 625 Hz, etc. The telephone eliminates frequencies below about 300 Hz, which in this case leaves frequencies of 375, 500, 625, etc. However, we hear a pitch corresponding to the fundamental of 125 Hz even though it is not transmitted. Male telephone speech does not appear to have a pitch different from that perceived in face to face conversation. Because of the limited frequency range transmitted, however, the quality of telephone speech may be different.

Differential Thresholds

In section 3.B we discussed the absolute threshold of hearing for pure tones of various frequencies. The threshold of hearing is the minimum level at which a sound stimulus creates the sensation of sound to a listener. We are now going to examine another threshold, the *differential threshold*. The differential threshold is the smallest change in stimulus which a listener can detect. For convenience, we will limit our discussion to pure tones and consider the minimum detectable change in intensity (when frequency is constant) and the minimum detectable change in frequency (when intensity is constant). The differential threshold is often called a *just noticeable difference* (or JND) and, like the other subjective characteristics of sound, it is not a constant. The JND depends upon both the frequency and the intensity of the tone for which it is determined.

Suppose two tones having the same frequencies but slightly different intensities are sounded one after the other. If the intensity difference is large enough the tones can be distinguished, but if it is sufficiently small the tones will sound identical. JNDs for sound level are shown in Table 3.C-2. Note that the JNDs range from as little as 0.2 dB for tones of 2,000 Hz at 80 dB to 4-6 dB for low-frequency tones presented at low levels.

Table 3.C-2. JNDs for intensity level (dB) of pure tones as a function of frequency (Hz) and level above threshold (dB).

Level (dB)	Frequency (Hz)							
	40	100	200	400	1000	2000	4000	10,000
10	5.5	4	3.4	2.8	2.4	1.9	1.8	3.5
20	3.8	2.5	1.8	1.4	1.3	1.2	1.2	1.8
40	1.7	1	.7	.6	.5	.5	.6	.9
80	—	—	.3	.2	.2	.2	.3	.5

JNDs for frequency of pure tones are shown in Table 3.C-3. Note that the JNDs for frequency range from as little as 0.2% for tones of 2,000 Hz at 60 dB to about 4% at low frequencies and low levels. JNDs for frequency of complex tones are generally smaller than those shown for pure tones.

Table 3.C-3. JNDs for frequency (percent) of pure tones as a function of frequency (Hz) and level above threshold (dB).

Level (dB)	Frequency (Hz)						
	100	200	400	1000	2000	4000	10,000
10	—	2.7	1.4	.7	.5	.5	.6
20	3.4	1.8	.9	.4	.3	.3	.5
40	2.9	1.5	.8	.35	.25	.25	.4
60	2.7	1.4	.7	.3	.2	.2	—

Even though JNDs for frequency are of the order of one percent, a listener is unable to "hear out" each of two simultaneously occurring tones unless they are separated by about 30 JNDs. The ability to hear out two components of a complex tone that lie close to each other in frequency seems to be more nearly related to critical bandwidth than to JND. It is possible to estimate the number of pure tones a normal listener can distinguish from the tables of JNDs. If the loudness is kept at 40 phons the average person, under ideal conditions, can distinguish approximately 1,400 different frequencies. If the frequency is kept constant at 1,000 Hz, about 280 different sound levels (between 0 and 120 dB) can be distinguished. If we multiply these two numbers, to account for all frequency and intensity changes, we compute that the total number of distinguishable tones (under ideal conditions) is almost 400,000! The ear indeed has amazing powers of discrimination.

Other JNDs of interest in speech and music are 0.3-5% for fundamental frequency, 3-5% for formant frequency, 20-40% for formant bandwith, 3% for formant amplitude, and 0.4-1.5 dB for overall intensity of speech signals (Flanagan, chapter 7).

Temporal Aspects of Hearing

In many respects the ear might be considered a time sensing device much as the eye might be considered a space sensing device. The ear is sensitive to the time duration of a tone and to the time lapse between successive tones. The duration of a tone can affect both its perceived loudness and pitch. Increasing the duration of a tone increases its perceived loudness up to a point in time (about 0.50 sec) beyond which the loudness decreases slightly due to a reduced firing rate of the hair cells because of their increased recovery times under continuous stimulation. It is much like placing one's hand on a hot stove: the extent of the burn is determined by the temperature of the stove and the length of time the hand is in contact, i.e., on the total heat absorbed by the hand.

There is an interesting unertainty relationship that appears in the description of many physical systems. It is stated as $\Delta f \cdot \Delta t \geq 1$ and means that the uncertainty in frequency times the uncertainty in time of a system is greater than or equal to one. It suggests that if we have a signal that is too short in duration (Δt small) the frequency (pitch) will be poorly defined (Δf large). If we listen to a sinusoid of a few milliseconds duration we will not be able to ascribe a pitch to it; as the duration is increased to 10 milliseconds or so we will be able to ascribe a pitch ("click pitch") to it which, however, is lower than the pitch ("tone pitch") we would ascribe to a longer duration of a tone required to ascribe a definite pitch. (Note that to some extent a constant number of cycles are required at low frequencies and a constant duration is required at high frequencies to make the judgment.)

Table 3.C-4. Number of cycles and duration of a pure tone required for a listener to ascribe a definite pitch.

Frequency (Hz)	80	100	200	400	1000	2000	4000
Number of cycles	3.5	4.0	5.5	7	12	23	56
Duration (msec)	44	40	28	18	12	11	14

Another temporal question of interest is related to the perception of successive tones. When two tones presented in time succession are separated in time by less than 0.002 sec they are perceived as being simultaneous; when their time separation exceeds 0.002 sec they are perceived as successive, although their order of presentation is not perceived until their time separation is about 0.02 sec. A listener is able to "count" the number of events when they occur at rates of 10 per second or less.

Echoes occur when a reflected (or delayed) sound is heard by a listener as separate and distinct from the direct sound. This typically occurs when time of arrival of the reflected sound is at least 0.05 sec later than that of the direct sound. If the difference in the distances traveled by the direct and reflected sounds is about 20 meters or more, then a time delay of 0.05 sec or more will exist between the direct and reflected sound and echoes will be heard. This must be accounted for to achieve proper design in buildings.

Perceptual Effects of Complex Stimuli

The preceding discussion has suggested that the ear has an "integration time" of about 0.01 sec and a "recognition time" of about 0.05 sec. This suggests that transients of musical instruments and of speech production may play a substantial role in the perception of such signals. Phoneme rates in speech production are of the order of 7-15 per second. Transient times in speech are of the order of 0.005-0.015 sec (considering plosive bursts or turn-on times for vowels). Musical instrument transients vary all the way from a few milliseconds (for percussives) to 0.02-0.04 sec (for an oboe) to 0.3-0.4 sec (for bowed strings). The consonants (basically transients) of speech and the attack transients of musical instruments play an important role in the identification and recognition of their respective sounds. (See section 6.A for further details on transients.) This may be due in part to the spectral changes that occur in the transients. The perception of one part of a changing sound is influenced by other parts of the sound—particularly those adjacent in time. This again offers us a partial explanation as to why the transients are so important in the identification of musical instruments and the intelligibility of speech.

In some listening situations, such as spoken communication, the perception of sound may be very context dependent. Our perception may be conditioned by constraints imposed on the situation. If unknown vocabulary or topical items are suddenly introduced in a conversation we may hear them as something we expected rather than as what was "said." In extreme cases it is a matter of perceiving what we anticipate even in the absence of what might be considered adequate stimuli.

An interesting auditory illusion can be constructed from an especially contrived set of complex tones. The theory for this effect (Shepard, 1964) is based on constructing a tone with many sinusoidal components, each consecutive partial having twice the frequency of the previous partial. A careful reading of the prior statement combined with some thoughtful reflection will lead you to conclude that these tones would be very unnatural. Musical tones produced by natural means have harmonics which are integer multiples of the fundamental (e.g., 100, 200, 300, 400, 500 Hz, etc.), while the above tones would have partials something like 100, 200, 400, 800 Hz, etc. Further- more, the amplitudes are large only for components of intermediate frequency. The amplitudes of the highest and lowest extremes taper off to subthreshold levels. The spectrum for such a tone is shown in Fig. 3.C-4. The lowest-frequency component is about 8 Hz, well below the lower limit of hearing. Note that several of the low-fre- quency components are also below the threshold of hearing (horizontal dashed line). Imagine now that an envelope is drawn over the frequency components, as repre- sented by the light curve in the figure. We now generate a second tone, in which all components are shifted upward in frequency by a constant percentage but where the amplitude of each component is still shaped by the same spectral envelope. The spectral composition of this second tone will differ from that of the original tone in the manner in which the dashed vertical lines of the figure differ from the original solid vertical lines. Note that the upward frequency shift of the components has been offset, to some extent at least, by the increased contribution of the lower com- ponents and the decreased contribution of the higher components. Indeed, if the second tone is shifted up by an entire octave (double the fundamental frequency) it becomes identical to the original tone, for at the octave the highest component (which has already faded below threshold) is introduced an octave below the pre- vious lowest component. By generating a set of 12 tones in this manner, each tone representing one note of the equally tempered scale, we can generate a chromatic scale.

Because these tones are very unnatural, they were constructed on a digital com- puter with an associate D/A converter. By having the 12 tones follow each other

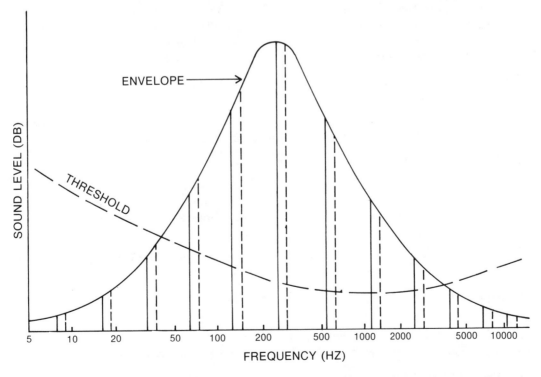

Fig. 3.C-4. Spectrum for producing auditory illusion. Sinusoidal components are spaced at octave intervals. (After Shepard, 1964.)

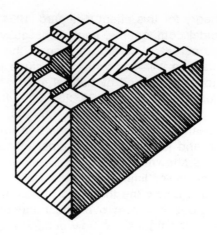

Fig. 3.C-5. "Circular" staircase illusion.

consecutively and then repeat, an illusion of constantly increasing pitch is produced. (This illusion is available on the tape *Descriptive Acoustics Demonstration,* section 3.C.) Each consecutive tone is heard as being higher in pitch than the preceding, but after several minutes of listening the tones do not seem to be getting any "higher." In a similar way, going through the tone sequence in reverse gives the illusion of a constantly decreasing pitch in which traversing the "stairs" in one direction produces an illusion of constant ascending while traversing in the other direction results in constant descending. (See Fig. 3.C-5.)

Exercises

1. What two mechanisms are at work in sound localization? Which is best for low frequencies? Which is best for high frequencies? Explain.

2. Can a person listening with one ear only tell much difference between mono and stereo sound? To get a more realistic stereo effect, a person should listen over earphones rather than through loudspeakers. Why? Even with earphone listening to stereo there is a lack of realism. What factors make this so?

3. Offer some explanations as to why we can localize sounds much more accurately in the horizontal plane than in the vertical.

4. What are two important results of masking experiments? How does the behavior of the basilar membrane at different frequencies explain these results?

5. How is the "critical bandwidth" concept explained in terms of the behavior of the basilar membrane?

6. What is the approximate critical bandwidth at each of the following frequencies (in Hz)? 100; 300; 1000, 3000; 10,000. (Use Table 3.C-1 and interpolate where necessary.)

7. What frequencies may emerge from a nonlinear loudspeaker when frequencies of 700 and 1,000 Hz are input to the speaker?

8. What frequencies may be heard if we put two sinusoids of frequencies 700 and 1,000 Hz into the ear at fairly high levels?

9. Suppose a complex tone having harmonic frequencies of 400, 600, 800, 1,000, and 1,200 Hz enters the ear. What fundamental frequency is perceived?

10. How can you explain hearing the above fundamental frequency when it is not present in the complex tone?

11. A tone is made up of the following frequencies (in Hz): 200, 300, 400, 500, 600, 900.
    a. Which frequencies are harmonics?
    b. What is the frequency of the fundamental?
    c. What "pitch frequency" is heard?

12. Suppose you build a room with hard walls. What are typical largest dimensions that you might use so that echoes would not be produced?

13. In the table below, in the column labeled "Perceptible?" indicate whether or not the two tones are noticeably different from each other.

| Tone 1 | | Tone 2 | | |
Frequency	Level	Frequency	Level	Perceptible?
50	60	55	60	
50	60	51	60	
2000	60	2002	60	
2000	60	2020	60	
40	20	40	43	
2000	80	2000	80.5	
2000	80	2000	84	

14. Suppose that the relationship $\Delta f \cdot \Delta t = 1$ held true. If a listener wanted to ascertain the frequency of a tone within 10 Hz, what length of tone would he need to listen to?

15. What minimum change of a second formant frequency of 1,500 Hz for speech would be just perceptible? Of a fundamental frequency of 130 Hz of a male talker? Or a fundamental frequency of 260 Hz of a female talker?

*16. Missing fundamental and perceived pitch
    Two sine wave generators producing approximately equal outputs were connected to an amplifier-speaker system. Explain the following listener judgments to the various complex tones.
    a. When sinusoids of 260 Hz and 520 Hz were added the resulting tone was judged to be in tune with C4 (about 260 Hz) on a piano. Why?
    b. When sinusoids of 260 Hz and 390 Hz were added the resulting tone was judged to be in tune with C3 (about 130 Hz) on a piano. Why?
    c. When sinusoids of 130 Hz and 195 Hz were added the resulting tone was in tune with C2 (65 Hz). Why?

Further Reading
Backus, chapter 7
Benade, chapters 12, 14
Denes and Pinson, chapters 5, 6
Flanagan, chapters 4, 7
Fletcher, chapter 9
Gerber, chapters 5, 8
Olson, chapter 7
Roederer, chapter 2
Stevens and Warshofsky, chapters 5, 6
Von Bekesy
White, chapters 5, 6
Winckel, chapters 3, 5, 6
Goldstein, J. L. 1966. "Auditory Nonlinearity," J. Acoust. Soc. Am. 41, 676–89.
Goldstein, J. L. 1970. "Aural Combination Tones," in Frequency Analysis and Periodicity Detection in Hearing R. Plomp and G. F. Smoorenburg, eds. (A. W. Suithoff, Leiden).

Hawkins, J., and S. Stevens. 1950. "The Masking of Pure Tones and of Speech by White Noise," J. Acoust. Soc. Amer. *22,* 6.

Hebrank, J., and D. Wright. 1974. 'Are Two Ears Necessary for Localization of Sound Sources on the Median Plane?" J. Acoust. Soc. Am. *56,* 935–38.

Plenge, G. 1974. "On the Differences between Localization and Lateralization," J. Acoust. Soc. Am. *56,* 944–51.

Shepard, R. 1964. "Circularity in Judgments of Relative Pitch," J. Acoust. Soc. Am. *36,* 2,346–53.

Stevens, S., and E. Newman. 1934. 'The Localization of Pure Tones," Proc. Nat Acad. Sci. *20,* 593.

Audiovisual

1. *The Science of Sound* (Album #FX6007, 1959, FRS)
2. *Auditory Illusions and Experiments* (Cassette #72232, 1977, ES)
3. *Descriptive Acoustics Demonstrations* (Tape Sect. 3.C, 1977, GRP)

Demonstrations

1. Two sine wave generators operating into mixer—amplifier—speaker.
2. Auditory illusions (pitch-up)

3.D. SCALES, TEMPERAMENTS, AND HARMONY

In the discussion of JNDs (section 3.C) we discovered that, at a constant intensity level, the ear can discriminate several thousand frequencies. However, this many frequencies represents a far too numerous set to be useful in music where a few hundred frequencies are found to be sufficient. The actual set of frequencies used in creating music has evolved over many centuries. Melody (homophony), or tones played in sequence, probably exerted some influence on the notes (or frequencies) chosen. Multiple concurrent melodies (polyphony) may have exerted further influences on the tones selected. However, the appearance of harmony, in which two or more tones are sounded concurrently to form chords or other harmonic musical structures, has probably had the most significant influence on the selection of notes to be used in the scales of Western music. In fact, in parts of the world where harmony has played a less important role in music, the selection of tones to be used differs from that of Western music. After some basic definitions we will consider some of the temperaments that have been given serious consideration over the years. This is followed by a discussion of the physical basis (in part, at least) for these systems and for harmony in terms of theories of consonance.

Intervals and Scales

To the musician it is not single tones which are interesting or important but rather a change in pitch, called a musical interval. When tones are played consecutively the interval is *melodic,* while tones played together (as in a chord) are *harmonic.* The interval of two tones whose fundamental frequencies have a ratio of 2:1 is termed an *octave.* A *scale* defines the collection of tones to be used in a system of music. An individual frequency of a scale is called a *note,* or tone, of the scale. (The word "note" is also used to connote a blob of ink on a sheet of music paper, which is the symbol of a musical tone.) Since the octave appears in virtually every system, we will limit our discussion of scales to a description of the way in which the notes within an octave are defined. The notation used to specify which octave is being referred to will be that of the USA Standards Association. In this notation the octave numbering starts on C, with C4 being middle C. A4 is the A above C4; C5 is the C an octave above middle C; C3 is an octave below middle C; and so on. An *interval* is the "distance" between two notes in a scale and can be expressed as a ratio of frequencies (e.g., 3:2) or by other means such as special names (octave, fifth, etc.). Musical intervals can generally be classified into two groups: consonant and dissonant. Consonant intervals result when the two tones sounded together give a "smooth" or pleasant sensation. Dissonant intervals are judged to be "harsh" or "rough" sounding. (Although modern music has made this distinction somewhat vague, there is still substantial agreement among average listeners.)

In the sixth century B.C., Pythagoras of Samos used a monochord (a one-stringed instrument in which a movable bridge allows the string to vibrate in two adjustable segments) to demonstrate consonance. He found a consonant sound when the two vibrating segments had lengths in the ratios of 1:1, 1:2, 2:3, and 3:4. We learned in chapter 2 that for a vibrating string the frequency is inversely proportional to the length of the string. The frequency ratios for the strings were then 1:1 (unison), 2:1 (an octave), and two new intervals having ratios of 3:2 and 4:3. A musician listening to these new intervals could immediately identify them as the "fifth" and the "fourth." Since this terminology has reference to a piano keyboard, let us digress momentarily and briefly consider the evolution of the piano keyboard.

The piano keyboard was borrowed from a much older instrument, the organ. In 1361 in the Saxon city of Halberstadt an organ builder, Nicholas Faber, completed a three-manual instrument which was somehow destined to exert a substantial influence on all future organs. The upper two manuals of the Halberstadt organ had a series of 9 front keys and 5 raised rear keys in groups of two and three, as shown in

Fig. 3.D-1a. Even though these keys were made to be struck by the fists, this was the prototype of the now well-known 7 white and 5 black keys per octave. ("Prototype" does not imply color, as black and white keys were not used until 1475, and even at that time the convention was reversed, a practice that prevailed for the next 300 years.) The modern piano keyboard is shown in Fig. 3.D-1b. If we number the notes, starting with C as one, we find that F is the fourth note, or "fourth," and G is the "fifth." The octave ends on the eighth note, which is C again. You will notice that when numbering the notes we counted only the white keys. Consecutive white keys, however, actually contain two different types of interval, the *whole tone* and the *semitone*. The interval between any two consecutive notes (including white and black keys) is a semitone, regardless of whether we go from white to black, white to white, or black to white. A whole tone is arrived at by passing through two semitones. The interval from C to D is thus seen to be a whole tone, while the interval from E to F is a semitone.

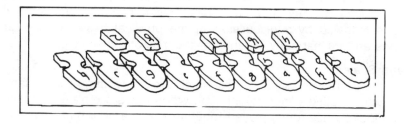

Fig. 3.D-1a. Third manual of the Halberstadt organ. (From Praetorius' *Syntagma Musicum,* 1619.)

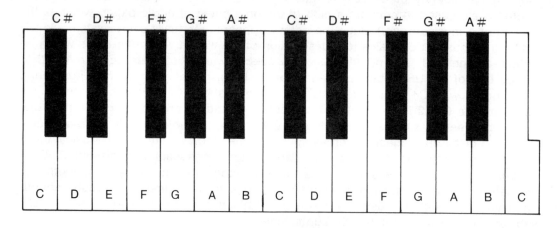

Fig. 3.D-1b. Section of modern piano keyboard.

Let us now repeat Pythagoras's experiment, but this time using two violins. The first plays C4 while the other produces a gradually increasing pitch from C4 to C5. The "degree of consonance" perceived as the pitch increases is represented in Fig. 3.D-2. At two additional intervals (besides the fourth, fifth, and octave) there is relatively little beating. These intervals correspond to E (the "third") and A (the "sixth"). If we add two somewhat dissonant intervals, D (the "second") and B (the "seventh"), to the collection we already have, we have one example of a scale. This particular scale (which utilizes only the white keys of a piano) is known as C major. An examination of Fig. 3.D-1b shows that the interval sequence for the scale of C major is whole, whole, semi, whole, whole, whole, semi. This sequence of intervals defines any *major scale,* the name of the scale being the starting note. A *minor scale* (harmonic form) is defined by the following sequence of intervals: whole, semi, whole,

whole, semi, whole + semi, semi. Again, the starting note determines the name of the scale.

Musical intervals can be defined from any starting note. Fig. 3.D-1b shows that the interval from C to G (a fifth) consists of 7 semitones, while a third (C to E) consists of 4 semitones. The third of G is thus seen to be B, while the fifth of G is the D above. But what is the fifth of B? If we go up by 7 semitones we end on a black key, F#, which is the fifth of B.

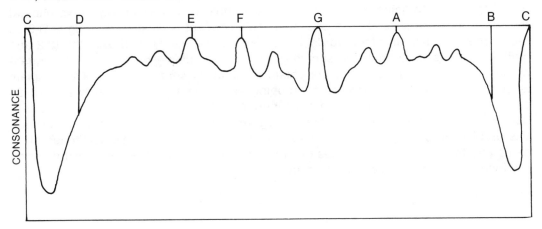

Fig. 3.D-2. Two violin tones sounding together create the degree of consonance shown here. (After Jeans, 1961.)

Three Systems of Temperament

We now consider three different ways that a keyboard instrument (such as a piano or an organ) can be tuned. Many other ways have been and still are being used, but three systems of temperament (tuning) will provide adequate illustrations for our purposes. They are the equally tempered, the Pythagorean, and the just systems of temperament.

Equal temperament is the system which is currently most widely adopted. It is based on compromises of the "perfect" intervals of other systems. Although the mathematics of equal temperament had been solved some fifty years earlier, it was not until 1688, three years after J. S. Bach's birth, that the first organ was tuned in accordance with this system. Bach himself favored equal temperament, and his many compositions (written in almost all keys) for harpsichord and organ undoubtedly had tremendous influence in moving the musical world to the almost universal adoption of this system. The *equal temperament* system divides the octave into twelve equal intervals, or semitones, any two consecutive notes having the same frequency ratios. The frequency ratio of two consecutive semitones is given by the twelfth root of two, which is 1.059463. In an equally tempered octave a whole tone is exactly equal to two semitones. In order to provide a finer division of the octave, we further divide the semitone into 100 equal parts, termed *cents.* The frequency ratio associated with the cent is 1.0005778. We can discuss any note relative to another note in terms of names (such as octaves), semitones, or cents. An octave consists of 12 semitones or 1200 cents, while the fifth, being 7 semitones, is 700 cents. The cent represents an interval somewhat smaller than the frequency discrimination ability of the ear. Although equal temperament is very convenient for keyboard instruments, the fifths are about 2 cents flat compared to Pythagoras's perfectly consonant fifths, while the equally tempered fourth is 2 cents sharper than a Pythagorean fourth.

Let us look at the exact fifths and fourths obtained earlier with Pythagoras's monochord. Suppose we now construct a C major scale where the frequency of each note is an exact fifth or fourth of one of the other notes. In other words, the frequency of G is exactly 1.5 times the frequency of C, and the frequency of F is exactly

1.333... (4/3) times the frequency of C. A frequency of 1.5 times that of G4 gives an exact fifth, D5, but an octave too high. Taking half the frequency of D5, then, yields the frequency of D4, which is back in the octave where we started. The frequency 1.5 times that of D4 gives A4, and so on. This system of temperament, based on exact fourths and fifths, is known as the *Pythagorean temperament.*

Although the idea of having a system of tuning with exact fourths and fifths looks appealing, there is one severe discrepancy. If we assume an octave has twelve notes, then by advancing frequencies by exact fifths we will eventually arrive back at the same note, but several octaves higher. This concept is illustrated in Fig. 3.D-3 as the "circle of fifths." In other words, if we take the frequency ratio of a fifth (3:2 and multiply it by itself 12 times we should get an exact multiple of the original frequency. In actual fact, the frequency we obtain is about 24 cents too high. This problem is avoided in equal temperament by robbing each exact fifth by one twelfth of 24 cents (2 cents) so that the twelve quasi-perfect fifths do indeed give an exact multiple of the original frequency. In the Pythagorean temperament the only way around this problem is to have two different sizes of semitones: the diatonic semitones of 90 cents and the chromatic semitones of 114 cents. (Note that these semitones differ by the Pythagorean discrepancy of 24 cents.)

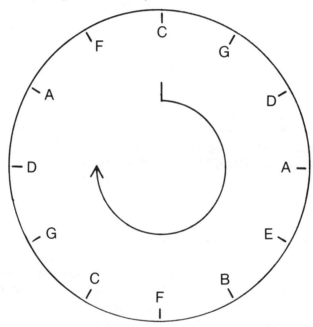

Fig. 3.D-3. Circle of fifths.

From the information given in Table 3.D-1 we can compare the Pythagorean and equal temperaments. The intervals (in cents) of a C major scale in Pythagorean temperament are C-D (204), D-E (204), E-F (90), F-G (204), G-A (204), A-B (204), and B-C (90). Notice that a whole tone (204 cents) is not equal to twice the diatonic (90 cent) or twice the chromatic (114 cent) semitone, but is equal to the sum of a chromatic and diatonic semitone.

With the coming into use of thirds as consonant intervals, the Pythagorean scale was somewhat deficient because its major third (386 cents) and minor third (294 cents) sounded sharp and flat respectively. Mean tone tuning attempted to compensate for these and similar difficulties by making upward or downward adjustments of certain intervals. However, mean tone tuning proved too restrictive as modulations from one key to another developed, and ultimately equal temperament prevailed.

Another system of some considerable interest from a theoretical point of view is *just temperament,* in which the intervals are defined in terms of the ratios of small whole numbers. Again, these intervals are different from those in equal temperament, being (in cents) C-D (204), D- (182), E-F (112), F-G (204), G-A (182), A-B (204), and B-C (112). There are two sizes of whole tones (182 and 204 cents) and neither is equal to twice a semitone (112 cents). This temperament has never been used extensively because to do it justice over many different keys (i.e., scales starting on different notes) would require the inclusion of more than twelve notes per octave, as will be discussed later. Its appeal lies in the number of "perfect" intervals (expressed as ratios of whole numbers) it includes and the resulting consonances of its diads and triads.

Other methods for dividing the octave have from time to time been devised. A quarter tone scale has been proposed that would involve 24 notes per octave, each differing by 50 cents from its neighbor. This system would include all of the equally tempered notes, but would have an additional note between each pair of notes of the tempered system. The division of the octave into nineteen equal parts has been suggested, and a nineteen-tone harmonium was actually constructed over 100 years ago. This system gives very good thirds, but the fifths are 5 cents lower than the fifths in a twelve-toned system. A division of the octave into 53 equal parts has been proposed, and a harmonium having 53 keys for each octave has even been constructed. Of all the various proposed octave divisions, this one is the most nearly "perfect" in that most of the important intervals are within 2 cents of being exact. It is thus possible to play with just temperament in any key, provided one is willing to invest the time and effort that would be required to master the keyboard. A detailed discussion (from a musician's viewpoint) of the history and development of various temperaments, as well as proposals for new systems, is provided in *Genesis of a Music,* by Henry Partch.

Table 3.D-1. Equally tempered, just, and Pythagorean scales and their frequency ratios and intervals. (Bracketed notes are identical in equal temperament.)

Note	Musical interval	Frequency ratio to first note			Interval from first note (cents)		
		Equal	Just	Pythagorean	Equal	Just	Pythagorean
C	unison	1.000	1.000	1.000	0	0	0
C#	semitone (chromatic)	1.059	1.042	1.068	100	71	114
D^b	minor second	1.059	1.067	1.053	100	112	90
D	major second	1.122	1.125	1.125	200	204	204
D#	augmented second	1.189	1.172	1.201	300	275	317
E^b	minor third	1.189	1.200	1.185	300	316	294
E	major third	1.260	1.250	1.266	400	386	408
F^b	diminished fourth	1.260	1.280	1.249	400	427	385
E#	augmented third	1.335	1.302	1.352	500	457	522
F	perfect fourth	1.335	1.333	1.333	500	498	498
F#	augmented fourth	1.414	1.406	1.424	600	490	612
G^b	diminished fifth	1.414	1.440	1.405	600	631	589
G	perfect fifth	1.498	1.500	1.500	700	702	702
G#	augmented fifth	1.587	1.563	1.602	800	773	816
A^b	minor sixth	1.587	1.600	1.580	800	814	792
A	major sixth	1.682	1.667	1.688	900	884	906
A#	augmented sixth	1.782	1.758	1.802	1000	977	1020
B^b	minor seventh	1.782	1.800	1.778	1000	1018	996
B	major seventh	1.888	1.875	1.898	1100	1088	1110
C^b	diminished octave	1.888	1920	1.873	1100	1.129	1086
B#	augmented seventh	2.000	1.953	2.027	1200	1159	1223
C	octave	2.000	2.000	2.000	1200	1200	1200

Beats and Perceived Roughness

Beats (see section 2.C) are due to alternate constructive and destructive interference of two sinusoids having slightly different frequencies. The ear is sensitive to these beats, which give the sound a certain perceived roughness. At a particular average frequency (average of the two frequencies involved) the degree of roughness depends on the number of beats per second. A certain number of beats per second at a certain average frequency produces maximum roughness. Fewer beats or more beats per second will decrease the perceived roughness. Table 3.D-2 gives representative values for the beat frequency that causes maximum roughness.

Table 3.D-2. Maximum roughness as a function of beat frequency. (After Plomp and Levelt, 1965.)

Average frequency of pair (Hz)	Beat frequency giving maximum roughness (Hz)
125	5
250	15
500	28
1000	66
2000	76

Critical bandwith (see section 3.C) plays a rather important role in determining whether a particular pair of simple tones sounds rough or smooth. If two tones are more than a critical bandwidth apart, they are approximately noninteracting and little roughness is produced. As the two tones come closer in frequency than a critical bandwidth, roughness is produced, with the maximum roughness occurring when the two tones differ by a frequency approximately equal to one-fourth the critical bandwidth. Then, as the frequency difference grows still smaller, the roughness decreases (or the smoothness increases). The diagram in Fig. 3.D-4 shows approximately how these relationships vary.

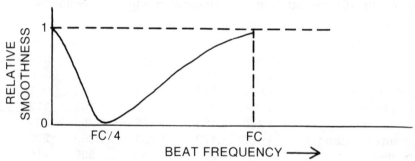

Fig. 3.D-4. Relative smoothness of simple tone pair. (After Plomp and Levelt, 1965.)

The above results can be applied to combinations of complex tones (tones having many partials). To do this, beat combinations between all possible pairs of partials must be considered. Such an analysis predicts that a trio sonata for organ by J. S. Bach should sound less rough than a Dvorak string quartet.

If we equate musical dissonance to roughness and consonance to smoothness, we have a scheme for determining how dissonant or consonant particular tone pairs or compositions will be. The consonance or dissonance depends not only on the notes written but also on the combination of instruments producing the tones. Obviously, instruments having many higher partials are more likely to produce dissonant combinations than instruments having few partials, because there are more chances of the partials interacting with each other. Procedures have been developed for calculating relative consonance of arbitrary complex tone combinations when the resulting spectrum is known; the predictions agree well with perceptual experiments. However, the equating of consonance to smoothness and dissonance to roughness leaves much to be desired, particularly for musicians. What was termed dissonant some years ago could very well be considered consonant today, even though the amount of roughness is the same. Consonance in this sense is a function of conditioning and is learned from the culture. Consonance may still depend on the relative roughness, but the relationship is neither simple nor fixed.

Consonance and Musical Temperaments

The relative roughness or smoothness of complex tone pairs plays a role in determining the relative frequencies of the tones in a scale. Suppose we play a complex tone having harmonics of 100, 200, 300, and 400 Hz. If a second complex tone is played at the same time, the resulting smoothness (or roughness) depends on the relative harmonic frequencies. Suppose the second tone has harmonics of 100, 200, 300, and 400 Hz also. The result will be smooth because no beats are produced. If the second tone is the octave and has harmonics of 200 and 400 Hz, again no beats will be produced and the result will be maximally smooth. As a general rule, we can state that the more harmonics of two complex tones are the same, the more likely the result is to be smooth.

If the ratio of fundamental frequencies is a ratio of small whole numbers, more of the harmonics of the tones will coincide, and the result should produce consonance. Tables 3.D-3 and 3.D-4 for the equally tempered and just temperaments show the relationships of fundamental frequencies between the first note in the scale and each additional note. (Observe that the tuning is to C4 = 260 Hz instead of A4 = 440 Hz. This is done as a convenience to reduce the use of decimal numbers in the example.) Even though it can be shown (see exercises) that the just system produces the smoothest tone pairs, equal temperament is used much more often because it is more practical. When a scale is constructed for each note in the just system (i.e., each scale starts on a different key), the notes have different frequencies in different keys. The equally tempered system attempts to average out these problems by making the frequency of each note the same, regardless of what key is used. Thus, equal temperament makes it possible to use only 12 notes per octave (including sharps and flats), whereas at least 53 are required in order to obtain reasonable accuracy in all keys with just temperament. Even though the equally tempered system is "out of tune" relative to just temperament, it is still of considerable advantage to keyboard instruments.

Table 3.D-3. Equally tempered scale showing frequency relationships among notes in the scale and higher partials.

Musical interval	Note	Ratio to C4	f_1	f_2	f_3	f_4	f_5	f_6
Unison	C4	1.000	260	520	780	1040	1300	1560
Second	D4	1.122	291.7	583.4	875.1	1166.8	1458.5	1750.2
Third	E4	1.260	327.6	655.2	982.8	1310.4	1638	1965.6
Fourth	F4	1.335	347.1	694.2	1041.3	1388.4	1735.5	2082.6
Fifth	G4	1.498	389.5	779	1168.5	1558.0	1947.5	2337
Sixth	A4	1.682	437.3	874.6	1311.9	1749.2	2186.5	2623.8
Seventh	B4	1.888	490.9	981.8	1472.7	1963.6	2454.5	2945.5
Octave	C5	2.000	520	1040	1560	2080	2600	3120

Table 3.D-4. Just scale of intonation showing frequency relationships among notes in the scale and higher partials.

Musical interval	Note	Ratio to C4	f_1	f_2	f_3	f_4	f_5	f_6
Unison	C4	1.000 1/1	260	520	780	1040	1300	1560
Second	D4	1.125 9/8	292.5	585	878	1170	1463	1755
Third	E4	1.250 5/4	325	650	975	1300	1625	1950
Fourth	F4	1.333 4/3	346.7	693.4	1040	1390	1735	2042
Fifth	G4	1.500 3/2	390	780	1170	1560	1950	2340
Sixth	A4	1.667 5/3	433.3	866.7	1300	1733	2167	2600
Seventh	B4	1.875 15/8	485	970	1455	1940	2425	2910
Octave	C5	2.000 2/1	520	1040	1560	2080	2600	3120

If a periodic musical tone has a fundamental frequency of 260 Hz, the higher partials will be 520, 780, 1040, etc., as seen in Tables 3.D-3 and 3.D-4. Doubling of frequency produces a pitch change of one octave. For instance, if a fundamental of 260 Hz corresponds to C4, then a tone with a fundamental of 520 Hz corresponds to a C one octave higher (C5), and the 1040-Hz tone is still another octave higher (C6). With this in mind, note that the tone having a fundamental of 260 Hz has a second partial of 520 Hz, which exactly corresponds to the fundamental frequency of the C one octave higher. Furthermore the fourth partial (1040 Hz) corresponds to the fundamental frequency of the next octave. Likewise, the eighth partial (2,080 Hz) corresponds to the next octave, C7. To what note then does the third partial (780 Hz) correspond? Consideration of Table 3.D-4 shows that in just temperament an interval of a fifth corresponds to a frequency ratio of 1.5. Since the frequency ratio of the third to the second partial is 780/520 = 1.5, the note corresponds to a fifth above C5, or the note G5. The sixth partial (1560 Hz) would then be the octave of G5, or G6. To what note does the fifth partial (1300 Hz) correspond? The chart of just intervals shows that the interval of a third has a frequency ratio of 5/4. Since the ratio of the fifth to the fourth partial is 1300/1040 = 5/4, the fifth partial is a major third above the fourth partial. Since the fourth partial is C6, the fifth partial must be E6 (in the just system).

Let us now consider how the intervals work out in the equally tempered system. Again, assume a tone having a fundamental frequency of 200 Hz. The harmonics of this tone are all multiples of 200 Hz, as shown in Table 3.D-3. The second harmonic has a frequency of 520 Hz, which corresponds to the octave C5. The third harmonic (780 Hz) corresponds to a just fifth above C5, or G5. The fourth harmonic (1040 Hz) is the superoctave, C6, while the fifth harmonic is the third above this, or E6. The sixth harmonic is the octave of the fifth, or G6. The seventh harmonic (1820 Hz) is approximately (even in just temperament) a minor third above the sixth harmonic, thus making the seventh harmonic a minor seventh of C6, which is B6[b]. The eighth harmonic is again the octave, C7. It is readily seen by comparing Tables 3.D-3 and 3.D-4 that in most cases the frequencies of the notes in the equally tempered system are quite close to the corresponding frequencies in the just system.

Harmony and Its Evolution

As we have seen above, when several tones are heard simultaneously the amount of dissonance present is proportional to the number of beats between partials. Pairs of tones whose fundamental frequencies are related by ratios of small integers have more of their harmonics in unison and fewer harmonics that can beat and create dissonance. A combination of tones also produces subjective difference tones equal to the difference between each frequency of the constituent tones. It is generally recognized that Western harmony is based on the *major triad,* which is a chord composed of the following intervals: the unison, a major third, and a perfect fifth. Why was this particular combination of intervals chosen, rather than some other set? While we cannot answer this question with complete certainty, there are several interesting hypotheses which are worth considering. First, consider a major triad consisting of A3, C4#, and E4, with respective frequencies (in just temperament) of 220 Hz, 275 Hz, and 330 Hz. Table 3.D-5 gives the first six harmonics of each of these tones.

Notice that a number of frequencies (indicated by an *) are common to two of the tones. Because of the duplication of a number of partials, and because there are no instances of any two partials being separated by the frequency which would produce maximum roughness, the tone of this triad is perceived as having a "smooth" sound. Perhaps, then, because the major triad sounded so consonant it was first adopted as the basis of musical harmony.

Table 3.D-5. Harmonics of tones of major triad A3, C4#, and E4.

	Harmonic					
Note	1	2	3	4	5	6
A3	220	440	660*	880	1100*	1320*
C4#	275	550	825	1100*	1375	1650*
E4	330	660*	990	1320*	1650*	1980

An alternate, and more tenuous, explanation of the role of major triads considers the difference tones produced when a chord is sounded. It has been hypothesized (Hirsch, 1967) that a combination of tones is perceived as "pleasant" if the primary and subjective tones are harmonically related to the difference tone of lowest frequency. The major triad sounds sonorous or full because the ear creates subjective tones whose pitch is a subharmonic of the frequencies in the primary chord, but which are above the threshold of tone perception. (The threshold of tone perception is where tones begin to have a definite pitch associated with them: it is usually considered to be about 40 Hz.) As an example, consider again the major triad consisting of A3, C4#, and E4. The frequency ratios to A3 are C#/A = 275/220 = 1.25 and E/A = 330/220 = 1.5. The differences of the fundamental frequencies are C# – A = 275 – 220 = 55 Hz; E – A = 330 – 220 = 110 Hz; and E – C# = 330 – 275 = 55 Hz. Note that the difference tone of 110 Hz is the frequency of A2 (the A one octave below A3). The difference tone of 55 Hz (which occurs twice) is the frequency of A1, the A two octaves below A3. You will notice that our three-tone chord has now become a five-tone chord which is deeper (and richer) than the original chord. If we perform a similar analysis on the upper partials of each of the original notes we obtain many more difference tones, all of them being some harmonic of 55 Hz. One problem with this hypothesis is that the difference tones computed above are not readily audible every time a major triad is played. To perceive the difference tones requires a trained and sensitive ear and a fairly intense tone. Even then, the difference tones are quite soft. The importance of the major triad seems to be well founded on considerations of roughness and smoothness, but the Hirsch hypothesis may have exerted an additional subtle effect. In either case, it appears as though the cornerstone of Western harmony has been laid by the basic physiological structure of the inner ear.

Some people have gone beyond merely considering the foundations of Western harmony and argued that the entire superstructure evolved in accordance with the physical nature of the harmonic series. Although their case is usually presented in an erudite manner, there are also some serious problems with such a viewpoint. Consider first the following information in support of this hypothesis. If a harmonic series is constructed based on C4 (such as shown in Table 3.D-6), using just intervals we find the following notes: C4, C5, G5, C6, E6, G6, B6^b, C7, D7, E7, F7#, G7, A7^b, B7^b, B7#, and C8. The perfect intervals (the octave, the fifth, and the fourth) occur quite early in the series. That is, the interval between the first two harmonics is an octave, the interval between the second and third harmonic is a fifth, and the interval between the third and fourth harmonic is a perfect fourth. (Some notes, namely B6^b, F7#, A7^b, and B7^b, would be out of tune, even in the just system.) Harmonics 1, 2, 3, and 4 formed the basis of medieval harmony (which consisted of chords in octaves, fifths, and fourths). After the year 1200, harmonic 5 (the major third) was added, giving the major triad, which was the basis of music for the next three hundred years. Around the year 1000, the seventh harmonic (B6^b) was added, thus contributing the dominant seventh chord. In the eighteenth century harmonic nine provided the ninth chord, and in the late nineteenth century the eleventh harmonic gave the eleventh chord. In the twentieth century music has been characterized by much dissonance, and the high harmonics give rise to dissonant sounds such as are found in the minor second.

Table 3.D-6. Harmonic series starting on C.

Harmonic	1	2	3	4	5	6	7	8	9	10	11	12
Note	C	C	G	C	E	G	B^b	C	D	E	F#	G
		oct	5th	4th	maj 3rd	min 3rd			maj 2nd		min 2nd	

Although the above argument presents a convincing view of the evolution of harmony from the harmonic series, there are several serious problems with such a point of view. One problem is, if nature is the basic source of harmony, why are certain notes (such as the seventh harmonic) out of tune, even in just temperament? Why are the fourth and sixth notes of the scale (F and A) missing from the series, while they have played an important part in harmony? Furthermore, why are the sharp fourth and flatted sevenths present in the harmonic series but not in the scale? Also, if the harmonic series is the basis of music, why did modal scales precede the major scales by many centuries, and why does the minor mode not appear in the harmonic series? Finally, the interval of a fourth was considered for many centuries to be a dissonant interval, and yet it appears quite early in the harmonic series. Obviously then, the viewpoint that all of Western harmony developed and evolved on the basis of the harmonic series is greatly oversimplified, if not inaccurate. We can probably say, however, that the natural structure of the harmonic series exerted some considerable influence on the development of harmony, but other nonscientific factors were probably equally important.

Intonation and Standard Pitch

Intonation is the extent to which the frequency of an instrumental tone corresponds to the expected value in the desired scale. In practice the accuracy of intonation of bowed strings is limited primarily by the skill of the performer. Instruments employing an air column are subject to temperature-caused variation in intonation as well as to intonation problems due to instrument design. Woodwind players can compensate to some extent and brass players to a much greater extent for instrumental intonation problems. The human voice, like the bowed strings, has intonation precision limited primarily by the skill of the performer.

Many of the more subtle differences among different scales are probably not realized in musical performance. The inability of a performer to play "perfectly in tune," as well as aesthetic considerations and ornamentations like vibrato, help to obscure perceptual differences. Sustained tones would be most likely to show such differences, but sustained tones are not common in musical performance. Just temperament has often been considered to be the most "natural" system of temperament. However, recent experiments have shown that musicians not limited to equal temperament (such as violinists and vocalists) make more use of the Pythagorean intervals than the just intervals.

It is necessary to have some standard pitch to tune against and then to tune all other notes relative to this standard within the intonation accuracies of an instrument and performer. It is important that the standard pitch be retained at a constant value over a long period of time because many instruments (both string and wind) are designed to perform optimally at a particular tuning and may perform less well at higher or lower tunings. Over the years the tuning of A4 has varied from as low as 415 Hz to as high as 460 Hz, an interval of almost two semitones. The currently accepted standard of pitch is the tuning of A4 to 440 Hz, although not all practicing musicians adhere to this value.

Exercises

1. In the following chart, indicate on a scale from 1 to 5 (where 1 = very smooth and 5 = very rough) the degree of roughness of the following tone pairs.

f_1	f_2	Smoothness/Roughness
100	100	
100	102	
100	104	
100	106	
100	110	
100	120	

2. Which pairs of instruments are most likely to produce dissonant combinations? Which are least likely? Why?

3. Complete the table below by supplying a tone 2 frequency that will approximately produce maximum roughness with tone 1 in each case.

	f (tone 1)	f (tone 2)	f (beat)
100	100		
	225		
	490		
	1,010		
	1,950		

4. When the following pairs of complex tones are played together, indicate on a scale from 1 to 10 (10 = maximally rough) the roughness of the combinations.
   a. Tone 1 = 100, 200, 300, 400 Hz; tone 2 = 200, 600 Hz.
   b. Tone 1 = 100, 200, 300, 400 Hz; tone 2 = 150, 300, 450 Hz.
   c. Tone 1 = 100, 200, 500, 1000 Hz; tone 2 = 110, 220, 520, 1020 Hz.
   d. Tone 1 = 100, 220, 490 Hz; tone 2 = 105, 240, 520 Hz.

5. Tables 3.D-3 and 3.D-4 give the ratio of fundamental frequencies between the first tone and each additional tone for the equally tempered and just scales. The frequencies of the first six partials are also given for each tone. Assume that each tone is played with C4 and determine the partials that will produce roughness, and order the pairs of tones in terms of relative roughness and smoothness. Work with each table separately. Note that the tuning is to C4 = 260 Hz.

6. Which of the two scales, just or equally tempered, produces the smoothest tone pairs? Why do we use the other scale more commonly?

7. Try constructing a just scale starting on G4. What is the fundamental frequency of A4 in this scale? What is the frequency of A4 for the just scale starting on C4? Since the two are not the same, does this present any difficulty?

8. Fill in the table below with the correct fundamental frequencies for equal temperament. (Note that in this case the tuning is to A4 = 440 Hz.)

n	C_n	D_n	E_n	F_n	G_n	A_n	B_n
4						440	
5							
6							
7							

9. Consider the discussion of the A major triad given in the subsection entitled "Harmony and Its Evolution." Assume that each note of the chord has the first six harmonics present. Calculate the frequency of every difference tone that will

10. For exercise 9, calclate all the frequencies that will be present in the A chord if summation tones (such as $f_1 + f_2$) are also present when the chord is sounded.

*11. Beat frequency and perceived roughness

A group of five listeners were asked to listen to two sinusoids presented simultaneously at the same level but at slightly different frequencies. A sine wave generator was set to the frequencies shown in the table. The listeners were asked to adjust the second generator until a maximally rough sound was produced. The resulting setting for each of the five listeners is shown in the table.

a. Calculate the difference between the frequency of generator 1 and the average frequency of generator 2 for each case and record in the table.

b. How do these differences compare with those in Table 3.D-2?

Generator 1 (Hz)	Generator 2 (Hz)	Difference (Hz)
125	129, 130, 131, 132, 130	
250	262, 266, 269, 268, 265	
500	527, 522, 532, 529, 524	
1000	1060, 1074, 1072, 1070, 1066	
2000	2081, 2083, 2071, 2077, 2085	

Further Reading

Backus, chapter 8
Benade, chapters 15, 16
Bartholomew, chapter 4
Helmholtz, chapters 8–19
Jeans, chapter 5
Olson, chapter 3
Roederer, chapters 4, 5

Seashore, chapter 10

Winckel, chapter 8

White, chapter 15

Wood, chapters 10, 11

Hirsch, C. 1967. "Some Aspects of Binaural Sound," IEEE Spectrum, Feb. 1967, 80–85.

Kameoka, A., and M. Kuriyagawa. 1969. "Consonance Theory Part I: Consonance of Dyads" and "Consonance Theory Part II: Consonance of Complex Tones and Its Calculation Method," J. Acoust. Soc. Am. *45,* 1451–69.

Partch, Harvy. 1974. *Genesis of a Music,* 2nd ed. (Da Caps Press, New York), chapters 15, 16, 17, 18.

Plomp, R., and W. J. M. Levelt. 1965. "Tonal Consonance and Critical Bandwidth," J. Acoust. Soc. Am. *38,* 548–60.

Young, R. W. 1939. "Terminology for Logarithmic Frequency Units," J. Acoust. Soc. Am. *11,* 134–39.

Audiovisual

1. *Temperaments: Equal, Meantone, Pythagorean* (Tape available from R. C. Nicklin, Physics Dept., Appalachian State Univ., Boone, N.C. 28608)
2. *Descriptive Acoustics Demonstrations* (Sect. 3.D., 1977, GRP)
3. *The Science of Sound* (Album #FX6007, 1959, FRS)
4. *Auditory Illusions and Experiments* (1977, Cassette #72232, ES)
5. *Science of the Musical Scale* (30 min, color, 1957, EBE)

Demonstrations

1. Two function generators operating into a mixer-amplifier-speaker system
2. The monochord

3.E. HEARING IMPAIRMENTS, HAZARDS, AND ANNOYANCES

Impairments and Corrections

The ear is a delicate instrument and, although well protected against normal environmental exposure, is subject to impairment from various causes. These include blockage of the ear canal, infection in the middle ear, "freezing" of the middle ear bones, and damage to the hair cells from high fevers incident with disease or from sudden or prolonged exposure to very intense sounds. A loss occurring because of problems in the outer or middle ear is termed a transmision loss because it is caused by a breakdown of the transmission (or mechanical coupling) of sound to the sensing elements of the ear. A loss occurring because of problems of the inner ear is termed a nerve loss because it is caused by a partial or complete breakdown of the nerve-sensing (or mechanical-to-nerve-pulse-converting) elements of the ear. If a person can hear bone-conducted sound but is deaf to air-conducted sounds, we can diagnose the problem as being in the middle ear. But if the person perceives no sound by bone conduction, then the auditory nerves are probably damaged.

Blockage of the ear canal can occur because of infections or because of a buildup of waxy secretions in the canal. The blockage simply prevents effective transmission of the sound wave from the external ear to the eardrum. Removal of the blocking material (preferably by a physician so that care will be taken not to damage any part of the ear structure) is the most usual solution to this problem.

Disablement of the bone chain in the middle ear can occur because of infections which destroy the mechanical link or because of a calcifying, or "freezing" (otosclerosis), of the stapes. In either case the mechanical link between the eardrum and the cochlea is partly or wholly impaired, and the vibrations arising from a sound are not passed on to the cochlea with their normal strength. In some instances the infection can be eliminated and the bone chain will repair itself, or the stapes can be freed with surgical techniques. In other instances it is necessary to remove the damaged parts and complete the mechanical link with a prosthetic.

Damage to the hair cells in the cochlea is the most serious of the various types of hearing impairments. At present there is no known means for repairing this kind of damage, and the hair cells do not regenerate.

A certain kind of nerve-deafness, presbycusis, seems to affect everyone as he/she grows older. The ear gradually becomes insensitive to high-frequency sounds, presumably because of a deterioration of the hair cells that are on the portion of the basilar membrane responding to high frequencies. In industrial nations most of the population 30 years and older cannot hear above 15,000 Hz. At 50 years the limit is 12,000 Hz; at 60 it is 10,000; and at 70 it is 6000 on the average. We will see later that there is evidence that loss of hearing with age may be due to our overall noisy environment. Permanent hearing loss usually occurs in people who work in particularly noisy industrial environments. In addition, tumors, infections, or blows on the head can produce loss of hearing. As a matter of fact, before World War II ear infections were the leading cause of deafness.

Hearing losses typically result in an inability to perceive some of the more subtle aspects of sounds. When the loss is more than a mild one, enjoyment of music may be reduced. The most critical situation occurs when the loss progresses to the point that speech reception is impaired. Then an individual's ability to function effectively with everyday speech communication is reduced. For this reason most laws dealing with hazards to hearing are concerned with the hearing loss that will impair speech communication. The first sounds to be lost by nerve deafness are the high-frequency sounds of speech. The most common complaint of people suffering from noise-induced deafness is "I can hear you, but I don't understand what you say." These people confuse words, because some similar speech sounds are distinguished by their upper frequencies.

The measurement of hearing acuity (or hearing loss) is of practical importance in clinical situations involving the fitting of hearing aids or the prescripton of therapy. An *audiometer* is a device used to test hearing acuity. It generates sinusoids at different frequencies and varies their intensity levels. A person listening to a tone produced by the audiometer responds when he can just barely perceive the tone. In this manner his individual threshold of audibility can be established and compared with the average threshold curve (0 phon level). If the subject has sustained a hearing loss, the degree of loss, expressed in dB above the normal threshold, is determined at each of the measured frequencies. (Clinical audiometers typically perform the subtraction between measured threshold and an average normal threshold and give a reading directly in terms of hearing loss.) The graph showing hearing acuity at different frequencies is called an *audiogram.*

An *electronic hearing aid* consists of a microphone to convert sound to electrical energy,an amplifier to increase its level, and a receiver to convert it back to sound. Hearing aids are useful for people with a hearing loss which is not too severe. A hearing aid is in some sense a "private public address system," which attempts to compensate for the decreased sensitivity of the ear by increasing the intensity of the signal sent to it. Hearing aids are of several types, including body-worn aids, ear-level aids, and in-the-ear aids. The latter two types are usually cosmetically more acceptable to the user, but body-worn aids can supply greater amplification when it is needed. A hearing aid should be prescribed by a qualified clinician on the basis of the patient's audiogram and other pertinent information. The clinician can also make an assessment as to whether a monaural or binaural aid should be used.

The Hazard of Noise

In primitive civilizations noise was, most likely, not a cause of damage to hearing, although it probably always was an annoyance. With the birth of the industrial revolution, however, a new invisible effluent of technology—noise—was produced by factories, and many workmen were exposed to noise levels which caused hearing damage. Today, with jet aircraft and highly amplified rock bands, we are all being subjected to an ever increasing hazard of noise.

Like the quantities we considered in section 3.B, noise can be defined either physically, in terms of frequencies and sound levels, or in terms of its effect upon humans. We will begin with the physical definition. *Noise* is considered to be any vibration which lacks the regularity which, we have seen, characterizes musical sounds. In other words, noise is an erratic, intermittent, or statistically random oscillation. As we have seen, most musical and speech sounds, even though complex, display a regularity: the wave pattern repeats itself periodically. If we plot the time-varying presssure for a noise, however, as illustrated in Fig. 3.E-1, the irregular nature of the pattern is apparent. We have shown that a periodic complex wave can always be broken down into sinusoidal components whose frequencies are harmonically related. Waves which are not periodic, like noise, can also be broken down into sinusoids, but their frequencies are not harmonically related.

Fig. 3.E-1. Presssure wave of noise.

The subjective definition of noise considers only the nuisance effect of sound. To a gardener, any plant growing in the wrong place is a weed. Using this analogy, *noise* can be defined as any unwanted sound (at a particular place and time). Although the 1812 Overture may be music (barely) to some people, if you are trying to concentrate on a problem while it is being played it becomes noise to you.

Short-term hearing loss is an experience that almost everyone has had. When exposed to intense sounds we become "partially deaf" for a period of time until the ears recover. This temporary hearing loss means there is a temporary upward shift of our threshold of hearing. Since the shift is not permanent it is called a *temporary threshold shift* (TTS). A loud hand clap close to the ear will produce a short-term hearing loss, as will several hours of exposure to very loud sounds. Recovery from a TTS of 30-40 dB usually requires several hours. For a shift of 50-60 dB even several days may be insufficient for full recovery. These temporary losses may not be too critical if they are not incurred often. However, studies show a positive correlation between TTS acquired on a regular basis and a PTS (permanent threshold shift). Fig. 3.E-2 illustrates temporary and permanent losses of band members and dancers exposed to very intense rock music. The vertical axis gives the threshold shift in dB, 0 dB being the average normal threshold. Note that the dancers suffer a smaller loss than the band members, but that the nature of the permanent loss (solid curve) is similar to the temporary loss (dashed curve) for both dancers and band members.

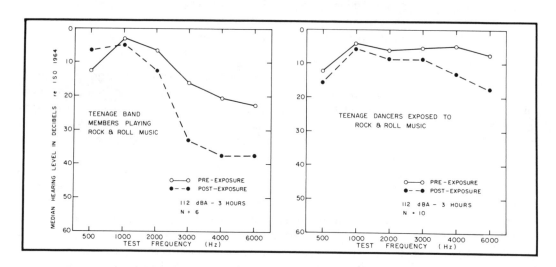

Fig. 3.E-2. Hearing levels of teen-age rock-and-roll musicians and dancers, measured just before and 5-11 minutes after a three-hour "rock session" with average sound levels of 112 dB(A). (From Cohen, et al., 1970. Data from PHS sample observations.)

Despite the data just presented, it should not be assumed that a permanent threshold shift caused by nerve deafness depends only upon the overall noise level. Whether or not a TTS will become permanent depends upon a person's total exposure to noise. Total noise includes the overall noise level (average dB level), the time distribution of the noise (continuous or intermittent), and the person's total lifetime exposure. The hazard of noise is the result of these three factors working in concert.

Exposure to a short duration of an extremely intense sound, such as an explosion, can produce a sudden (and usually permanent) hearing loss. Hearing loss can be an occupational hazard for people who work in very noisy environments, such as found in certain factories or around jet aircraft. The losses in these latter cases are not as sudden as for explosions, but they can be appreciable over comparatively short periods of time. In either case, exposure to adverse environmental conditions can give rise to a premature deterioration of the hair cells and thus to an early hearing impairment. Fig. 3.E-3 illustrates a hearing loss due to exposure to a firecracker explosion. The loss appears permanent because little recovery is noted after five months. Fig. 3.E-4 illustrates hearing losses due to use of a lawnmower on a regular basis without adequate ear protection. Fig. 3.E-5 illustrates hearing losses due to use of firearms on a regular basis without adequate ear protection. Note that the left ear, which has greater exposure to sounds from the muzzle, shows an appreciably greater loss than the right ear.

As a means of controlling or preventing noise-induced hearing loss, a set of curves called the "damage risk criteria" have been developed. These curves, shown in Fig. 3.E-6, show the duration of a given combination of sound pressure level and frequency range that the average adult can tolerate without danger.

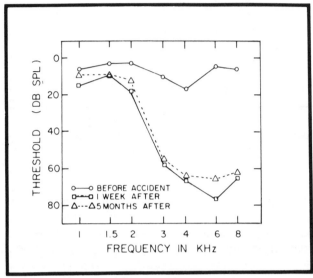

Fig. 3.E-3. Traumatic unilateral hearing loss caused by accidental exposure to a firecracker. (From Cohen, et al., 1970. Data from Ward and Glorig.)

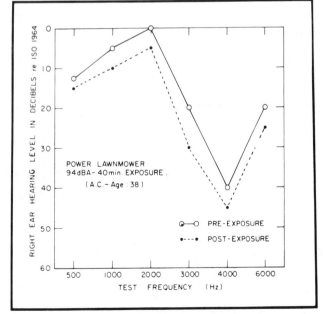

Fig. 3.E-4. Differences in pre-exposure and 2-8 minutes post exposure hearing levels. (From Cohen, et al., 1970. Data from PHS observations.)

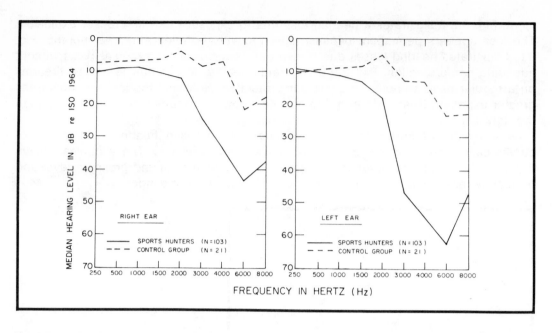

Fig. 3.E-5. Comparison of hearing levels of sports hunters with those of control group. (From Cohen, et al., 1970. Data from Taylor and Williams.)

An examination of the graph yields the following information: (1) in any frequency region the tolerable exposure period decreases as the decibel level increases; (2) at a constant sound pessure level the exposure period decreases as frequency increases (up to 4,800 Hz); and (3) for a constant exposure time the sound pressure level must decrease as frequency increases. Briefly, then, the damage-risk criteria specify the maximum period of time to which a person can be exposed to sound of various pressure levels and frequencies with minimal risk of hearing loss.

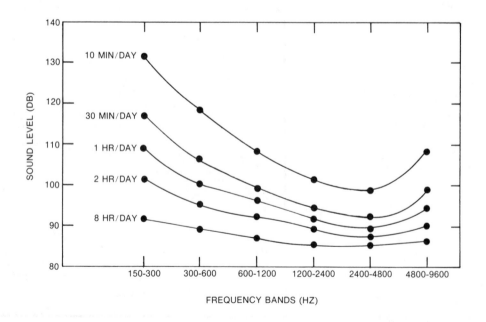

Fig. 3.E-6. Damage risk criteria for a single daily exposure. (After Kryter, et al., 1966)

Although industrial noise is still the principal cause of occupational hearing loss, many teenagers are being subjected to undue hazard when they listen to or dance to loud rock music. Charles Speaks and David Nelson reported the results of a study on the sound level produced by ten bands in the Minneapolis area. As a part of their research, they discovered that the sound levels ranged from 105 to 120 dB, which is about the same level of noise one would encounter in a boiler shop! Other experimenters have investigated the effects of loud rock music on the ears of guinea pigs (Lipscomb, 1969). After ninety hours of intermittent exposure to this music, the cells of the cochlea were examined. It was discovered that they "had collapsed and shriveled up like peas." A similar study (Dey, 1970) tested the effects of rock music on teenagers. The music was recorded in a local discotheque and played back to various subjects at the 100- and 110-dB levels for various time periods. Because the TTS was so great for those exposed to the 110-dB level, the researcher scrapped plans to play the music at the 120-dB level. In concluding his report Dr. Dey stated: "It is not likely that society could insist that our young restrict themselves to so mild a sound as 100 dB for two hours, so we shall have to reconcile ourselves to damaging the 14 percent most susceptible and later providing a variety of social rehabilitative and medicare support to which these persons certainly will eventually turn."

Because the hazard of intense sounds is beginning to be recognized by goverment officials, legislation has been enacted in an attempt to define conditions of noise hazard and methods of measurement. The Occupational Safety and Health Act of 1970 is a federal law which attempts to govern the amount of exposure to intense sounds. The law applies specifically to industry, but the sound levels recommended by the law are probably liberal guidelines for any exposure to intense sound. Table 3.E-1 shows maximum permissible exposure times for occupational noise measured with the A network of a sound level meter. Table 3.E-1 also shows some recommended maximum allowed exposure times for nonoccupational noise.

Table 3.E-1. Maximum allowed exposure times for occupational noise and recommended maximum exposure times for nonoccupational noise per 24-hour day at the sound levels shown. (After Cohen, et al., 1970.)

Sound level (dBA)	Occupational exposure	Nonoccupational exposure
80		4 hours
85		2 hours
90	8 hours	1 hour
95	4 hours	30 minutes
100	2 hours	15 minutes
105	1 hour	8 minutes or less
110	30 minutes	4 minutes or less
115	15 minutes or less	2 minutes or less

In a remote part of the Egyptian Sudan live the Mabaans, a tribe whose life is essentially free of the noises of modern civilization. Although the Mabaans are subject to presbycusis like everyone else, even Mabaans 70 years old have hearing acuity. similar to that of young boys (Rosen, 1962). Furthermore, there is no noticeable difference in hearing acuity between men and women. This result is in sharp contrast to the industrialized nations, where men show greater hearing losses than women of the same age. (Presumably this result is due to the higher noise levels encountered by men working in industry.) This effect is summarized in a striking manner by Fig.

3.E-7. Dr. Rosen's research suggests rather strongly that a substantial hearing loss with age is not a necessary part of growing older. Rather, it seems to be due to our total lifetime exposure to unnecessarily loud sounds.

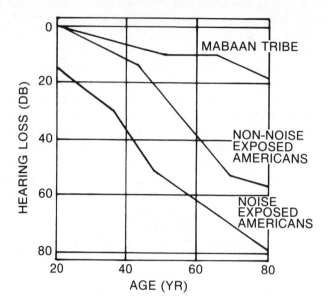

Fig. 3.E-7. Hearing loss with age for Mabaans, non-noise and noise exposed Americans. (After Rosen, 1962.)

Hearing Annoyances

Many noises do not constitute a hearing hazard, but these less intense sounds may still constitute a nuisance, leading to annoyance or frustration on the part of the recipient. Even mild noise can have a strong psychological effect on humans. Thousands of years ago the Chinese realized that to a man in quiet isolation the sound of slowly but steadily dripping water assumes the proportions of a loudly beaten drum. Such noise was used to drive a man mad or to break his will. Today, extremely quiet conditions are very difficult for city people to tolerate. Noise has become such an intimate part of their lives that silence is a condition to be avoided.

Community noises which qualify as annoying can be traffic, aircraft, factories, music halls, stereophonic equipment, radios, and television. The most common source of annoying community noise, however, is automotive traffic. In some cities the problem of aircraft noise is quite severe, although the actual effect of this noise on health is still a source of controversy. In order to minimize the effects of aircraft noise, standards and limits are often imposed on aircraft landing and taking off.

One can only wonder about the possible effects that our noisy modern appliances are having upon housewives. During the day their ears are bombarded by the sounds of mixers, blenders, garbage disposals, dish washers, clothes washers, dryers, and vacuum cleaners—not to mention radio, television, stereo, and crying babies!

Noise, like other types of pollution, has tended to increase with the advances of technology and with population increase. In the future, considerable effort is going to be necessary just to prevent a further increase in community noise levels. Lack of progress toward noise abatement in the past has been due to widespread apathy of the public and of government agencies toward this problem. The federal government, however, entered the arena wih the enactment of the Noise Control Act of 1972. This law deals with noise emission standards for products distributed in commerce and noise standards for aircraft. Provisions of this law are enforced by the Environmental Protection Agency. This act is aimed at dealing with major noise sources in commerce, including transportation, construction, motors and engines, and electrical and electronic equipment. The intention is that manufacturers be required to produce quieter products. The Environmental Protection Agency also was given the authority to require environmental impact studies (including noise impacts) of new highways, industrial installations, etc. If it is determined that the impact on the environment will be too great, approval of a project may be withheld.

Typical local noise ordinances deal with the annoyance aspect of sound. These ordinances usually specify a sound level in dB(A) that is not to be exceeded. These are often established for different areas—such as residential, business, and industrial—and for different times of the day. The actual values established depend on the particular community. (Some levels have been set so low that the local crickets were in violation.) Table 3.E-2 gives representative values. Intermittent noises (10% of any hour) are often permitted to be 5 dB(A) higher and impulsive noises (short bursts) 10 dB(A) higher than the values in the table. Current noise emission levels for cars and cycles is generally set around 80 dB(A) or less at a distance of 50 feet and for trucks around 85 dB(A). The trend is toward a further reduction of these allowed levels over the next several years.

Table 3.E-2. Representative values of allowed sound levels for continuous noise in a typical local noise ordinance for annoyance.

Area class	Day (dBA)	Night (dBA)
Residential	50	45
Business	55	50
Industrial	60	55

One difficulty in writing noise ordinances to control annoyance is that the reaction of different individuals to particular noises varies so widely. Most ordinances are written to satisfy some "average" person and, therefore, may fail completely to satisfy some individuals or groups. A noise must exceed the overall background noise to be considered annoying. If it has special qualities, such as being musical, having a rhythm, or being speechlike, it may be particularly noticeable. The past experience of the individual and his conditioning to noise will have a significant effect on his response to noise; different cultures will even respond differently to the same noise. Emotional factors also can influence response to noise. Once a person has been "emotionally involved" with a sound he will be more sensitive to it, either positively or negatively. People are usually more annoyed by the sounds that someone else creates than by the sounds they create themselves. For many people, producing loud noises is emotionally satisfying because the noises represent the accomplishment of something. To those listening, however, the effect may be unpleasant.

The manner in which a recipient of noise responds is often determined by his mood. The same sound at different times of the day may evoke different responses, depending upon whether we are tired, angry, busy, or relaxed. Another factor which can influence our reaction to sound is whether the noise is essential or not. We may accept the hum and buzz of an air conditioner or a refrigerator since we deem them necessary, but the sounds of hard rock music coming from the stereo may be perceived as annoying because we believe they are nonessential. Also, sounds with visual information are often more acceptable to the listener than sounds without an accompanying visual stimulus. The sounds of a shoot-out on T.V. could be quite noisy and bothersome unless accompanied by the visual part of the program. Likewise the sound of loud laughter from the apartment next door could be very annoying; but in conjunction with the funny antics of two puppies at play could be quite enjoyable.

Other factors which influence our reaction to sounds are their frequency of occurrence and their predictability. An annoying sound becomes increasingly bothersome the more often it is repeated. Also a very loud, unpredictable noise, such as a car backfiring or a sonic boom, causes a startle reaction and thus is extremely annoying.

Exercises

1. Considering hearing impairments due to problems in the outer, middle, or inner ear, which one is likely to be most serious? Least serious? Why?

2. Indicate what type of loss is associated with the following parts of the ear: outer, middle, and inner. What are the possible causes of the loss and means for correcting it?

3. Which frequencies are affected most in hearing loss due to aging (presbycusis)? Which frequencies are affected most by exposure to intense sounds?

4. Hearing loss due to exposure to adverse environmental conditions might be termed "premature presbycusis." In what sense is this so?

5. Determine the amount of hearing loss at frequencies of 100, 200, 1000, 2000, 5000, and 10,000 Hz if the measured thresholds are 40, 25, 5, 5, 20, and 50 dB respectively. (Use the threshold curve in Fig. 3.B-1 as the reference).

6. After attending weekly rock sessions for a period of three years a person is observed to have a hearing threshold of 20 dB at 2000 Hz. What is the hearing loss (in dB)?

7. After a day at the rifle range, a person's right ear is found to have a threshold of 20 dB at 2000 Hz. What is the temporary hearing loss (in dB), assuming that the hearing was normal at the beginning of the day?

8. What occupations are most likely to produce some hearing impairment?

9. What precautions might be taken to reduce the extent of hearing impairment in the occupations of exercise 8?

10. Farmers who spend a great deal of time driving tractors are observed to have greater hearing loss on the average than people who live in urban areas. Why?

11. What noise hazards exist in a typical house or yard? What noise annoyances?

12. Which feature, hazard or annoyance, is most important in typical noise ordinances?

13. What criteria would you use if you were responsible for writing a noise ordinance?

14. We have defined noise two different ways. If you are enjoying the experience of very loud rock music, does it qualify as noise by either of these definitions? How could you interpret the word "undesirable" in the second definition so that very loud, but enjoyable musical sounds are included in this definition?

*15. Noise levels in residential, business, and industrial areas
   a. Use the sound level meter to measure the A-scale readings in three different residential areas. Describe the areas and record the readings.
   b. Repeat 15a for three business areas.
   c. Repeat 15a for three industrial areas.
   d. How do your results compare wih those from a typical noise ordinance in Table 3.E-2?
   e. Do any of the levels you measured represent a hazard to hearing?

*16. Personal audiogram
   Schools that provide training in audiology are often willing to run audiorams free of charge (or at a very nominal charge) in order to provide experience for their students in audiometry. Check your locale and make arrangements to get an audiogram run on yourself.

a. Get a copy of the audiogram. If your audiogram shows losses at any frequencies, try to account for how the losses may have come about.
b. What is the reference level relative to which losses are shown in an audiogram? Is it the same at all frequencies?
c. Interpret the reference line in a typical audiogram in terms of an ''average'' threshold of hearing curve.

Further Reading

Baron, chapters 1, 2

Chedd, chapter 7

Fletcher, chapters 19, 20

Gerber, chapter 4

Stevens and Warshofsky, chapters 7, 8

White, chapters 7, 8, 9, 10, 11, 12, 13, 17

Angevine, O. L. 1975. ''Individual Differences in the Annoyance of Noise,'' Sound and Vibration 9 (11), 40–42.

Bragdon, C. R. 1974. ''Quiet Product Emphasis in Consumer Advertising,'' Sound and Vibration 8 (9), 33–36.

Cohen, A., J. Anticaglia, and H. H. Jones. 1970. ''Sociocusis—Hearing Loss from Non-occupational Noise Exposure,'' Sound and Vibration 4 (11), 12–20.

Dey, F. L. 1972. ''Auditory Fatigue and Predicted Permanent Hearing Defects from Rock-and-Roll Music,'' New England J. of Medicine 282, 467.

Fletcher, D. H., and C. W. Gross. 1977. ''Effects on Hearing of Sports-Related Noise or Trauma,'' Sound and Vibration 11 (1), 26-27.

Kryter, Ward, Miller, and Eldridge. 1966. ''Hazardous Exposure to Intermittent and Steady State Noise,'' J. Acoust. Soc. Am. 39, 451–64.

Lipscomb, D. M. 1969. ''High-Intensity Sounds in the Recreational Environment,'' Clinical Pediatrics 8, 63.

Miller, J. D. 1974. ''Effects of Noise on People,'' J. Acoust. Soc. Am. 56, 729–64.

Noise Control Programs and Ordinances. 1974. Sound and Vibration 8 (12), 10–30.

Rosen, S., et al. 1962. ''Presbycusis Study of Relatively Noise-Free Population in Sudan,'' Annals of Otolaryngology, Rhinology, and Laryngology 71, 727.

Schwartz, J. M., W. A. Yost, and A. E. S. Green. 1974. ''Community Noise Ordinance in Gainsville, Florida,'' Sound and Vibration 8 (12), 24–27.

Shaw, E. A. G. 1975. ''Noise Pollution—What can be Done?'' Physics Today 28 (1), 46–58.

Simmons, R. A., and R. C. Chanaud. 1974. ''The 'Soft Fuzz' Approach to Noise Ordinance Enforcement,'' Sound and Vibration 8 (9), 24–32.

Sound and Vibration News. 1972. ''Summary of Noise Control Act of 1972,'' Sound and Vibration 6 (11).

Audiovisual

1. *Death Be Not Loud* (26 min, bw, 1970, MGHT)
2. *Ears and Hearing,* 2nd edition (22 min, color, 1969, EBE)
3. *Noise and Its Effects on Health* (20 min, color, 1973, FLMFR)
4. *Quiet Please* (21 min, color, 1971, JACBMC)

Demonstrations

1. Sound level meter and measurements of sound levels, that can be created in a classroom

Chapter 4

Acoustical Environments

4.A. LISTENING ENVIRONMENTS

An optimum listening environment should provide for the elimination of noise and the enhancement and proper control of the sounds of interest. We consider first the problem of noise and then several different listening environments.

Insulation from Noise

We have defined *noise* as any unwanted sound or as any sound that competes with a desired sound. We are concerned here with noises that annoy and distract from listening to some performance. Noise can come from sources that are external to a listening enclosure (e.g., aircraft and traffic noise in the vicinity of a concert hall). Noise can also arise from sources inside a listening enclosure (e.g., air conditioning and the people noise within a concert hall).

Noise from either source can ruin an otherwise well-designed auditorium. Nevertheless, examples abound where adequately insulating against noise was ignored or where fallacious principles were used. One of the most common misconceptions about sound insulation is that materials which are effective as heat insulators are also useful for sound insulation. This is, in almost all cases, untrue. Porous materials which are good heat insulators are good absorbers of sound, but they are usually poor insulators. Acoustical absorbers can reduce the noise level within a room, but absorbing material will have little effect on noises transmitted into a room from outside. In this section we will consider the sources of noise which can be transmitted into a room and how the transmitted noise can best be attenuated.

Some of the ways by which noise can be transmitted from a source to a listener are shown in Fig. 4.A-1. The noises are produced in three ways: (1) directly in the air (the voice or a musical instrument), (2) by impact (slamming a door or banging on the floor), and (3) by vibrating machinery. The noise can then travel throughout the building in two paths: (1) through air only (airborne noise), or (2) by vibrations of the solid structure of the building (structure-borne noise), such as those caused by motors and fans. These two types of noise transmission differ in several important ways. Impacts are generally short but powerful. Consequently, they can be propagated great distances, often with little attenuation. Sometimes the radiated noise is even increased when a larger surface area is set into vibration. Airborne noises are usually of smaller power but longer duration. The disturbance they cause is usually confined to the immediate vicinity of their origin. Furthermore, a boundary which provides good insulation for airborne noise may be very poor at attenuating impact noises. Airborne noise is generally transmitted in two ways: (1) through openings and (2) by means of forced vibration. Open windows and doors are the most common culprits, but even the crack under a closed door can transmit a substantial percentage of the incident noise. Heat ducts are the next most common opening which transmits unwanted sounds, but gaps around water pipes and electrical fixtures also provide paths for noise. Forced vibration occurs when a noise in one room is transmitted through the wall to the wall of an adjacent room. The wall in the adjacent room is caused to vibrate, thus transmitting the sound. A listener in the adjacent room hears a "reproduction" of the original sound provided by the partition. Reradiated airborne noise can best be attenuated by a more massive partition which offers greater resistance to vibration. Noise from internal sources, such as that generated by an air conditioner, or sounds from other rooms are often carried by the ductwork. To reduce this transmitted noise, the ducts can be lined with sound-absorbing materials.

Fig. 4.A-1. Some of the many sources of sound and paths by which it reaches a room.

Structure-borne noise and vibration, however, should be suppressed at the source. Impact noises can be greatly reduced by resilient flooring, such as carpeting or foam-backed tile. To prevent the transmission of machinery vibration, flexible mountings or antivibration pads are necessary. If the noise cannot be adequately suppressed at the source, the various structures should be isolated from each other. That is, a floor can be acoustically isolated from the ceiling below if the ceiling supports are connected to the floor joists by resilient mounts.

External noises are dealt with in two basic ways. One is to reduce the amount of sound emitted by any particular device, such as an aircraft, an automobile, or a truck. However, there are practical limits below which sound emission cannot be reduced, especially near an airport or a heavily traveled highway. Often zoning regulations are imposed in these circumstances, so that certain types of buildings (e.g., homes, schools, hospitals) are not allowed within particularly noisy environs. Unfortunately, zoning is not always practiced with enough consistency to produce satisfactory results.

Under the best of circumstances, with zoning and control of noise emissions there still remain significant amounts of noise, even in residential areas. These can be prevented from entering a building by surrounding the building with a sound barrier. Two basic conditions must be satisfied by this barrier: (1) it should be massive, and (2) it should be airtight. When sound waves in air strike a massive barrier, they are mostly reflected because of the mismatch in sound speed and density between the air and the barrier. Even though a barrier is massive, however, it will not be effective unless it is made fairly airtight. For example, suppose a window does not fit well, so that the cracks around the edge are equal in area to 1% of the area of the window. About 4% of the total sound energy striking the window will leak into the dwelling, which corresponds to a loss of only about 14 dB, even though an airtight window may provide a loss of 30 dB or so. As poor-quality construction can thus negate the value of a good barrier, care in construction to provide tight fits is in many cases more important than the selection of proper sound barriers. Typical sound loss characteristics of various barriers are shown in Table 4.A-1. Note that, in general, the greater the mass of the partition, the more the transmitted noise will be attenuated.

Table 4.A-1. Typical sound transmission losses for various acoustical barriers. (The losses expressed in dB are for the frequencies shown and assume no leakage around the barrier.)

	Frequency—Hz		
Barrier	125	500	2000
Wall—solid			
.25 kg/m²		23 ⎫ average	
1.0 kg/m²		29 ⎪ for 100	
5.0 kg/m²		38 ⎬ to 3200	
25 kg/m²		50 ⎭ Hz	
Wall—double			
with air core			
(increase in loss			
over solid wall			
of same mass)			
4 cm air	1	2	8
15 cm air	4	11	18
Doors			
4 cm hollow core			
(14 kg)	11	16	22
Solid (27 kg)	15	14	25
4 cm solid core			
(42 kg)	20	14	26
Wall—double and			
filled with			
17 cm foam	28	51	61
6 cm foam and			
1 cm sound board	27	45	58
10 cm mineral wool			
and 1 cm sound board	28	46	60
Windows			
.3 cm glass	12	20	25
1 cm insulating glass	17	19	27
same with storm sash	16	27	35

As an example of a structure where the external environment noise was a particular problem, consider the J. F. Kennedy Center for the Performing Arts in Washington, D.C. The site chosen for the center is on the Potomac River, close to the National Airport. Aircraft often fly as low as 200 feet above the roof, and low-flying helicopters are often seen in the immediate vicinity of this building. In addition, there are the usual noises of automotive traffic around the structure. The design utilized to suppress the external noise is that of a box-within-a-box. The three auditoriums are completely enclosed within an exterior shell. Furthermore, the columns suppoting each auditorium have been designed to isolate both airborne noise and mechanical vibrations from the interior surfaces. A double-walled construction with enclosed air space is used for the exterior shell. The noise of interior sources is controlled with resilient mounts, flexible connectors, and acoustically lined ductwork. At all outside entrances special doors are used, and there is a "sound lock" region between the foyer and the interior of each auditorium. Because of this special construction it is possible to enjoy concerts without undue interference from internal or external noise sources.

Sounds in the Open

The auditorium (which literally means "hearing place") developed from the Greek open-air theater (which can be loosely translated as "seeing place"). Although the Greeks did not consider acoustical principles when constructing the open-air theaters, they did sometimes use a sound-reflecting enclosure around the actors, and they arranged the seats in a semicircular pattern on a steep embankment. It is well known that outdoor listening conditions are generally poor, and by modern standards the typical Greek outdoor theater would be unacceptable for speech or music. Today many open or semiopen structures are being designed to provide good listening conditions for large audiences. In comparing open and enclosed structures, the open structure has certain definite advantages. It is less expensive to construct, it can easily accommodate very large audiences, and it is usually located in a rural area, thus lending an aura of informality. The main problem associated with unenclosed structures is the difficulty of hearing, due to the sound absorption of the audience, the loss of all sound energy directed upward (over the audience), and the interfering noises from external sources. To overcome these difficulties, unenclosed structures should be designed so that the concert areas are as far from sources of noise as possible. Placement should also utilize any natural sound barriers (e.g., hills, vegetation) between sources of noise and the performers and listeners. Not much can be done beyond this to control external noises. Sound in the open spreads out and becomes less intense (see section 2.F) as it travels, and this can create listening levels that are too low to be acceptable. Desirable acoustics for an outdoor arena are achieved first by having a large reflecting surface, called the band shell, behind the stage. The band shell directs the sound toward the audience and helps the members of a musical group to stay in tune and in time with one another. Some of the better classical Greek theaters provided a reflective surface to reinforce the direct sound. When large masks were also used to amplify their voices, an adequate environment for drama was created.

Electronic reinforcement of sound is also an important means of overcoming the problem of inadequate sound power. However, even with sound reinforcement the listeners close to the performers will hear much louder sound than those farther away. As we will see later, this is not the situation for a well-designed concert hall.

Sound in Enclosures

Sound waves outdoors are quickly attenuated as the distance from their source increases. In an enclosed space, however, very little sound is lost by transmission through the wall; so the energy remains within the room. Fig. 4.A-2 illustrates how sound waves, in striking various surfaces, could be reflected, absorbed, dispersed, diffracted, or transmitted elsewhere. Let us now consider each of these processes in more detail.

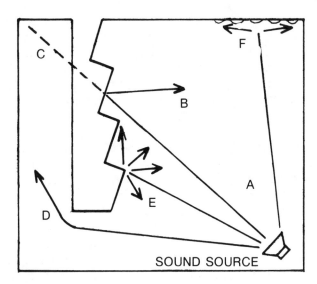

SOUND SOURCE

Fig. 4.A-2. Paths of sound in an enclosure: (A) direct from a source; (B) reflected from a wall; (C) transmitted through a wall; (D) diffracted around a corner; (E) diffused from an irregular object; (F) absorbed at a surface.

Sound reflection was discussed in section 2.B, where the law of reflection was mentioned. This relationship holds only when the wavelength is small compared to the reflecting surface. Hard, rigid, flat surfaces (concrete, plaster, etc.) reflect most of the sound energy which strikes them. Concave surfaces tend to focus (or concentrate) reflected sound waves, while convex surfaces tend to disperse the waves. When sound energy is incident on a soft, porous material some of the sound energy is changed into heat, leaving less energy to be reflected. The change of sound energy into another form is known as *absorption.* Sound absorption can be realized by treated walls, floors, ceilings, or by objects within the room (draperies, people, etc.).

Sound diffusion occurs when there is an approximately equal distribution of sound energy in a room. Adequate diffusion is highly desirable in concert halls, as it helps accentuate the natural qualities of music. Diffusion of sound is usually acquired by irregularly shaped objects which scatter the sound. Diffraction, which was discussed in section 2.B., helps to spread the low-frequency sounds (below 200 Hz) beneath balconies and into other places where there may be no direct sound path.

As pointed out previously, in addition to modifying the sound, an enclosure also serves the important function of blocking out external noise. Once a location has been selected for an enclosure, typical noise conditions at the site should be measured and then construction materials should be selected to provide adequate internal noise levels. Table 4.A-2 shows typical tolerable levels. One can select the building materials after determining the required transmission loss (in dB) as external sound level (dB) minus the tolerable internal sound level (dB).

Table 4.A-2. Typical tolerable noise levels (in dB). (After Olson.)

Studio	20-30
Theater	25-35
Auditorium	25-40
House	35-45
Hotel	40-50
Offices	40-50
Eating places	45-55

Achieving an appropriate reverberation time (see next section) is a prime consideration for the internal design of an enclosure. Short reverberation times provide greater clarity for individual sounds, while long ones provide more blending of the sounds. In general, a shorter reverberation time is desirable when the room is to be used for speech, so that speech sounds being produced at any instant will not have too much blending with previously spoken sounds. The object with speech is intelligibility, or understanding of what is said. Some kinds of music (e.g., chamber music) are better presented with short reverberation times to provide clarity. Other kinds of music (e.g., organ music) are better presented with long reverberation times to provide blending.

The reverberation time alone gives an insufficient description of a room for optimum results. If the room focuses the sound, dead spots may be produced and the sound will not be of approximately uniform intensity throughout. Problems of this nature can generally be avoided by having convex reflecting surfaces on the walls and/or by having nonparallel walls and by placing absorbing materials properly so that the sound is made diffuse.

There should be an appropriate mixture of direct sound (sound arriving first with no reflections), early reflected sound (sound next and within 0.05 seconds of the direct sound), and reverberant sound (all other reflected sound). If there is strong reverberant sound reaching the listener more than about 0.05 seconds after the direct sound, it is heard as an echo and is generally objectionable. Echoes from distant side walls or from rear walls can be avoided by tilting the walls so that they do not reflect sound back to the vicinity of the source. Early reflected sound (within about 0.05 seconds of the direct sound) blends with the direct sound to produce a single acoustical image. The best direction for early reflected sounds is from the side walls of the auditorium or from structures enclosing the performing group. The reflected sounds should generally be no more than about 10-12 dB more intense than the direct sound for best results. Many early reflections are desirable to provide adequate diffusion.

Another feature of interest in concert hall and auditorium design is the effect the rows of seats have on the direct sound as it travels from the stage to the back of the hall. When the rows of seats are placed on a level floor so that the direct sound passes over front rows to get to rows in the back, an abnormally high reduction in energy occurs at frequencies of about 100-300 Hz. This problem can be dealt with by providing a floor whose elevation increases from front to back so that the direct sound can reach all of the seats unobstructed by the seats in front.

In closing, we note that as much consideration should be given to acoustical aesthetics as to visual aesthetics in the design of enclosures in which sound is to play a significant role.

Electronic Reinforcement

In a well-designed auditorium of medium size a speaker on stage can be understood at the rear of the room without the aid of electronic amplification equipment. In poorly designed or very large halls, however, a sound amplification system is required to insure adequate loudness and good sound distribution. Sound-reinforcing systems are also used to provide desired directional properties of radiated sound, to provide desired time delays, and to correct improper spectral balance in enclosures. The basic elements of electronic sound reinforcement systems are microphones, amplifiers, and loudspeakers, with optional electronic delays and equalizers.

A well-designed sound system should do all of the following: (1) it should transmit a wide range of frequencies (30 Hz to 15,000 Hz) without distortion; (2) it should be capable of a wide dynamic range without distortion; (3) it should be free from echoes or feedback; (4) it should remain undetected, giving the illusion that the amplified sound is the actual sound source. The three basic components of any sound-reinforcing system are microphones, amplifiers, and loudspeakers. It is important that each component of the system be properly matched and that adequate fidelity be maintained in each component.

138

Generally speaking, public address systems require the least fidelity and systems for reinforcing music the highest fidelity.

Microphones and loudspeakers should be placed properly to provide best results. For a speaker at a podium, placement of the mike some 30-50 cm from the speaker should be adequate. Use of a unidirectional mike is desirable to reduce pickup of audience noise. Several microphones may be required for use with large performing groups. They should be placed several meters from the nearest performer (except in circumstances where one performer is to be emphasized above the group, as in the case of a soloist), so that a fairly uniform response is obtained over the whole performing group. Microphones used with theatrical productions are usually placed in the footlight trough, although other placements to achieve special effects may be employed. The mikes should be placed every 3 or 4 meters and may have selected directional characteristics. Loudspeakers should be placed so as not to produce much sound that is picked up by the microphones; otherwise the system squeals in a very undesirable manner. Loudspeakers placed at the front of the hall tend to produce a better illusion of the sound coming from the original source. Because sound travels much faster in the electrical part of the system than in air, loudspeakers may produce sound that will arrive at the listener before the sounds going directly from the source to the listener. This may create the effect of an apparent source at the loudspeakers rather than at the actual source. This problem can be reduced by delaying the sound that goes into the loudspeakers, by means of electrical or mechanical delays. It is possible to create loudspeaker arrays that will beam the sound into selected parts of the hall, which may be useful in concentrating the sound on the audience and avoiding undesirable reflections from back walls or ceilings in the hall. The general rule is that a loudspeaker array with a large width will produce a beam of sound with a narrow width (see section 2.F).

If a hall responds nonuniformly at different frequencies, it is often possible to improve the situation by inserting a *spectrum equalizer* into the sound reinforcement system. A spectrum equalizer is a collection of variable-gain filters that cover some frequency range of interest. Frequencies emphasized by the room can be deemphasized in the equalizer before the sound is radiated by the loudspeakers, and in this way the reinforced sound in the room can be made to have greater spectral uniformity than the original.

Exercises

1. List several external sources of noise that may have a significant effect in a classroom, a living room, a bedroom, or a concert hall. Describe the features of the source and the path the sound travels. Describe steps that might be taken to eliminate or reduce the problem.

2. Repeat exercise 1 for internal sources of noise.

3. Why are massive partitions needed to eliminate external sound? Why are airtight partitions needed to eliminate external sound?

4. How successful will a massive partition be in eliminating external sound if it is not airtight? How successful will an airtight partition be if it is not massive? Give examples.

5. What units in a typical house are most likely to permit external sounds to enter? What practical steps can be taken to correct the problem?

6. A person attempting to sleep in a hot hotel room opens the window to get some cool air. The window provides a transmission loss of 30 dB when closed. When open it allows 25% of outside sound energy to get in. What is its transmission loss when open? How effective is this transmission loss for providing quiet sleeping conditions?

7. What role does zoning play in dealing with external noise sources?

8. Reverberation times for optimum results vary between speech and music. Why should the reverberation time for speech generally be shorter than for music?

9. What type of music requires the longest reverberation time for optimum results? What type of music requires the shortest reverberation time for optimum results?

10. Suppose people in the front row receive strong reflections from the back of a room 20 meters away. Will it be perceived as an echo?

11. Suppose a time delay of 0.2 seconds produces maximum interference for a speaker. How far away from the speaker is a rear wall that will produce reflected waves giving rise to maximum interference? How does this distance compare with the distance between stage and back wall in typical theaters and concert halls?

12. How are the potential problems of delayed feedback avoided in drama theaters and concert halls?

*13. A sound level of 90 dB is measured through an open door. When the open door is draped, the measured level is 88 dB. When the door is closed, the level is 76 dB; and when the door is closed and the cracks around the opening stuffed, it is 72 dB. What is the transmission loss in each case? How well do these values agree with those in the text? How can you account for the discrepancies?

*14. A sound level of 100 dB is measured through an open door. When a 60-cm layer of dense mineral wool fills the opening, the level is reduced to 70 dB. When a 15-cm-thick solid-core door fills the opening, the level is reduced to 65 dB. When both are used, the level is 55 dB. How do these values agree with those in the text? How do you account for the discrepancies?

Further Reading

Backus, chapters 9, 15
Beranek, chapters 1–12
Culver, chapter 17
Doelle, chapters 4, 13, 16
Olson, chapters 8, 9
Stevens and Warshofsky, chapter 8
U.S. Gypsum, chapters 2, 3
Winckel, chapter 4
Beranek, L. L. 1975. "The Changing Role of the 'Expert'," J. Acoust. Soc. Am. *58*, 547–55.
Bishop, D. E., and P. W. Hirtle 1968. "Notes on the Sound-Transmission Loss of Residential-Type Windows and Doors," J. Acoust. Soc. Am. *43*, 880–82.
Harris, C. M. 1972. "Acoustical Design of the John F. Kennedy Center for the Performing Arts," J. Acoust. Soc. Am. *51*, 1113-26.
Heebink, T. B. 1970. "Effectiveness of Sound Absorptive Material in Drywalls," Sound and Vibration *4* (5), 16-18.
Schultz, T. J., and B. G. Watters. 1964. "Propagation of Sound Across Audience Seating," J. Acoust. Soc. Am. *36*, 885-96.
Sessler, G. M., and J. E. West. 1964. "Sound Transmission over Theatre Seats," J. Acoust. Soc. Am. *36*, 1725-32.
Shankland, R. S. 1972. "The Development of Architectural Acoustics," Am. Scientist *60*, 201-9.
Shankland, R. S. 1973. "Acoustics of Greek Theatres," Physics Today *26* (10), 30-35.
Wetherill, E. A. 1975. "Noise Control in Buildings," Sound and Vibration *9* (7), 20-26.
Yerges, L. F. 1971. "Windows—The Weak Link?" Sound and Vibration *5* (6), 19-21.

Audiovisual
1. The Science of Sound (33 1/3 rpm, 4 sides, FRSC)
2. Acoustics of the Classroom (19 min, color, USNAC)

Demonstrations
Squeal with too much gain or poorly placed components in sound reinforcement systems

4.B. REVERBERATION TIME

As noted in the previous section, the manner in which an enclosed volume can affect an interior source of sound is very complex. The volume, the shape, the configuration, and even the furnishings will all have a profound influence on the sound perceived by a listener. The single most important property of any listening environment, however, is probably the reverberation characteristics of the room. Reverberation is the prolongation of a perceived sound, after the source is turned off, due to successive reflections in an enclosed space. In this section we discuss some of the properties of a room which determine the reverberation time and the optimum reverberation times for various auditoriums.

Absorption and Reverberation

When a sound wave strikes a wall in a room, part of the energy is reflected, but part of the energy is absorbed by the wall and converted into heat or other non-sound energy. An extreme for lack of absorption might be a concrete-walled room in which little energy is absorbed each time a wave encounters a wall. A long time is then required for the energy in the room to die out because the sound wave must run into the walls many times. (Losses of energy to the air through which the sound travels may be important when losses to the walls are small.) Another extreme for absorption is the *anechoic chamber* (or outside away from buildings and other obstacles), in which almost all of the sound energy is absorbed the first time a wave strikes the highly absorbing walls of the chamber; the sound dies out almost immediately. The amount of time required for a sound to die out is termed reverberation time. Obviously, the less the absorption in a room, the greater will be the reverberation time, and vice versa.

Consider a hard-walled tube 34 m in length, as shown in Fig. 4.B-1. Imagine that a pulse of sound is somehow introduced into the tube so that it bounces back and forth between opposite ends of the tube. Assume that no sound energy is lost to the air in the tube or to the sides of the tube, but that all losses are at the ends of the tube. (In many practical cases, such as a wind instrument or the vocal tract, this assumption does not hold true, because as a wave moves from a reed or the vocal cords part of the sound energy is absorbed by the walls of the instrument or the vocal tract and a weaker pulse reaches the end of the instrument than left the reed or the vocal cords.) If the ends of the tube were completely nonabsorbing, the pulse would bounce back and forth forever. If the ends of the tube were moderately absorbing, the pulse would bounce back and forth for a long time, but would eventually die out. If the ends of the tube were highly absorbing, the pulse would die out quickly. We could place a microphone at some point in the tube (see Fig. 4.B-1) and measure how quickly the pulse died out. The microphone would detect the pulse once every trip along the tube, or, since $t = l/v$, once every $t = (34 \text{ m}/[340 \text{ m/sec}])$ sec. The plots in Fig. 4.B-2 illustrate microphone measurements for the cases of no absorption, small absorption, and large absorption. The energy level of each reflected

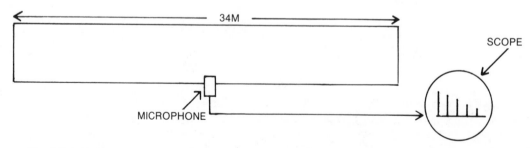

Fig. 4.B-1. Hard-walled tube containing a pulse bouncing back and forth between opposite ends.

142

pulse is plotted in dB for each of the three cases. The energy level decrease (in dB) after each reflection can be determined by dividing one by the fraction of energy reflected and then taking 10 times the log of the ratio as energy level loss = 10 log (1.0/fraction reflected). Case 1 assumes that none of the energy is absorbed and that all (1.0) is reflected, or a 0-dB loss per reflection; case 2: 0.27 absorbed, 0.63 reflected, 2-dB loss; case 3: 0.75 absorbed, 0.25 reflected, 6-dB loss.

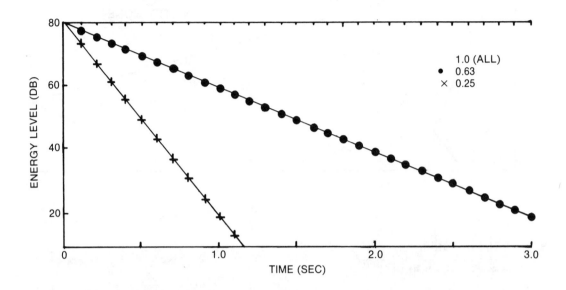

Fig. 4.B-2. Energy levels of successive reflected pulses in a tube 34 m long assuming no loss at the tube walls. Three cases are considered: energy is all reflected at each end, 0.63 reflected, and 0.25 reflected.

Now suppose we are in a room with perfectly reflecting walls. At another place in the room a toy balloon is popped. The first sound we hear travels directly from the source. After a short time we will hear the first reflected sound, followed by more and more reflected sounds. This is illustrated in Fig. 4.B-3. After a while the original wave will be traveling in all directions, and so it will be spread around the room in a fairly uniform manner. Fig. 4.B-3b illustrates what we would hear in the above situation. After the direct sound and the first reflection, the reflections would be so close in time that we would perceive a continuous diffuse mixture of sound. This is known as reverberation.

In this hypothetical room the reverberation would continue for a very long time, as the only absorption would be due to the person in the room and the absorption of the air. In actuality, the walls, ceilings, floors, and furnishings of rooms all absorb some sound energy and convert it to heat. The reverberation is thus reduced considerably.

Reverberation Time

Reverberation time (RT) is defined as the time required for the sound level to decrease by 60 dB. Note that for the case of complete reflection in Fig. 4.B-2 the RT is infinite, for 0.63 reflection RT is 3 seconds, and for 0.25 reflection RT is 1 second. The larger the reflection (or the smaller the absorption) the longer the RT. Materials are usually described in terms of their absorptive properties so that RT may be expressed as being inversely proportional to absorption. For the sound pulse in a tube the RT also depends on the length of the tube and on the speed of sound in the tube. (Clearly, if the pulses moved faster or if the tube were shorter the RT would be shorter because the pulse would strike the ends of the tube more times per second.)

143

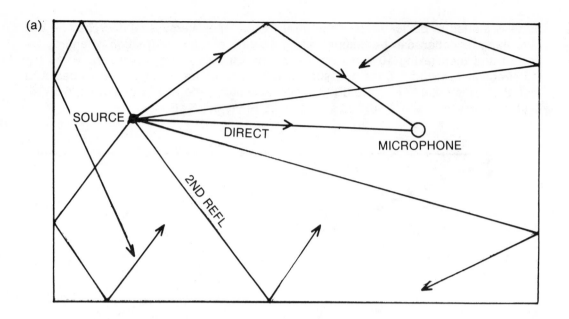

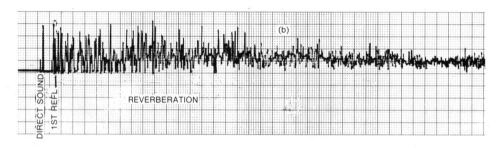

Fig. 4.B-3. (a) Paths of direct and reflected sound from source to "listener"; (b) simulated microphone response for sound in room. (Courtesy of S. E. Stewart.)

The RT in three-dimensional structures, such as rooms, is of primary interest in building design. As might be expected on the basis of the one-dimensional example, RT depends inversely on absorption, but it depends on the volume of the enclosure rather than on length. The expression (called the Sabine equation) is RT = 0.16 V/EAA, where V is the room volume in cubic meters and EAA is an equivalent absorbing area associated with the interior surfaces of the room. If all the surfaces of the room were completely absorbing, EAA would be the surface area of the room in m². The only perfect absorber, however, is an open window. If, instead of an open window, we have twice the area of some material which absorbs half of the incident sound energy, the resulting reverberation time will be exactly the same. Let us define the *absorption coefficient* (C) of any material as the fraction of energy absorbed on each reflection of a sound wave. Then we can see that for any surface area, A, EAA = A·C. When many different surfaces are present the total absorption is given by EAA = $A_1C_1 + A_2C_2 + A_3C_3 \ldots$, where the subscripts represent different surfaces. It will be noted that EAA has the dimensions of an area and that it is smaller in magnitude than the total wall area of the room.

Typical absorption coefficients are shown in Table 4.B-1. Note that the values vary with frequency for any given material and that most materials are more highly absorbing at higher frequencies.

The sabine equation will provide an estimate of reverberation time only when the sound is diffuse, that is, it dies away in a fairly smooth manner. A sound field is not diffuse when (1) acoustical absorbing materials are concentrated in one area, (2)

144

Table 4.B-1. Typical absorption coefficients for some building materials. (Actual values depend on backing and mounting of material. Entries for adult person and upholstered chair are in EAA values (m^2).)

| Material | Frequency—Hz | | |
	125	500	2000
Acoustical plaster	0.15	0.50	0.70
Acoustical tile	.20	.65	.65
Brick	.02	.03	.05
Carpeted floor			
heavy, on heavy pad	.10	.60	.65
light, without pad	.08	.20	.60
Concrete	.01	.01	.02
Draperies			
heavy	.15	.55	.70
light	.03	.15	.40
Fiberglass blanket			
2.5 cm thick	.3	.70	.80
7.5 cm thick	.6	.95	.80
Glazed tile	.01	.01	.02
Paneling—plywood supported at 1 m intervals and backed with 5-cm air space			
.15 cm thick	.10	.20	.06
.30 cm thick	.30	.10	.08
Plaster	.04	.05	.05
Vinyl floor on			
concrete	.02	.03	.04
Wood floor	.06	.06	.06
Adult person	$.30m^2$	$.45m^2$	$.55m^2$
Upholstered chair	$.20m^2$	$.35m^2$	$.45m^2$

there are curved surfaces which concentrate sound, and (3) when one dimension is substantially different from the other two. Since most listening environments do not have a totally diffuse sound field (such a thing would not even be desirable), there is often a considerable discrepancy between the measured and the calculated RT.

Ambient Energy Levels

We have to this point discussed the relationship of the RT of a room to the room volume and the room absorption. There are two related aspects of interest: (1) the buildup time and (2) the final intensity of a constant power sound source in a room. If a sound source having constant power output is placed in a room, the sound intensity in the room will increase until it reaches some final value (at which time the energy being absorbed is just equal to the energy being supplied by the sound source).

The relationships between absorption, power input, RT, buildup time, and final intensity level can be illustrated with the following example. Take a bucket and punch small holes (absorption) in its side from bottom to top. Now take a garden hose and turn it on to some steady flow rate (source of constant power). Let the hose run into the bucket and observe the time required (buildup time) for the water level in it to reach its final value (final intensity level). Now remove the hose and observe the time required for the water to drain from the bucket (RT). Note that RT and buildup time both tend to be of about the same length. If the holes in the bucket are small (small absorption), the times are long and the final water level (intensity level) is high. For large holes the times are short and the final level is low. The formula $I_{max} = P/EAA$

gives the relationship between I_{max}, the final intensity (in watts/m²); P, the power output (in watts) of the source, and EAA, the equivalent absorbing area (in m²). The final intensity of sound in a room can also be written by combining the two previous equations to give $I_{max} = (P) \cdot (RT)/(0.16)(V)$. From this it can be seen that the RT should be made longer for large rooms so that a sufficient energy level can be achieved. It can be seen that as the volume increases the RT must be increased to maintain a roughly constant intensity in the enclosure. Optimum RTs are listed in Table 4.B-2. (The RTs in Table 4.B-2 are generally valid at frequencies of 500 Hz but should increase slightly at frequencies above 4000 Hz. The RTs should also be increased at lower frequencies so that at 125 Hz the RT is about 1.4 times the values shown.)

Table 4.B-2. Optimum reverberation times (in seconds) at 500 Hz for different room sizes and uses. (Data from Knudsen, Olson, and U.S. Gypsum.)

Use	Room size—cubic meters			
	30	300	3000	30,000
Office—speech	.4	.6	—	—
Classroom—speech	.6	.9	1.0	—
Workroom—speech	.8	1.2	1.5	—
Rehearsal room—music	.8	.9	1.0	—
Studio—music	.4	.6	1.0	—
Chamber music	—	1.0	1.2	—
Classical music	—	—	1.5	1.5
Modern music	—	—	1.5	1.5
Opera	—	—	1.4	1.7
Organ music	—	1.3	1.8	2.2
Romantic music	—	—	2.1	2.1
Room in home—speech	.5	.8	—	—
Room in home—music	.7	1.2	—	—

Example of RT Calculation

We now consider as an example a living room 4 × 5 × 3 m, with heavy carpet on the floor, acoustical plaster on the ceiling, and 0.30 cm plywood paneling on the walls. We calculate the RT at 500 Hz by computing V and EAA and using the Sabine equation.

$$V = 4 \times 5 \times 3 = 60 \text{ m}^3$$

$$
\begin{aligned}
EAA = {}& 4 \times 5 \times .60 && \text{(floor)} \\
& + 4 \times 5 \times .50 && \text{(ceiling)} \\
& + 2 \times (4 \times 3 \times .10) && \text{(end walls)} \\
& + 2 \times (5 \times 3 \times .10) && \text{(side walls)} \\
EAA = {}& 27.4 \text{ m}^2.
\end{aligned}
$$

$$RT = (0.16V)/EAA = 0.35 \text{ sec}$$

When the sound power input is 0.01 watt the intensity is $I_{max} = .01/27.4 = .00036$ watt/m², which gives an intensity level of 76 dB.

RT in an actual auditorium can be measured by using a sound source at one location and a microphone at another location. The source can be used to provide quasi-steady sound (white noise or a warble tone) in the room and then the RT measured from the time the source is turned off. (A sine wave source should not be used, as complications may occur because of the natural modes of the room.) Often an impulsive sound, such as a pistol shot, is used as the source.

Exercises

1. Describe qualitatively the absorption characteristics and reverberation time you might expect to find in each of the following rooms: concrete-walled room, living room, bathroom, anechoic room, concert hall, recording room, and wood-paneled room.

2. Consider an auditorium 30 × 20 × 15 m high. Assume that the floor is concrete, with an average of one upholstered chair every square meter. Assume that the ceiling and walls are plastered. Calculate the RT. What is the intensity level in the hall if the orchestra produces a sound power of 2 watts?

3. A sound source having a power output of 0.001 watts radiates equally in all directions. It produces a measured level of 65 dB in an anechoic chamber. It produces a sound level of 77 dB in a concrete-walled room of 1000 m³ volume and a level of 70 dB in a 1000 m³ room with part of the walls covered with absorbing material. What are the approximate RT of the two 1000 m³ rooms? Do these results agree with what you might expect? How do RT and sound level relate to each other?

Further Reading

Backus, chapter 9
Benade, chapters 11, 12
Beranek
Culver, chapter 17
Doelle, chapter 5
Kinsler and Frey, chapter 14
Knudsen and Harris, chapters 6–8
Olson, chapter 8
White, chapter 4

4.C. AUDITORIUM DESIGN

Auditoriums as we know them evolved from the open-air theater mentioned previously. The first enclosed theater was called an odeion. This was basically an enclosed open-air theater (with a wooden roof) of moderate size. The odeion would seat between 200 and 2000 people. The enclosed theater not only provided shielding from extraneous noise, but the additional reflecting surfaces made it considerably easier to hear the actors. By the seventeenth century the theaters had become completely enclosed and the seating area had evolved to a U-shape with multiple balconies. Because of the relatively small size of the theater and the high absorptivity of the audience, reverberation times were short. Since members of the audience were also located in close proximity to the stage, many of these theaters provided almost ideal listening conditions. During the eighteenth century auditoriums became larger, multiple balconies disappeared, and hard plaster was used for the walls. All of this had a deleterious effect on the acoustics, but pleasing the eye seems to have been more important than pleasing the ear. During the nineteenth century scientists began to appreciate and to understand some of the acoustical difficulties of auditoriums, but it was not until the present century that any systematic acoustical research was performed. The principles of acoustics are now fairly well established, and it is possible to utilize them to achieve desirable results when new auditoriums are designed. In this section we will consider the differing acoustic requirements of various types of auditoriums. Then we will consider the acoustic requirements of a "good" concert hall in considerably more detail. Finally, we will present several examples of well-known concert halls.

Basic Acoustic Requirements

In any auditorium the hearing conditions are primarily determined by purely architectural considerations. Practically every detail within the enclosed space will contribute in some manner to the overall acoustical character of the room. Nevertheless, by considering several basic acoustical rules and by making judicious use of the variety of materials presently available, almost any structure can be designed so as to have adequate acoustics. The primary requisite for a good acoustical environment is that the structure be free from exterior and interior noises, as discussed in section 4.A. The second requisite is that the room have optimum reverberation for its volume and for its intended purpose. We have already discussed optimum reverberation time in section 4.B. After these two basic requirements have been satisfied, three additional matters must be considered if an auditorium is to provide good listening conditions: (1) there should be adequate loudness in all parts of the room; (2) the sound energy should be diffused (uniformly distributed) throughout the auditorium; (3) the room should be free of acoustical defects, such as echoes, sound concentration, flutter, and sound shadows. We now consider each of these requirements in more detail.

Adequate loudness is achieved in small or medium-sized rooms in two ways: (1) by directing as much of the sound energy as possible toward the audience and (2) by preventing excessive absorption of the sound. Sound can be directed toward the audience by several means, as illustrated in Fig. 4.C-1. First, it is important that the room be shaped so that the audience is as close as possible to the sound source. Generally this means that long, narrow rooms should be avoided and balconies should be used whenever possible. Second, there should be a raised stage to elevate the sound source. Third, the floor of the auditorium should be raked (constructed on a ramp). Since the audience is a very absorptive surface, sound waves which just graze their heads are greatly attenuated. Fourth, large sound-reflective surfaces (e.g., plaster or thick plywood) should be located as close to the sound as is feasible. Finally, the ceilings and walls need to be designed to provide favorable reflections of sound, especially for the seats located farthest from the stage. These measures, however, while extremely important, will not perform miracles. The first provision for adequate loudness must come from the performer. A speaker, for instance, must speak slowly and clearly and with a volume sufficient for the room. In very large auditoriums even the above conditions may prove inadequate and the installation of a sound amplification system would then be in order.

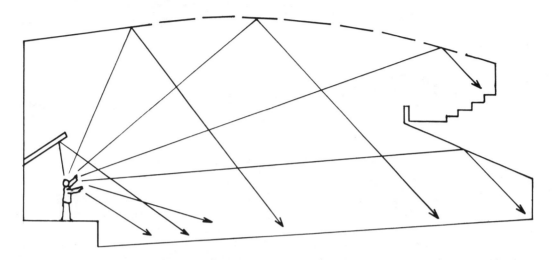

fig. 4.C-1. Direct and reflected sound for an auditorium.

The proper diffusion of sound is important for the achievement of a uniform distribution of sound in an auditorium. Diffusion can be created in several ways. First, by the use of many surface irregularities in the construction of the room. Irregularities such as exposed beams, coffered ceilings, sculptured balcony railings, and protruding boxes are effective if there are many of them and they are reasonably large. Second, for smaller rooms, diffusion can be achieved by using alternate regions of sound reflective and sound absorptive materials. Finally, an irregular or random arrangement of materials having different absorption coefficients will also produce diffusion. Even rooms with excessive reverberation times often show a considerable improvement in listening conditions when a number of properly sized diffusers are installed.

Although the acoustical attributes mentioned above are very important in auditorium design, it is equally important that all potential acoustical defects be minimized, if not entirely eliminated. The most common of these defects are echoes, flutter, focusing of sound, distortion, room resonances, and sound shadows. Each of these defects will be described briefly, and some of them are illustrated in Fig. 4.C-2.

Echoes are probably the most serious of the acoustic defects listed above. While reverberation is highly desirable in an auditorium, echoes are to be avoided at all costs. Echoes, like reverberation, are caused by reflections. But, whereas the reflections making up reverberant sound occur immediately after the direct sound, an echo is heard at least 0.05 sec after the direct sound is perceived. With this time interval between direct and reflected sound, the ear can resolve the reflected sound and it no longer sounds like a continuation of the original sound. A sound-reflective rear wall in an auditorium is often the source of an annoying echo on the stage. The echo can be eliminated by treating the rear wall with an absorptive material.

Flutter consists of a rapid succession of small, but noticeable, echoes. It occurs most commonly between two parallel but highly reflective surfaces. The easiest way to avoid this defect in auditoriums is to avoid parallel reflecting surfaces. If parallel surfaces cannot be avoided, the next best solution is to treat them with sound-absorbing materials. If acoustical treatment is not feasible, another solution would be to incorporate sound diffusing elements into the parallel walls or to tilt the walls slightly.

149

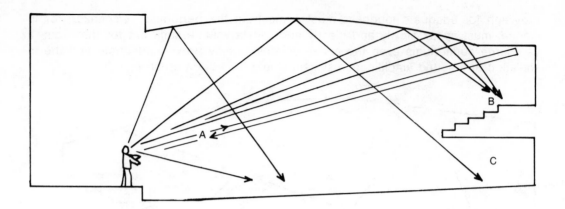

Fig. 4.C-2. Acoustical defects: (A) echo; (B) focusing of sound; (C) sound shadow.

Sound focusing is produced whenever there are concave surfaces. The effect of the focused sound is to produce areas where the intensity is unnaturally high at the expense of other areas where it is too low. Since we are attempting to achieve a uniform distribution of sound in the auditorium, focusing effects are highly undesirable. If concave surfaces are unavoidable they should be either treated with sound-absorbing materials or covered with a set of convex surfaces (which diffuse sound). This suggestion is illustrated in Fig. 4.C-3.

Distortion, in auditorium acoustics, is any undesirable change in the quality of a musical sound due to the uneven or excessive absorption of sound at certain frequencies. For instance, if an auditorium has an RT of 2.0 sec at 2000 Hz, but the RT is only 1.0 sec at 200 Hz, low-frequency sounds will attenuate much more rapidly than the treble. The listener will perceive a sound which seems deficient in bass, even though the original sound produced by the orchestra may have been balanced. This defect can be avoided if the interior surfaces and acoustical materials applied have absorption characteristics which are fairly uniform over the audio frequency range.

Room resonances are present when some of the natural frequencies of the room are excited by a source of sound. Room resonance problems are most acute in small

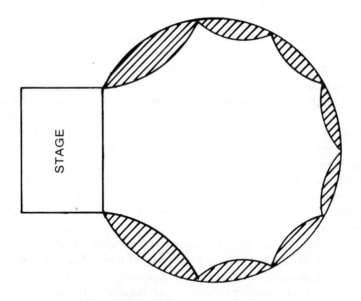

Fig. 4.C-3. A circular auditorium with conven diffusers.

150

rooms with hard interior surfaces, especially when several natural modes occur near the same frequency. The natural modes of a room are usually not uniformly distributed across all frequencies. Some frequency regions have more modes, or a greater response than other regions. Irregularities in the shape of a room or an irregular distribution of absorptive materials throughout a room will greatly improve the uniformity of spacing of natural modes. Also, just increasing the amount of absorptive material in a room will broaden and lower the natural mode peaks, thus giving a more uniform room response.

A *sound shadow* occurs in a region of an auditorium where there is direct sound energy, but little if any reflected sound. This defect is most often noticed under balconies, especially low and deep balconies. To avoid a noticeable under-balcony sound shadow, the depth of the balcony should not exceed twice the height of the opening. It is also desirable to utilize the underside of the balcony to reflect sound to the seats below.

Criteria for Speech and Music

Auditoriums that are used primarily for speech require that three criteria be met: (1) a low ambient noise level for clear syllable recognition, (2) a fairly short reverberation time for good speech intelligibility, and (3) a highly directional sound field to provide a direct communication link between speaker and listener. The criteria for rooms designed primarily for music are somewhat different and considerably more complex. Furthermore, during each of the major periods of music, the style of music was, to some extent, influenced by the type of hall where the music would be performed. There is therefore no one ideal type of concert hall, but rather an optimum hall for each period of music as shown in Table 4.C-1. The important variables are reverberation time (which determines fullness of tone), the definition (the degree to which successive sounds stand apart), intimacy (which indicates the size of the room), and dynamic range (the difference between the faintest and loudest musical sounds heard).

Table 4.C-1. Optimization features of concert halls.

Period	RT (sec)	Fullness of tone	Definition	Intimacy	Dynamic range
Baroque (1600–1750) (secular)	under 1.5	low	high	high	low
Baroque (sacred)	2.0	high	m. low	low	high
Classical (1750–1820)	1.5 – 1.7	medium	high	medium	medium
Romantic (1800–1900)	1.9 – 2.2	v. high	low	v. low	v. high
Modern (1900–)	varies	varies	varies	varies	varies
European (non-Wagnerian opera)	under 1.5	low	high	m. high	m. low
Wagnerian opera	1.6 – 1.8	high	m. low	low	m. high

Note that although there is a considerable variety in these characteristics, the trend has been toward larger RTs as time progressed, with the noticeable exception that the modern period is characterized by a complete lack of uniformity in any of these properties. (Perhaps this is because there seems to be no uniformity in modern music.)

In 1962 Leo Beranek, a well-known acoustician, published the result of an exhaustive study of 54 of the world's best concert halls. As part of the study he conducted elaborate acoustical measurements in each hall, he interviewed musicians and music critics familiar with each hall, and he listened to concerts in these halls. The results of his study indicate the important acoustical attributes of each hall and the correlation between the acoustical properties and the musicians' subjective analyses. In Table 4.C-2 the most important of these subjective qualities, their physical correlates, and optimum values are summarized.

151

Table 4.C-2. Important acoustical attributes of concert halls. (After Beranek.)

Subjective attribute	Physical correlate	Optimum
1. Liveness	Reverberation.	Depends on period of music.
2. Intimacy	Time delay gap between direct and first reflected sound.	Less than 0.02 sec.
3. Warmth	RT longer for low frequencies than for high.	ave. low freq. RT about 1.2 times avg high freq RT
4. Definition or clarity	Ratio of loudness of direct sound to reverberant sound.	Direct sound somewhat louder than reverberant sound at all locations.
5. Balance and blend of orchestra	Design of stage enclosure so sound mixes on stage.	If stage wider than 15 m ceiling should be low and irregular in shape. For ceiling higher than 10 m the width should be less than 16 m and the depth less than 10 m.
6. Ensemble	Design of stage enclosure so musicians can hear each other.	Ample reflecting surfaces on side of stage and above ochestra. Stage less than 20 m wide.
7. Texture	The pattern in which sound reflections arrive at a listener's ear.	Five or more reflections during first 0.06 sec.
8. Tonal quality	Absence of defects such as echoes, flutter, etc.	No defects.
9. Uniformity	Diffuse sound field, absence of focused sound or sound shadows.	As diffuse a sound field as possible.
10. Dynamic range	Loudness of fortissimo; relation of background noise to loudness of pianissimo.	Fortissimo passages should not exceed 120 dB(A). Background noise should be 30 dB(A) or less, so as not to obscure pianissimo passage.

The two most important criteria for a successful concert hall are liveness (reverberation) and intimacy (short initial time delay gap). When the initial time delay gap is too long the hall seems cold and impersonal, even though it may have optimum reverbation characteristics. Placing reflecting screens near the orchestra will provide early reflections to the audience and greatly improve the acoustical intimacy of a hall. This is illustrated in Fig. 4.C-4. The importance of having an adequate reverberation time can be seen by considering that most of the attributes of good concert halls, as tabulated above, involve sound reflection. One additional aspect of reverberation which is not obvious from the above considerations is that the reverberation curve (a plot of reverberant sound intensity vs. time) should have a uniform decay.

An ideal reverberation curve is shown in Fig. 4.C-5a: this figure shows how the reverberant energy dies away with time. Note that the curve is almost a straight line, which indicates that the room has a very uniform reverberation response; i.e., the sound is getting soft at an even rate. This is highly desirable. Fig. 4.C-5b and c indicate two highly undesirable reverberation responses which, by the way, have the same RT of 2.0 sec. Fig. 4.C-5b shows a highly irregular response: the sound seems

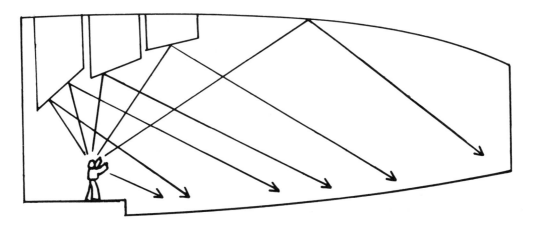

Fig. 4.C-4. The use of ceiling reflectors to shorten the initial time delay gap.

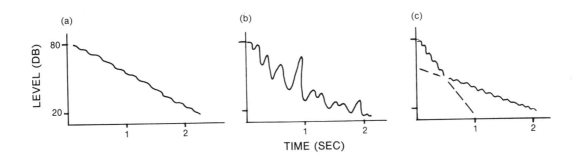

Fig. 4.C-5. Reverberation curves: (a) ideal; (b) irregular; (c) double sloped. (After Bliven, 1976.)

to get softer, then louder, then softer again in a type of erratic warble. Generally, the more diffuse the sound field, the less likely the chance of finding this type of irregular curve. Finally, Fig. 4.C-5c shows another situation which is very undesirable, even though the reverberation curve is smooth. In this illustration, the reverberation falls off rapidly during the first half second, thus telling the ear that the RT will be short. In actuality, the RT is 2.0 sec, and the resulting effect is a type of auditory confusion: we expect a short RT but actually perceive a substantially longer RT. Also, even though the RT is 2.0 sec, because of the rapid initial decay of sound the room will still seem somewhat dead. The best way to realize a uniform reverberation curve is to fill the concert hall with diffusing elements of all sizes so that the sound will bounce evenly throughout the room.

Examples of Concert Halls

We will now consider three well-known concert halls, ranging from an acoustical disaster to a recently constructed hall which was carefully designed to have the best possible acoustics. The acoustical disaster was the Royal Albert Hall in London, which opened in 1871. The principal acoustical problems which have plagued this hall during the past 100 years arise from three main causes. First, the reverberation time was excessive, even for a structure of its cavernous size. Second, there were very pronounced and annoying echoes, intensified in some areas by the dome-shaped ceiling. Finally, there was a decrease in loudness of the direct sound, as it travels such great distances in this hall. The echo has been recognized as being the most serious of the above defects, and attempts have been made over the years to minimize this problem. In 1961 substantial changes were made in an effort to control the echo and to reduce the mid-frequency reverberation time. This was done by hanging 109 fiberglass saucers (from 2 to 4 meters in diameter) 25 meters above the floor. The disks were made so that mid-range frequencies were absorbed while the lower and upper frequencies were largely reflected. They are hung so that all together they form a convex dome at the base of the true dome. A large 20-meter reflector was also installed behind the orchestra to help bounce the sound toward the rear of the hall. These two changes have greatly improved the hall; the worst of the notorious echoes are no longer present and the mid-frequency reverberation has been reduced substantially.

The second hall is the Mormon Tabernacle in Salt Lake City, completed in 1867. While this structure was not designed as a concert hall, it is often used for this purpose. The tabernacle has several features that often lead to acoustical problems. However, by a fortuitous set of circumstances the acoustics are not as bad as they could be. The focusing effects due to the elliptical shape cause serious disturbances at relatively few locations. The sound focusing has also been alleviated somewhat by the installation of an elaborate sound-reinforcing system. Since the interior of the tabernacle is mostly plaster, one would expect the reverberation time to be excessive. It is not, because at the time the building was constructed large quantities of cattle hair were mixed with the plaster, thus increasing the absorption of the plaster. Even so, the empty tabernacle has a mid-frequency RT in excess of 4.0 sec, which is about double the optimum time. When the room is about one-third full the RT is just about right, but when completely filled (8,000 people) the room becomes somewhat dead (RT = 1.0 sec).

The final concert hall which we consider is the Philharmonic Hall at Lincoln Center, New York City. This hall, which opened in the fall of 1962, was to be designed so that acoustic considerations were paramount. The acoustical consultant spent several years studying 54 of the world's best concert halls. As a result of this study he discovered the primary requisites of good concert halls and applied this knowledge to the design criteria of Philharmonic Hall. It was anticipated that Philharmonic Hall would rank with the world's greatest halls, thus verifying that good concert hall acoustics can be achieved by the application of scientific principles (Beranek, 1962).

Unfortunately, when the hall opened many musicians and music critics commented very unfavorably on the hall and its acoustics. Apparently the hall was so plagued with acoustical problems that three times over the next dozen years major changes were made to the interior in an effort to correct the problem. After each change the hall was slightly better, but the problems remained. What went wrong? While no one can be completely certain, we do know what the main acoustical problems were. We also realize that many of these defects were of the type which the consultants recognized and sought to avoid. Perhaps the unconventional design of the hall introduced more unknown variables than could be satisfactorily reckoned with. Apparently the "final" architectural plans the consultant had worked on and had expected to see constructed were modified and expanded into a larger, more modern design without his consent or that of the orchestra (Beranek, 1975). Also, plans for an adjustable ceiling and many sound-diffusing features were discarded as too expensive.

The main complaints about the hall were as follows. First, the orchestra sounded dry and lifeless. There was no effect of being surrounded by sound; rather, the sound seemed to come primarily from the stage area. Second, the hall was rated as being deficient in bass. Third, there were several annoying echoes. And fourth, the musicians on the stage could not hear each other well. In more scientific terms, there was not enough diffusion of sound, giving irregular fluctuations in the reverberation decay curve. Also, the initial reverberation decay was too fast, giving a double-sloped reverberation curve. Although the RT was about right for the size of room, the room seemed dead because of the rapid initial decay. Also, the sound of the orchestra was not well distributed, so that it did not seem well balanced in all parts of the hall.

In 1974 another acoustical consultant was engaged to see what could be done to correct the problem (Bliven, 1976). After some study of the matter he concluded that the only feasible way to make it a first-rate concert hall was to knock it down and build it over! He felt that it would be hopeless, as well as being extremely expensive, to make any further renovations to the structure. Early in 1975 money was found to build an entirely new concert hall within the concrete shell of the original hall. He insisted that three conditions were to be adhered to: (1) everything would be ripped out down to the girders, (2) acoustics would have priority over aesthetics, and (3) the acoustical consultant's views would have priority over those of the architect. The design philosophy was to use only the "tried and true" methods of acoustical design. The consultant attempted to construct an acoustical setting which would be as much like that of Boston's famous Symphony Hall as possible. In 1976 the rebuilding commenced. When the hall, renamed Avery Fisher Hall after its chief benefactor, opened during the fall of 1976 musicians and critics were lavish in their praise. We can see, then, that it is possible to use scientific principles to design a concert hall with excellent acoustics, provided that the well-established principles are adhered to and radical architectural considerations (which might prove to be acoustically disadvantageous) are avoided.

Exercises

1. What are the five basic requirements for a well-designed auditorium? Which of these is most important? Least important? Explain.

2. Suppose you are an acoustical consultant. An architect asks you for advice on the design of a high school auditorium to seat 1200. What specific advice would you give him? (Your recommendations should be fairly detailed, including rough dimensions and a top view and side view of your proposed design).

3. If you want the RT of an auditorium to be approximately independent of audience size, would you use cushioned or hard surfaced seats? Explain.

4. If you are designing a multipurpose auditorium, to be used for drama and musical concerts, which will seat 1000 people, what RT would you recommend? What volume should the room have in order to have this RT be optimum (see section 4.B)? What recommendations would you make concerning the ceiling of the room? The stage enclosure? The rear wall?

5. Suppose you are called upon to act as an acoustical consultant for a small church. Would you recommend hard interior surfaces (and no carpeting) or acoustically treated surfaces and the introduction of a sound system? Explain in detail which design is preferable and why.

6. How far must a rear wall be located from a stage in order that an echo be perceived on the stage?

7. How would you design a balcony for an auditorium in order to get the maximum number of seats, but without any acoustic defects under the balcony? Make several sketches and give a detailed explanation.

8. Which of the subjective attributes of Table 4.C-2 are most important for a good concert hall? Which are least important? Explain why.

9. Do the data of Table 4.C-2 imply that there might be a maximum size to a concert hall, which, if exceeded, would render the hall less desirable for orchestral concerts? Explain.

10. Is it desirable to use the same hall for orchestral performances and for opera? Why or why not?

11. Which of the attributes of Table 4.C-2 were probably present in the original Phil-harmonic Hall at Lincoln Center? Which were probably deficient? Why?

12. Can you imagine any type of music where the undesirable acoustical properties of the Royal Albert Hall could be used to advantage? Explain.

13. Is it better to design a concert hall with too high an RT or too low an RT? Which situation is easier to correct? Why?

14. A municipal auditorium is observed to sound cold and impersonal during orches-tral performances. Also, the RT for low frequencies is substantially larger than the high-frequency RT. What recommendations would you make to help rectify these situations?

Further Reading

Beranek, chapters 3, 4, 9, 10, 11, 12, 15

Doelle, chapter 6

Knudsen, chapter 9

Olson, chapter 8

Barron, M. 1971. "The Subjective Effects of First Reflections in Concert Halls—the need for lateral Reflections," J. Sound & Vibration *15*, 475–94.

Beranek, L. L. 1975. "Changing Role of the 'Expert'," J. Acoust. Soc. Am. *58*, 547–55.

Beranek, L. L., et al. 1964. "Acoustics of Philharmonic Hall, New York, during Its First Season," J. Acoust. Soc. Am. *36*, 1247–62.

Bliven, B. 1976. "Annals of Architecture (Avery Fisher Hall)," New Yorker, Nov. 8, pp. 51–135.

Clark, R. W. 1958. *The Royal Albert Hall* (Hamish Hamilton, London).

Doelle, L. L., 1967. "Auditorium Acoustics," Architecture Canada, Oct., pp. 35–44.

Hales, W. B. 1930. "Acoustics of the Salt Lake Tabernacle," J. Acoust. Soc. Am *1*, 280–292

Jordan, V. L. "Room Acoustics and Architectural Acoustics Development in Recent Years," Applied Acoustics, Jan., pp. 59–81.

McGuinnes, W. J. "Adjusting Auditoriums Acoustically," Progressive Architecture, March 1968, p. 166.

Northwood, T. D. 1967. *Room Acoustics: Design for Listening* (National Research Council, Ottawa), Canadian Building Digest 92, August.

Shankland, R. S. 1972. "The Development of Architectural Acoustics," Amer. Sci. *60*, 201–9

Chapter 5

Acoustics of Speech

5.A. ENERGY TYPES IN SPEECH PRODUCTION

In bare outline the human vocal mechanism can be viewed as consisting of an energy source, a mechanism for interrupting the air stream, and a variable resonator for creating different sound spectra, as shown in Fig. 5.A-1. The reservoir of air in the lungs is the energy source. The air can be expelled to provide an air stream which flows through the rest of the vocal system. The air flow from the lungs may be interrupted by the vocal cords or by constrictions in the vocal tract. The variable resonator is the vocal tract, the effects of which will be considered in the next section.

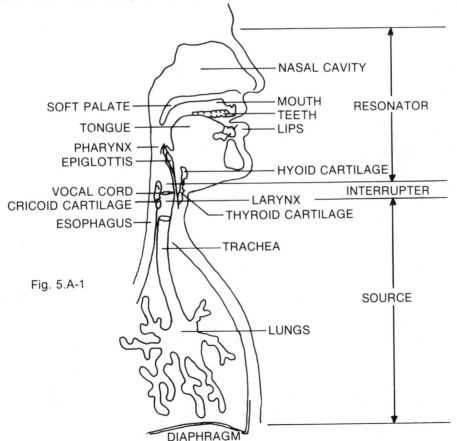

Fig. 5.A-1

Fig. 5.A-1. The speech production mechanism.

Anatomy and Function

Several parts of the speech production mechanism serve dual functions. When one eats, for example, food must pass through the mouth and pharynx into the esophagus, and it is important that the epiglottis be lowered to cover the larynx so that food will not enter the air pathway to the lungs. When one breathes, the epiglottis must be raised and the vocal cords must be wide open to permit a free flow of air between the lungs and the outside.

When a person speaks, air is forced from the lungs and flows through the trachea, then through the opening in the vocal cords, then through the larynx, pharynx, mouth, and optionally the nasal cavity to the outside. The energy required to produce speech sounds is supplied by exhaling air from the lungs. However, exhaling alone is not adequate to produce useful speech sounds. (You might try alternately inhaling and exhaling to convince yourself that the act of exhaling in and of itself is not sufficient to produce disturbances that are useful in the speech communication process.) To be audible, the otherwise steady flow of air must be converted into an oscillating one. The interruption of the air flow from the lungs is accomplished for speaking purposes in one of three ways: (1) by using the vibrating vocal cords to interrupt the airstream in a periodic manner; (2) by forming a constriction in the tract that causes the air stream to become turbulent and noisy as it flows through the constriction; (3) by completely closing off the vocal tract, thus stopping the air flow, and then releasing the air pressure suddenly. The energy types produced by the above means will be labeled respectively voice energy, noise energy, and burst energy. It is also possible to get a mixture of voice and noise energy by vibrating the vocal cords while constricting the vocal tract.

Vocal Cord Action

Voice energy is produced when the air flow from the lungs is alternately interrupted by the closing and opening of the vocal cords. The *vocal cords* can be thought of as small bands of muscle, each of which has approximately the same characteristics. The vocal cords are located within the larynx, a cartilaginous tube at the top of the trachea. The two cords meet each other and are attached to the thyroid cartilage at the front of the larynx, as shown in Fig. 5.A-2. (The thyroid cartilage produces the protrusion from the neck commonly called the Adam's apple, which is most apparent in adult males.) Each cord is attached to an arytenoid cartilage at the back of the larynx. The arytenoids are normally positioned widely apart from each other to permit breathing, as can be seen in Fig. 5.A-2a. However, they can be brought into contact with each other, which causes the vocal cords to come close together (touching each other in many cases) for the production of voice energy. When the air pressure in the trachea is caused to exceed the air pressure in the vocal tract, the vocal cords will tend to be blown upward along the tract and to be blown apart.

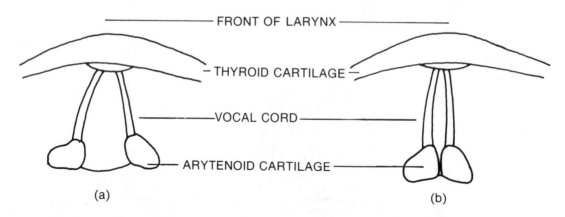

Fig. 5.A-2. Simplified top view of the vocal cords: (a) breathing position; (b) voicing position.

In a simplified way, each cord can be thought of as a simple vibrator with its associated mass, tension, and resistance. The cords are caused to vibrate by the air pushing on them or rushing between them. When the opening between the vocal cords is very small, the air tends to force them apart, but then as the opening gets larger the rushing air creates a decreased pressure between them (via the Bernoulli effect) and pulls them back together. The muscular tension in the cords also serves to pull them back toward their rest positions. The interaction of the inertia of the cords, the tension in the cords, and the pushing and pulling of the air stream serves to produce an oscillatory motion of the cords. The resistance and nonlinear tension of the cords tend to prevent the amplitude of the oscillatory motion from becoming too large.

Three primary variables serve to control the vibration frequency of the vocal cords: (1) tension in the vocal cords; (2) mass of the vocal cords; and (3) pressure difference between the trachea and the vocal tract. The muscular control of the vocal cords is very complicated and not completely understood at the present time. Researchers have developed mathematical models of the vocal cords that are capable of modeling various aspects of observed vocal cord behavior. However, from a simplified point of view, the tension in the vocal cords can be viewed as being increased when the muscles cause the cords to be stretched. Apparently this stretching also produces a reduced effective mass by "locking out" part of the vocal cord mass from the vibratory motion. In a simplified view of the cords as mass and spring vibrators, increased tension will increase the vibration frequency and decreased mass will also increase the vibration frequency. For any given values of vocal cord tension and mass, larger pressure differences produce higher vibration frequencies than do smaller pressure differences.

An interesting feature of the vocal cords that seems to contradict our general description of vibrating systems is that the cords tend to lengthen for the production of high-frequency sounds. For a normal male talker the cords are short and thick for the production of low frequencies (about 0.9 cm long and 0.5 cm thick for a frequency of 125 Hz) and comparatively long and thin for the production of high frequencies (about 1.8 cm long and 0.3 cm thick for a frequency of 250 Hz). As seen in section 2.D, longer strings produce lower frequencies than do identical shorter strings. However, in the case of the vocal cords the tension is increasing and the mass is decreasing along with the vocal cord lengthening. The tension and mass effects tend to produce higher frequencies and more than offset the tendency of the increased length to produce lower frequencies.

Certain of the fundamental features for simple vibrators do hold true when comparing the behavior of the vocal cords of adult males, adult females, and children. In general, male vocal cords are both longer and more massive than female vocal cords, which are in turn longer and more massive than the vocal cords of children. Furthermore, the range of fundamental voicing vibration frequencies produced by adult males (about 80-240 Hz) is lower by about a factor of two than the fundamental voicing vibration frequencies produced by adult females (about 140-500 Hz), which in turn is lower than the corresponding frequencies for children. (Ranges given are appropriate for speech. Ranges for singing would be more like 80-700 Hz and 140-1100 Hz for adult males and females, respectively.) In this sense the relationship that shows more massive simple vibrators to have lower vibration frequencies than less massive ones is consistent with the different masses and frequencies of vocal cords for males, females, and children. The relationship between frequency and length derived for strings also holds in a qualitative sense for vocal cords—the shorter female vocal cords have a higher frequency than do the longer male vocal cords.

The valving action of the vocal cords results in an approximately periodic flow of air through the glottis. This air flow is generally not sinusoidal, but contains energy at many harmonics of the fundamental frequency in addition to the energy of the fundamental. The harmonic energy arises from nonlinear behavior of the air flow through the glottis. To illustrate this nonlinear behavior, we make three assumptions, one or more of which turns out to be unrealistic: (1) the glottis is an opening whose area varies sinusoidally; (2) the pressure across the glottis (pressure in the trachea minus that in the larynx) is constant; (3) under the conditions assumed in (1) and (2) the air flow through the glottis varies directly as the area of the glottis.

If all three of the assumptions just stated hold, the air flow through the glottis will vary sinusoidally (and consist of a single frequency) because the area of the glottis varies sinusoidally. In reality, however, the air flow does not vary sinusoidally for one or more of the following reasons. (1) The airflow does not vary in direct proportion to the glottal area. This is because the resistance of the glottis to air flow changes rather drastically as the glottal area changes. In practice, doubling the area more than doubles the air flow (especially for small areas) if all other conditions are constant. (2) The pressure across the glottis is not constant. Even if we assume the pressure in the trachea is constant, the pressure in the larynx will be fluctuating because of standing waves in the vocal tract. The fluctuating pressure across the glottis gives rise to fluctuations in the air flow. (3) The area of the glottis does not vary sinusoidally. This occurs because the forces on the vocal cords are nonlinear and not sinusoidal. For example, when the cords collide on closing, the forces suddenly (and nonlinearly) become much larger. Furthermore, when the vocal cords open wide more force is required to stretch them than when they are less open. When the vocal cords are driven with higher pressures, they open and close more abruptly, the motions are more violent, and the vocal cords tend to remain closed for larger portions of the whole cycle. These features given rise to a glottal area that changes in a way which is very different from a sinusoid. A waveform representing a typical air flow through the glottis and its idealized spectrum appear in Fig. 5.A-3. In this example the air flow is shut off for part of a cycle and the resulting flow is pulselike. Note that the flow begins rather gradually after being shut off, but then is shut off rather abruptly after reaching some largest value. The spectrum shows a rather rich harmonic content due to the pulselike nature of the wave.

In summary, the air flow through the glottis is periodic, but not sinusoidal, and contains harmonic components because of one or more of the following conditions: (1) air flow does not vary directly with the glottal area; (2) pressure across the glottis does not remain constant; (3) the glottal area does not vary sinusoidally.

Other Sound Sources

The voicing source is located at the vocal cords, but noise sources occur at any one of several different locations in the vocal tract. Any time air is forced through a small or irregular constriction, the smooth flow of the air stream is disrupted and the flow becomes irregular and turbulent. The "hissy" sounds of speech are produced with a noise source. Typical points of constriction in the tract are teeth to teeth, teeth to tongue, and teeth to lips. When the vocal tract is constricted and the vocal cords are also caused to vibrate, a mixture of both voice and noise energy results. The voicing causes alternate puffs of air to pass through the constriction, thus giving rise to repetitive bursts of noise. This procedure is used for the production of the voiced fricatives.

It is possible to completely block off the vocal tract while trying to force air from the lungs, which gives rise to a large pressure on the lung side of the constriction. If the constriction is suddenly removed, the sudden release of pressure produces a burst of energy which is characteristic of many of the unvoiced plosive sounds. A more gentle release of a smaller pressure, which is typical of the voiced plosive sounds, usually gives rise to a negligible burst of energy.

You can discover what energy types are used in the production of the various speech sounds with the following techniques. Test for the presence of voicing by placing the fingertips gently on the Adam's apple; a slight vibration will be felt when voice energy is present. Test for the presence of noise energy simply by listening for

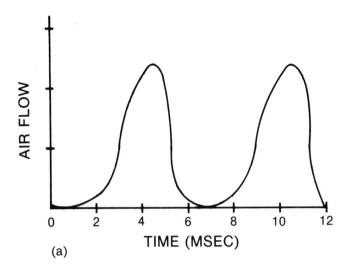

(a)

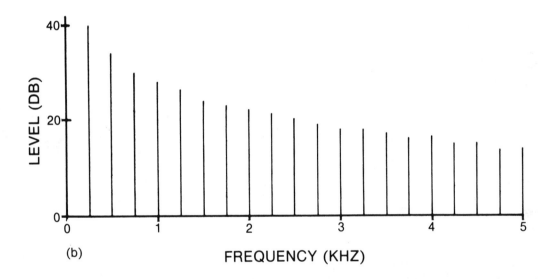

(b)

Fig. 5.A-3. (a) Typical glottal airflow; (b) idealized spectrum of glottal airflow.

a "hissy" character in the sounds produced. If both of these conditions are satisfied, we label the result a mixture of voice and noise energy. Look for burst energy by noting whether a large pressure is built up in a constricted tract and, if so, whether it is suddenly released.

Classification of Speech Sounds

A convenient concept in dealing with speech is the *phoneme,* which is defined as a distinguishable speech sound. The number of phonemes to be used depends to some extent on how finely one wishes to divide the world of speech sounds. The human vocal mechanism is capable of producing an almost infinite variety of different sounds; however, most of these are not readily distinguishable from one another. In the present context we will use the phonemes tabulated in Table 5.A-1. Speech sounds may be grouped as in Table 5.A-1, where the groupings are done partly on the basis of energy type. In later sections other ways of grouping and classifying will be discussed.

161

Table 5.A-1. Phoneme classification, giving examples and two sets of symbols for the phonemes.

	Text symbol	Example	IPA* Symbol
Vowels	EE	beet	i
	I	hit	I
	E	bed	ε
	A	had	ae
	O	hot	a
	U	put	U
	OO	cool	u
	UH	fun	∧
	AE	make	e
Nasals	M	me	m
	N	no	n
	NG	sing	η
Liquids	L	law	l
	R	red	r
Glides	W	we	w
	Y	you	j
Unvoiced fricatives	WH	when	hw
	H	he	h
	F	fin	f
	TH	thin	θ
	S	sin	s
	SH	shin	∫
	CH	chin	t∫
Voiced fricatives	V	view	v
	DH	then	∂
	Z	zoo	z
	ZH	mirage	ʒ
	JH	judge	dʒ
Unvoiced plosives	P	pea	p
	T	tea	t
	K	key	k
Voiced plosives	B	bee	b
	D	down	d
	G	go	g

*International Phonetic Alphabet

Different speech pressure waves result from the production of different phonemes. The differences are due in part to the energy types involved. (They are due also to vocal tract shape, which will be discussed in section 5.B.) When voicing energy is used, the wave is approximately periodic. When noise energy is used, the wave is irregular. A pressure waveform recorded with a microphone for the sentence "Joe took father's shoe bench out," as produced by one talker, is shown in Fig. 5.A-4.

Note the relatively high-level, periodic portions of the wave associated with the vowel sounds. Other periodic, but lower-level, portions of the wave are associated with voiced sounds produced with a partly closed vocal tract, such as /DH/ and /B/. Note the rather low-level, non-periodic portion of the wave associated with the /JH/, /SH/, and /CH/ sounds.

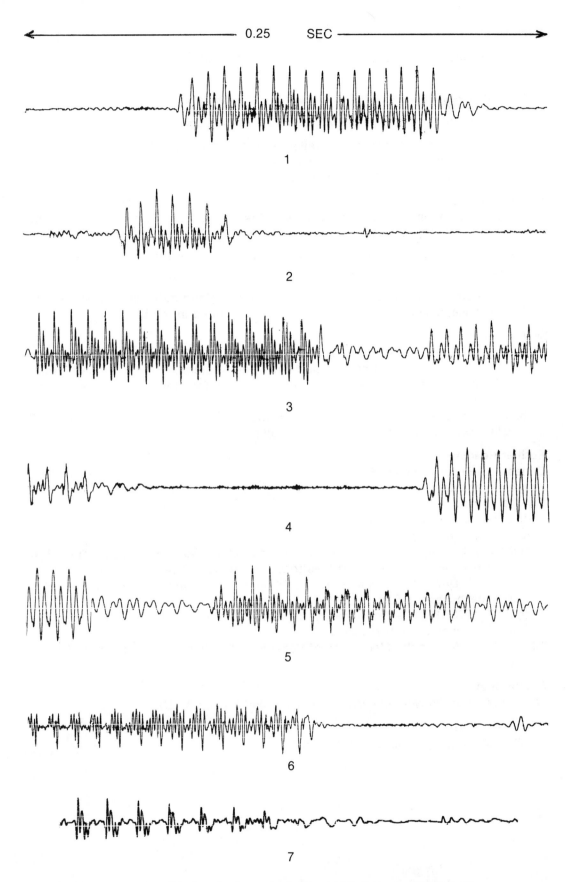

Fig. 5.A-4. Waveform for the sentence "Joe took father's shoe bench out."

Exercises

1. Describe the basic parts of the human vocal mechanism and the functions of each.

2. What is the primary force that causes the vocal cords to vibrate?

3. What controls the fundamental vibration rate of the vocal cords?

4. Tell why you would expect the vibration frequencies of adult female vocal cords to be higher than those for an adult male. Frame your description in terms of a simple vibrator model of the vocal cords.

5. What ranges of fundamental voicing frequency are associated with the speech of adult males, adult females, and children? Are these ranges comparable with the respective singing ranges?

6. What effect does a higher blowing pressure have on the behavior of the vocal cords? How does this affect the spectrum of the vocal cord waveform?

7. Produce each of the phonemes listed in Table 5.A-1 and determine its energy type (voice, noise, mixture, burst). Apply the tests described in the text.

*8. Analysis of energy types in speech
    a. Try to identify the portions of the wave in Fig. 5.A-3 associated with the phonemes in "Joe took father's shoe bench out." Determine the degree of periodicity in each part of the wave (much, some, none) for each of the 12 phonemes.
    b. Compare the "periodicity" results with the "energy type results" of exercise 7. What correlation is there between periodicity and voicing or noise?

Further Reading

Denes and Pinson, chapter 4
Flanagan, chapters 2, 3, 4
Fletcher, chapters 1, 2
Ladefoged, chapters 1, 2, 3, 4, 6, 7, 11, 12
Singh and Singh, chapters 1-4

Flanagan, J. L., and L. Landgraf. 1968. "Self-oscillating Source for Vocal Tract Synthesizers," IEEE Trans. Audio Electroacoust. *AU-16,* 57-64.

Hollien, H., D. Dew, and P. Philips. 1971. "Phonational Frequency Ranges of Adults," J. Speech Hearing Res. *14,* 755-60.

Titze, I. R. 1973; 1974. "The Human Vocal Cords: A Mathematical Model, Parts I and II," Phonetica *28,* 129-70; *29,* 1-21.

Titze, I. R. 1976. "On the Mechanics of Vocal-fold Vibration," J. Acoust. Soc. Am. *60,* 1366-1380.

Audiovisual

The Function of the Normal Larynx (20 min, color, 1956, ILVAD)

Demonstrations

Oscilloscope display of speech waveforms

5.B. VOCAL TRACT MODIFICATIONS OF SPEECH ENERGY

In the previous section we listed most of the phonemes (distinguishable speech sounds) used in English. A smaller number of phonemes than those listed would make English as we know it impossible to use. The energy produced at the vocal cords or at a constriction cannot be varied in enough different ways to produce this many useful and distinguishable speech sounds. However, the speech energy must pass through the vocal tract (or at least portions of the vocal tract) on its way to the outside, where it will be received by a listener. It is possible to vary the spectrum as the speech energy passes through the vocal tract by causing the vocal tract to assume different shapes. (You might at this point do a small experiment to convince yourself that vocal tract shaping plays an important part in helping to create a sufficient number of distinguishable speech sounds. Try opening your mouth to some comfortable position and then, keeping it in this position, see how many different and useful speech sounds you can produce by only controlling your vocal cords.)

Vocal Tract Features

The vocal tract is distinguished from most musical wind instruments by the manner in which it can be controlled. Most wind instruments make use of a tube that maintains an approximately constant shape but whose length can be varied through the use of finger holes or valves. The vocal tract length, on the other hand, is of a more or less fixed length (ignoring the effects of lip rounding, etc.), but it can be caused to assume a large number of different shapes. This variable shape capability of the vocal tract permits the selective modification of speech energy produced at the vocal cords or at a constriction and thus makes possible the creation of enough distinguishable speech sounds to make English a viable spoken language.

The *vocal tract* consists of three main sections of tube: the pharynx, the mouth, and the nasal cavity. These can be modified with certain vocal organs—the tongue, the teeth, the lips, and the soft palate—as shown in Fig. 5.B-1.

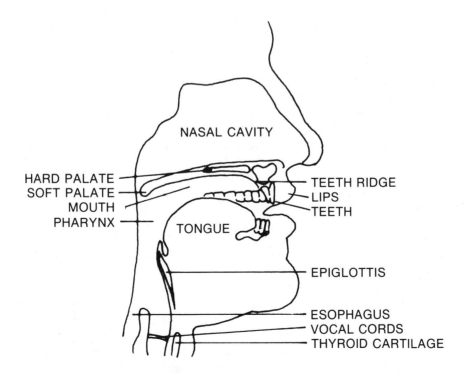

Fig. 5.B-1. Simplified cross-section of vocal tract configured for nonnasal sounds.

165

We will restrict our attention to modification of voice energy produced at the vocal cords. We consider a case in which the velum (or soft palate) is raised and brought into firm contact with the back of the pharynx so that the nasal tract is not coupled to the pharynx. This means that energy produced at the vocal cords must flow through the pharynx and then through the mouth to reach the outside world. (Cases that involve a lowered velum—thus coupling the nasal cavity to the pharynx—will be considered later.)

Neutral Vocal Tract

In order to study the way in which the vocal tract modifies speech energy, we begin with a greatly simplified tube and consider its effects. The tube is assumed to be a uniform cylinder closed at one end and open at the other end and is termed a *neutral tract* because it has constant area from one end to the other. This tract is probably most closely approximated in an actual vocal tract configuration by the vowel /UH/. The neutral tract also gives us something that is simple enough to consider in some detail. Fig. 5.B-2 shows a neutral tract where the closed end is at the vocal cords and the open end is at the mouth. In the figure the tract has been unbent to form a straight tract. (The unbending has no appreciable effect on the behavior of the waves which pass through it.)

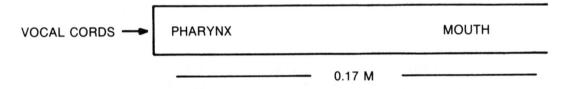

Fig. 5.B-2. Neutral vocal tract.

Now we come to the question of how the vocal tract modifies the speech energy and find that it treats energy at different frequencies in different manners. Figuratively, it prefers some frequencies and discriminates against other frequencies. If the vocal cords produce a "preferred" frequency, this frequency will be enhanced by the vocal tract. However, if the vocal cords produce a "nonpreferred" frequency, this frequency will be diminished by the vocal tract. In practice, of course, the vocal cords produce both "preferred" and "nonpreferred" frequencies.

The "preferred" frequencies we have been discussing turn out to be the natural frequencies of the tube. The methods developed in section 2.D can be used to gain information about what frequencies the vocal tract will emphasize. Our neutral vocal tract is equivalent to the closed-open tube that was considered in section 2.D., whose lowest natural frequency was determined as $f_1 = v/4\ell$, and whose higher natural frequencies were given by taking odd-integer multiples of f_1 as $f_n = (2n-1)f_1$. In speech research the first natural frequency is usually called the first formant (F1),and the higher natural frequencies the second formant (F2), the third formant (F3), and so on. A *formant* in this sense is just a resonance of the vocal tract (or a frequency that the vocal tract tends to enhance). If we assume that the tract in Fig. 5.B-2 is 0.17 m long and that the speed of sound in the tract is 340 m/sec, we calculate a value of 500 Hz for the first formant (F1) of the neutral tract. Similar calculations give values of 1500 Hz for F2 and 2500 Hz for F3. We can use the method described in section 2.D to measure the input impedance of the neutral tract or, for that matter, the input impedance of a tube having any shape. The loudspeaker in this case is placed at the vocal cord end of a simulated tract. The transmission (which is related to the input impedance) of the neutral tract is shown in Fig. 5.B-3. Recall that the input impedance (transmission) at a particular frequency tells us how the tract responds at that frequency. High values of input impedance occur for frequencies at which strong standing waves are produced; low values occur for weak standing waves. Note in Fig. 5.B-3 that the formant frequencies are enhanced by the neutral tract and are those frequencies at which the input impedance is high.

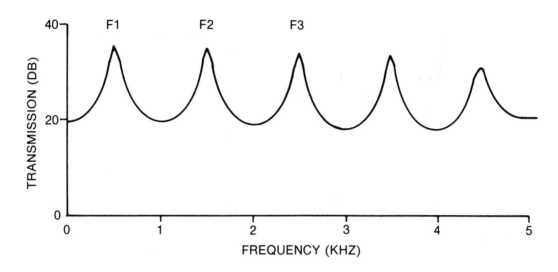

Fig. 5.B-3. Neutral vocal tract transmission at different frequencies.

Modifications to Neutral Tract

If the neutral vocal tract has some small changes made in its shape, it is not unreasonable to suppose that the formant frequencies will be changed by small amounts. If large changes are made in the vocal tract shape the changes to the formant frequencies will also be large. However, for any vocal tract shape there will be an accompanying set of formant frequencies that are related to the tract shape. In other words, the formant frequencies (peaks) that we observe in the spectrum of a speech signal coming from the vocal tract are determined by the vocal tract shape. (Additional peaks and valleys can occur in the spectrum if the waveform produced by the vocal cords is too "unusual.")

We have seen that the formant frequencies of the neutral tract occur when standing waves are set up in the tract; the standing waves depend on length of the tract, the speed of sound, and reflections from the ends of the tract. As a further approximation to a real vocal tract, we consider a "two-tube tract," each tube having a different cross-sectional area. In multi-tube tracts wave reflections occur where tubes of different areas join as well as at the ends of the tubes. As you can well imagine, the standing waves are more complicated and occur at different frequencies than for our single-tube neutral tract. As an example, we approximate an /EE/ tract shape in terms of the two-tube model shown in Fig. 5.B-4 with L1 = 9 cm, L2 = 6 cm, A1 = 8 cm², and A2 = 1 cm². The transmission for this tract is shown in Fig. 5.B-5a. The formant frequencies are seen to be about F1 = 250 Hz, F2 = 1875 Hz, and F3 = 2825 Hz.

We have to this point considered only how the vocal tract modifies different frequencies, but we must now consider how it modifies a particular spectrum of interest: the spectrum of the air flow from the glottis. The idealized spectrum of air flow through the glottis from Fig. 5.A-3 is repeated in Fig. 5.B-5b for convenience. The spectrum of the speech wave that emerges from the vocal tract is obtained by adding the spectra (when they are in dB) of the vocal tract transmission and the glottal air flow. The resulting spectrum of the output speech is shown in Fig. 5.B-5c. Spectra of several different speech sounds are shown in Fig. 5.B-6. Note that spectra of the sounds produced with voice energy show harmonic spectra, i.e., well-defined

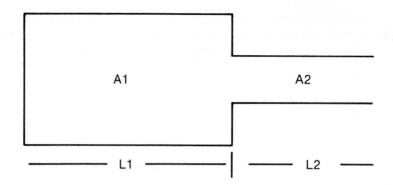

Fig. 5.B-4. Two-tube approximation to /EE/-shaped vocal tract.

spectral lines. Spectra of noise energy sounds (/S/ and /SH/) do not have well-defined lines; spectra of mixed energy sounds (/Z/ and /ZH/) show some lines at low frequencies.

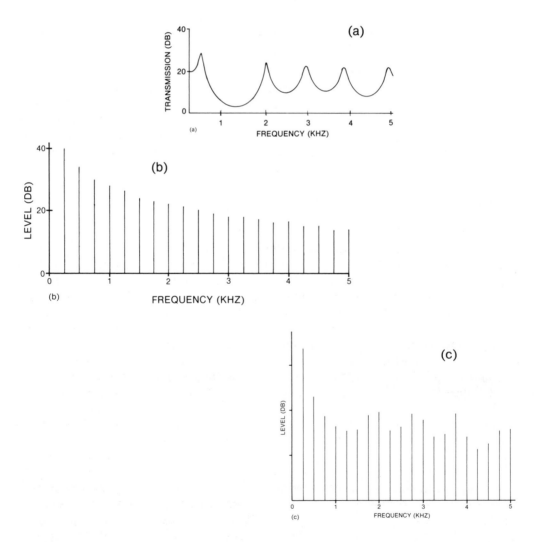

Fig. 5.B-5. (a) Transmission of two-tube /EE/-shaped tract; (b) spectrum of glottal wave; (c) resulting spectrum of output from tract.

168

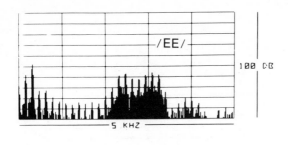

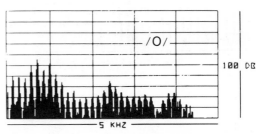

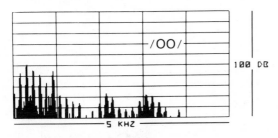

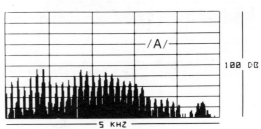

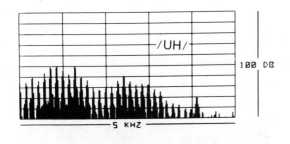

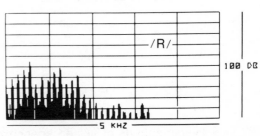

(Continued on next page.)

169

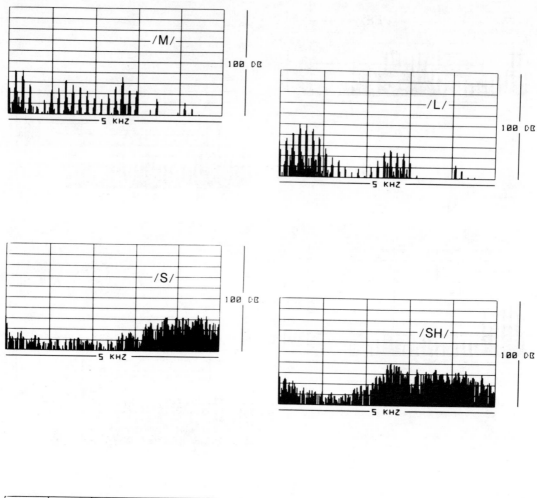

Fig. 5.B-6. Spectra of several speech sounds.

The diverse sounds of voiced speech can be produced by means of various vocal tract shapes. The particular spectrum produced is mostly independent of the voicing fundamental frequency; it depends primarily on the vocal tract shape and the resulting formant frequencies. (There must be some internal scaling in speech perception to account for our ability to recognize the same speech sounds produced by the shorter tracts and the resulting higher formant frequencies of adult females and children. It is interesting to note that the formant frequency ratios remain roughly the same as for male speech.) The mouth can be varied more extensively in shape than either the pharynx or the nasal cavity because of the great ability of the forepart of the tongue to be maneuvered. The nasal cavity is basically fixed in terms of internal dimensions and can only be coupled to the pharynx in varying degrees. The size of the pharynx is somewhat variable and is under control of the back of the tongue. Approximate average formant frequencies for men, women, and children are shown in Table 5.B-1.

Table 5.B-1. Approximate average formant frequencies for men, women, and children. (After Peterson and Barney, 1952.)

Phoneme	EE	I	E	A	O	U	OO	UH
F1								
Men	270	390	530	660	730	440	300	640
Women	310	430	610	860	850	470	370	760
Children	370	530	690	1010	1030	560	430	850
F2								
Men	2290	1990	1840	1720	1090	1020	870	1190
Women	2790	2480	2330	2050	1220	1160	950	1400
Children	3200	2730	2610	2320	1370	1410	1170	1590
F3								
Men	3010	2550	2480	2410	2440	2240	2240	2390
Women	3310	3070	2990	2850	2810	2680	2670	2780
Children	3730	3600	3570	3320	3170	3310	3260	3360

Exercises

1. What are the main components of the vocal tract?

2. What function does each component of the vocal tract serve?

3 Suppose a "neutral female tract" is defined to be a neutral tract that is 14 cm long. At what frequencies will the first three formants occur?

4. A "neutral child's tract" is 11 cm long. What are the first three formant frequencies?

5. The speed of sound in helium is greater than it is in air. Suppose a neutral tract is filled with helium (v = 970 m/sec). At what frequencies do the formants occur? Does this account for the "Donald Duck" speech quality of a diver who is breathing helium?

6. What are the formant frequencies of a neutral tract filled with krypton (v = 200 m/sec)?

7. What determines the spectrum of the output speech waveform?

8. An /0/-like tract might be configured with two tubes whose dimensions are shown below. If the vocal tract transmission is as shown below for this two-tube model and the spectrum of the vocal cords is as shown in Fig. 5.B-5b, sketch the spectrum of the resulting speech.

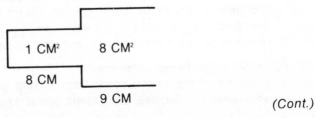

(Cont.)

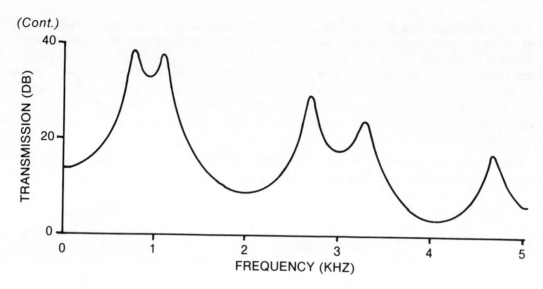

9. An alternative means can be used for determining which frequencies the neutral vocal tract will emphasize. Consider, for example, the effect of the neutral tract upon air pulses produced at two different rates: one pulse every 0.002 sec or one every 0.001 sec. The time required for a single pulse to travel the length of a 0.17-m tract is equal to 0.0005 sec. At the end of 0.0005 sec the first pulse in each case will be at the open end of the tract and will be reflected back toward the closed end as a negative pulse. After a total elapsed time of 0.001 sec (0.005 sec to return) the negative pulses will be back at the closed end of the tube. In the case of the pulse train with pulses every 0.001 sec, a new positive pulse is produced at this instant and is partly canceled by the negative pulse. At the closed end of the tube the pulse reflects without flipping over and begins its journey to the open end again. After reflection from the closed end of the tube we have a negative pulse moving toward the open end of the tube for the one case (0.001 sec) and a positive pulse moving toward the open end of the tube for the other case (0.002 sec). After 0.0005 sec the pulses arrive at the open end of the tube, where the negative pulse is reflected as a positive pulse returning to the closed end and the positive pulse is reflected as a negative pulse returning to the closed end. The positive pulse and the negative pulse arrive back at the closed end just as new positive pulses are produced by each pulse train. The two positive pulses add together to produce a larger positive pulse in the one case (0.002 sec), whereas the positive and negative pulses partly cancel again in the other case (0.001 sec). From this example we deduce that the tube serves to enhance the pulses produced at a rate of one new pulse every 0.002 sec by means of constructive interference, and to diminish the pulses produced at a rate of one new pulse every 0.001 sec by means of destructive interference. To what frequency does the pulse period of 0.002 sec correspond? How does this relate to the formant frequencies? To what frequency does the the pulse period of 0.001 sec correspond? How does this relate to the formant frequencies? What of other pulsing rates?

10. Spectra of speech sounds
   a. Use the spectra in Fig. 5.B-6 to determine the approximate frequencies for the first three formants for each speech sound by looking for the broad peaks in the spectrum. Record the results.
   b. How well do your results for vowel sounds agree with those in Table 5.B-1?

*11. Construction of model vocal tracts
   a. Construct three model vocal tracts using paper and masking tape (or other appropriate materials). Make them about 17 cm long, one with constant cross-

sectional area and the other two with cross-sectional areas appropriate for the vowels /EE/ and /O/.

b. Use an artificial larynx to excite the tracts. Do the sounds from the /EE/ and /O/ tracts shift perceptually from the neutral tract in a way similar to real speech?

c. Use a spectrum analyzer to determine the formant frequencies. What effect does vocal tract shape have on the spectra?

d. Do the formant frequencies for /EE/ and /O/ shift in the same direction from the neutral values as occurs in actual speech?

Further Reading

Denes and Pinson, chapter 4

Flanagan, chapters 3, 5

Fletcher, chapter 3

Ladefoged, chapters 8, 9

Singh and Singh, chapters 3-5

Peterson, G. E., and H. L. Barney. 1952. "Control Methods Used in a Study of the Vowels." J. Acoust. Soc. Am. *24*, 175-84.

Audiovisual

1. *Your Voice* (10 min, b&w, 1947, EBE)

Demonstrations

1. Artificial larynx operating into model vocal tracts
2. Sine wave driver operating into 17-cm "brass tube tract" with microphone and meter to show relative output

5.C. DISTINGUISHING CHARACTERISTICS OF SPEECH SOUNDS

Phonemes are the individual or distinguishable sounds of speech and can be thought of as the building blocks of speech from which syllables, words, and sentences are constructed. In sections 5.A and 5.B we discussed how it is possible for distinguishable phonemes to be produced by the human vocal mechanism. We now consider phoneme categorization schemes first and then analysis and synthesis methods for examining speech characteristics.

Categorization of Phonemes

Phonemes can be categorized in several different ways. In section 5.A we considered a categorization by energy types. Other categorizations might be by the steady-transitory dichotomy, or the manner of articulation (see Table 5.A-1), or the place of articulation, etc.

It is possible to produce some phonemes in a steady state; for instance, a vowel can be produced as long as we have the breath to drive the vocal mechanism. However, there is no such thing as a steady-state production (even in an idealized sense) for some other sounds, such as the plosives. We will label sounds that can be produced in a steady state as *steady phonemes* and sounds that are inherently transitory as *transitory phonemes*. (In running speech almost all of the phonemes are transitory and very seldom is a steady state achieved. However, the fact that some sounds can be produced in a steady state and others cannot is interesting. Furthermore, perception of the transitory phonemes may be more dependent on transitions than for steady phonemes.) The category of steady phonemes includes the vowels, the liquids, the fricatives, and the nasals, whereas the category of transitory phonemes includes the plosives, the semivowels, and the diphthongs. (You can determine into which category a particular phoneme falls by performing the simple experiment of trying to produce the phoneme in a steady state; if you are able to do so it is a steady phoneme, and if not it is a transitory phoneme.)

Steady sounds are produced when the vocal tract maintains a constant shape; transitory sounds are produced when the shape of the vocal tract varies in time. Since the formant frequencies are related to the vocal tract shape, we expect a constant vocal tract shape to produce formant frequencies that are constant in time, and a changing vocal tract shape to produce formant frequencies that change in time. We will find when we study spectrograms that these features appear in the spectrograms as formant trajectories that change with time, thus indicating the changing nature of the vocal tract shape with time.

There are several different manners of speech production: plosive production involves the sudden release of pressure; fricative production involves forcing an air stream through a constriction; nasal production involves coupling of the nasal cavity with the pharynx; production of liquids involves a fairly small constriction in the tract; the semivowels are produced by starting from a vowellike configuration and then moving to the following vowel configuration. Plosives and fricatives can be either voiced or unvoiced. Note that the manner of articulation is closely related in many cases to the energy types discussed in section 5.A.

The various articulators used in speech production determine the place of articulation. Places of articulation (or constriction) are between the lips (labial), between the teeth (dental), between lips and teeth (labio-dental), between the tongue and gum ridge (alveolar), between the tongue and hard palate (palatal), between the tongue and soft palate (velar), and in the vicinity of the glottis (glottal). These various places of articulation produce different vocal tract shapes, which in turn give rise to various formant patterns.

Speech Analysis

Section 5.B described how the vocal tract produces characteristic formant patterns, depending on its shape. If it were possible to perform many spectral analyses, of the kind described in section 5.B, at different times in a changing speech signal and then to place these speech spectra contiguously, we would be able to see a three-dimensional representation showing the amount of energy at each of many different frequencies and to see how the energy changes in time. Fig. 5.C-1 illustrates how a three-dimensional display of sound level, frequency, and time might appear. A roughly equivalent display can be produced by using relative lightness and darkness to represent the third dimension; in this way we can get a three-dimensional display on a two-dimensional page. The *sound spectrograph* is a device that creates this latter type of display and is shown schematically in Fig. 5.C-2. The important elements of the spectrograph are the bandpass filters that divide the incoming speech signal

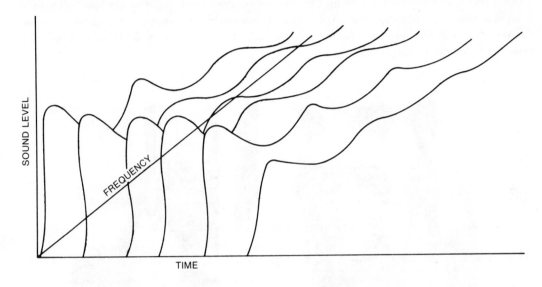

Fig. 5.C-1. Three-dimensional display of contiguous speech spectra.

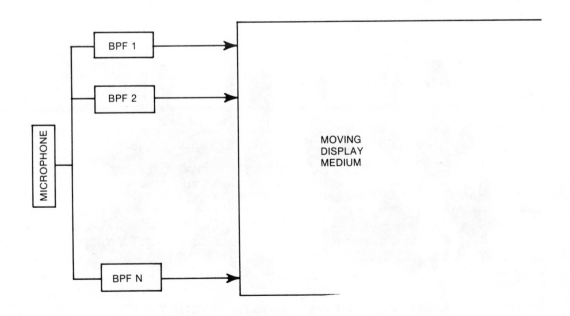

Fig. 5.C-2. Elements of a sound spectrograph.

175

into many different frequency bands(from 50 to about 250 bands, depending on the application). The amount of energy that comes through each filter is measured and used to control a marking mechanism. This mechanism may be a voltage-controlled sparking wire in contact with heat-sensitive paper, as in a conventional analog sound spectrograph; when a computer is used as a sound spectrograph the output signal from each channel is typically used to control the brightness of a display point on an oscilloscope screen.

A typical analog spectrograph produces spectrograms in which sound level is indicated by relative darkness, as shown in Figs. 5.C-3 and 5.C-4. A typical computer-generated spectrogram indicates the intensity level by relative brightness (see Fig. 5.C-5); the fundamental frequency (in Hz) and sound level (in dB) are also shown. In a typical spectrogram the time axis is along the horizontal direction and the frequency axis is along the vertical. As already noted, the sound level is shown in terms of relative darkness or brightness.

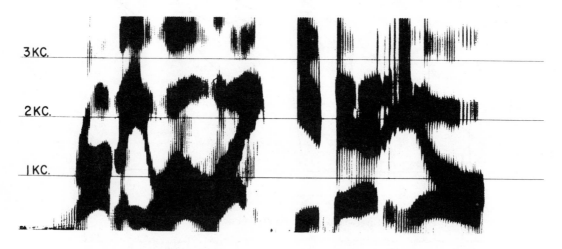

ROBBY WILL LIKE YOU DADDY-OH (NATURAL)

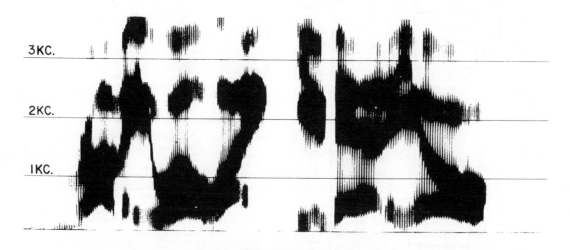

ROBBY WILL LIKE YOU DADDY-OH (SYNTHETIC)

Fig. 5.C-3. Spectrogram of the sentence ''Robby will like you, daddy-oh.'' (From Strong, 1967.)

176

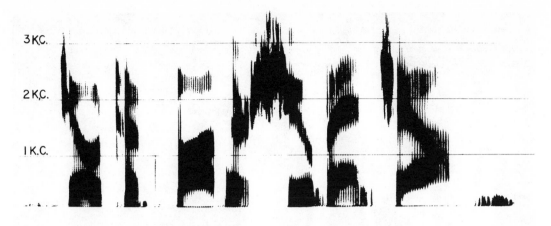

JOE TOOK FATHER'S SHOE BENCH OUT (NATURAL)

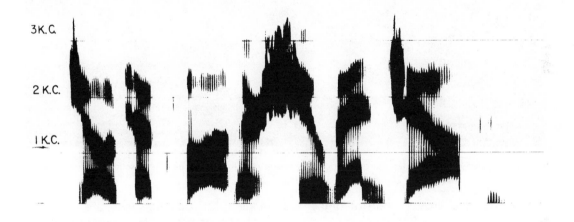

JOE TOOK FATHER'S SHOE BENCH OUT (SYNTHETIC)

Fig. 5.C-4. Spectrogram of the sentence "Joe took father's shoe bench out." (From Strong, 1967.)

The sentences spoken for the spectrograms of Figs. 5.C-3 and 5.C-4 were, respectively, "Robby will like you, daddy-oh" and "Joe took father's shoe bench out." The sentence spoken for the spectrogram of Fig. 5.C-5 was "The second planet was inhabited by a conceited man." The various distinguishing spectrographic features for each of the several phoneme categories can be seen in the spectrograms of these figures. Keep in mind that we are discussing idealizations here; it is not usually possible to sit down with a spectrogram and determine unambiguously what was spoken. Many subtleties occur when the phonemes blend together in running speech. A feature commonly observed for vowel sounds is the presence of three or more fairly distinct formant bands of energy, which may become approximately constant in frequency if the vowel is of fairly long duration. The diphthongs are also typically characterized by three or more distinct formant bands that are seen to change their frequency positions quite markedly from the beginning of the diphthong to the end. Nasals are typically characterized by a low-frequency formant, with much smaller amounts of energy at high frequencies. (It should be noted that even for nasals there

177

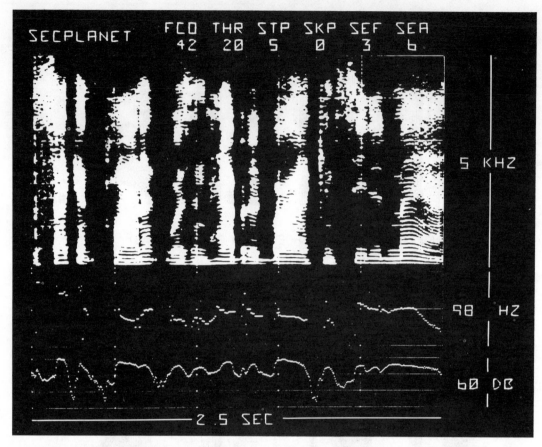

Fig. 5.C-5. Computer-generated narrow-band spectrogram of the sentence "The secont planet was inhabited by a conceited man." (From Strong and Palmer, 1975.)

are bands of energy present at higher frequencies, but because they are so much weaker than the low-frequency band they do not appear very dark in sound spectrograms.)

Voiced plosives typically start with a low-frequency formant present (during the vocal tract closure for the plosive production), and then higher-frequency formants appear (following the vocal tract opening); these formants usually exhibit transitions into the following speech sound. The voiced plosives usually show very little if any "burst energy" because the release of the vocal tract closure involves the release of much smaller pressures than is often the case with unvoiced plosives. The unvoiced plosives typically give rise to a very short burst of energy that is spread across many different frequencies. The vocal cord vibrations usually begin later after the release of the vocal tract constriction than is the case with the voiced plosives. An example can be seen in the /T/ of *took* in "Joe took father's shoe bench out," in Fig. 5.C-4.

The unvoiced fricatives (of which the /SH/ of *shoe* in "Joe took father's shoe bench out" is a good example) typically give rise to high-frequency bands of energy in a spectrogram. These bands of energy are more ragged in appearance than the formants for voiced speech, because the vocal tract is excited with noise energy, whereas voice excitation is used for vowel sounds. Again, it should be noted that unvoiced fricatives produce energy at low frequencies but that it is weaker in comparison to the high-frequency energy than for voiced sounds and so it often does not appear on the spectrogram. The voiced fricatives (of which the /Z/ in *was in* of "The second planet was inhabited by a conceited man" is an example) have spectrographic characteristics similar to the unvoiced fricatives; however, their spectrograms also typically exhibit a low-frequency formant, due to the presence of voice energy in their production.

Note in the spectrograms of Figs. 5.C-3, 5.C-4, and 5.C-5 that our idealizations for the spectra of phonemes spoken in isolation or for simple consonant-vowel syllables break down when the phonemes are combined in running speech. It is not possible to determine in a spectrogram explicitly where one phoneme ends and another begins. The vowel formant frequencies very often do not achieve their steady-state values in running speech, and everything seems to be in a constant state of transition. Unusual groupings of sounds also occur. For example, consider the behavior of the /CH/ sound in *bench.* In the written version we think of the /CH/ as belonging to the word *bench,* but in the spectrogram it appears, rather, to belong to the word *out.* The utterance actually produced seems more like *ben chout* than *bench out.* The same phoneme may also sound different when it occurs in different contexts. Note in the spectrogram of "Robby will like you, daddy-oh" that there was a much tighter closure for the first /D/ than for the second /D/ of *daddy-oh,* because the first formant is the only one apparent in the first case, whereas the higher formants are also apparent in the second. Further study of the spectrograms will reveal other interesting features of actual speech.

One general feature of the spectrograms just discussed should be mentioned in passing. The spectrograms in Figs. 5.C-3 and 5.C-4 are termed wide-band spectrograms because the spectrograph used wide filters and individual voicing harmonics are not distinguishable. The spectrogram in Fig. 5.C-5 is a narrow-band spectrogram; the spectrograph used narrow filters, and individual voicing harmonics can be seen. An uncertainty relation similar to that described in section 3.C is also apparent here. The wide-band spectrogram has poor frequency resolution and fails to show harmonics, but it has good time resolution and shows transients well. The narrow-band spectrogram has good frequency resolution and shows harmonics, but it has poor time resolution and shows transients poorly.

Another analysis method has been made possible with computers. It might be termed a "formant spectrograph" because it produces "formant spectrograms" in which only the spectral peaks (and not wide bands) are plotted. Fig. 5.C-6 illustrates such a "formant spectrogram" for "Joe took father's shoe bench out." Note the similarities and differences between Figs. 5.C-4 and 5.C-6; both were produced from the same utterance, but each emphasizes different features.

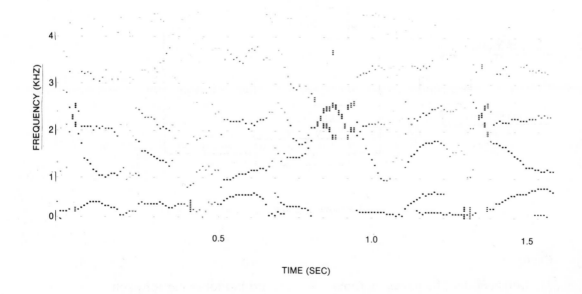

Fig. 5.C-6. "Formant spectrogram" of the sentence "Joe took father's shoe bench out."

Speech Synthesis

In addition to analysis, we can also discover distinguishing characteristics of speech through speech synthesis. We simply synthesize speech from a set of parameters that we think are significant for speech production, and then, by asking listeners to identify the "speech sounds" produced, we can ascertain which of the parameters are in fact the significant ones. The Pattern-Playback machine, designed and used at the Haskins Laboratories, provided much of the early information regarding the significant parameters for speech synthesis. Basically, it involved using either spectrograph-generated spectrograms or hand-drawn spectrograms to produce the synthetic speech. By modifying an existing spectrogram, or by hand-drawing a spectrogram, it was possible to eliminate or add features, thus helping to determine which features of the spectrogram were most significant.

Another device that is proving useful for speech synthesis is the *formant synthesizer*. The basic elements of one type of formant synthesizer are shown in Fig. 5.C-7. Each of the band-pass filters represents a different formant. The frequency and amplitude of each filter can be controlled independently, thus making it possible to simulate formant frequencies and amplitudes. Two types of excitation are used as inputs to the synthesizer: (1) a pulse train and (2) noise. The pulse train input is used to simulate voicing energy, and the noise input is used to synthesize phonemes requiring noise energy. A switch controls which of the two types of energy goes into the band-pass filters. Both the pulse train and the noise have large amounts of energy at many different frequencies. The band-pass filters selectively eliminate most of the energy outside their own band and are thus able to simulate formants by permitting only a selected range of frequencies to pass through. The outputs of the filters are added together, and the resulting signal, when played through a loudspeaker or a set of headphones, produces the synthetic speech. The parameters used to control a formant synthesizer can be varied in many different ways to assess the relative importance of various features for synthesizing speech. Different versions of formant synthesizers have been successfully used to produce synthetic speech of good quality. Output from formant synthesizers is being used in many practical applications, such as telephone answering systems and audio output from computers.

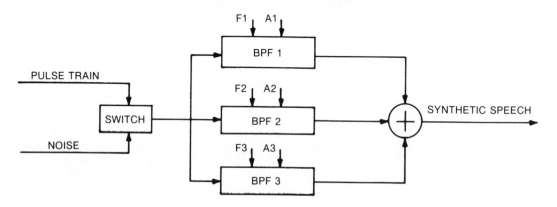

Fig. 5.C-7. Elements of a formant speech synthesizer.

Exercises

1. Consider the phonemes in Table 5.A-1 and do the following for each.
    a. Indicate whether the phoneme is steady or transitory.
    b. Indicate the place of articulation (where appropriate) as one of the following: labial, labio-dental, alveolar, palatal, velar, glottal.

c. Indicate the manner of articulation as one of the following: vowel, semivowel, liquid, nasal, voiced fricative, unvoiced fricative, voiced plosive, unvoiced plosive.

2. a. List the distinguishing features typically seen in a spectrogram for the following categories of sounds: nasals, voiced fricatives, unvoiced fricatives, diphthongs, voiced plosives, unvoiced plosives. (For example, vowels are typically characterized by three or more fairly well defined energy bands or formants.)

b. Sketch basic spectrographic features for each category, as illustrated in the diagram for a vowel.

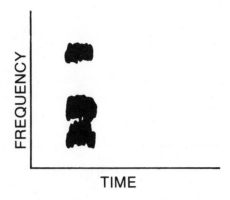

c. Describe what the vocal cords and the vocal tract are doing to produce the observed spectrographic features.

3. Compare the spectrogram in Fig. 5.C-4 to the "formant spectrogram" in Fig. 5.C-6. Both are for the same utterance, "Joe took father's shoe bench out." Describe similarities and differences between the two spectrograms.

4. Compare the sound pressure wave in Fig. 5.A-3 with the spectrogram in Fig. 5.C-4. Both were obtained for the utterance "Joe took father's shoe bench out." Describe the corresponding features of the two.

5. Compare formant frequencies obtained from approximately steady portions of the spectrograms in Figs. 5.C-3, 5.C-4, and 5.C-5 with values obtained from the spectra of Fig. 5.B-7. How well do they agree? Name some items that might account for the discrepancies.

6. Fundamental frequency can be determined by measuring the frequency of some higher harmonic (such as the 10th) and then dividing by the harmonic number (10 in this case). Use this technique to determine the fundamental at various points in the narrow-band spectrogram of Fig. 5.C-5.

7. Suppose a two-formant speech synthesizer is to be controlled by means of a lap board. The lap board is shown in the diagram below and is used by placing a pointer on the board in a position appropriate for the first two formant frequencies of the desired sound. The first formant frequency is controlled by the horizontal position of the pointer (increasing from left to right), and the second formant frequency is controlled by the vertical position of the pointer (increasing from down to up). Show on the diagram of the lap board where the pointer should be placed to produce the following vowel sounds: /EE/, /I/, /E/, /A/, /O/, /U/, and /OO/. Show also the approximate paths that should be traced out by the pointer to produce the diphthongs /OI/ and /OW/, and to produce the following syllables: /D/O/, /G/O/, /B/EE/, /D/EE/, and /G/EE/.

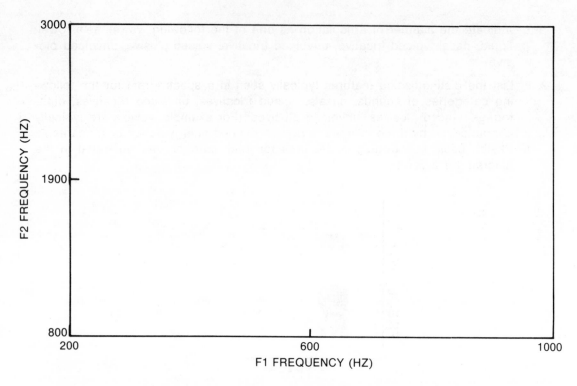

8. Interpret the spectrograms in Figs. 5.C-3, 5.C-4, and 5.C-5. Label each of the phonemes in the spectrograms with an appropriate phoneme symbol and show the approximate time duration of each by bounding each with vertical slashes. Note transition regions and the influences of one phoneme on another.

* 9. If you have access to a sound spectrograph, produce a sound spectrogram for your own voice while speaking a sentence that is about 2 seconds in length. Interpret the sound spectrogram by labeling different parts of the spectrogram with appropriate phoneme symbols. Also, indicate points of high and low pitch.

*10. If you have access to a speech synthesizer, synthesize some steady vowel sounds and some consonant-vowel syllables.

Further Reading

Denes and Pinson, chapters 4, 7

Gerber, chapter 10

Flanagan, chapters 5, 6

Fletcher, chapters 3, 4, 5

Ladefoged, chapters 6-9

Singh and Singh, chapters 5-10

Cooper, F. S. 1977. "Observations on Speech Research: Objectives, Strategies and Some Partial Answers," Languages and Linguistic Symposium 1977, Brigham Young University.

Potter, R. K., G. A. Kopp, and H. C. Green. 1947. *Visible Speech* (D. Van Nostrand Co).

Strong, W. J., and E. P. Palmer. 1975. "Computer Based Sound Spectrograph System," J. Acoust. Soc. Am. *58,* 899-904.

Strong, W. J. 1967. "Machine-aided Formant Determination for Speech Synthesis," J. Acoust. Soc. Am. *41,* 1434-42.

Audiovisual

Tapes of synthetic speech

Demonstrations

Overhead of speech spectrograms

5.D. PROSODIC FEATURES OF SPEECH

Prosodic features are information-bearing features of speech other than those features that are specifically associated with phonemes; they extend over more than one phoneme or speech segment. The more common prosodic features are pitch, stress, and rhythm. In spoken English these prosodic features play a secondary role to that of the phonemes in terms of communicating meaning. However, they are often used to clarify otherwise ambiguous sentences and to add emphasis to selected words within a sentence. In some spoken languages, such as tone languages, prosodic features play a more significant role, in which they can actually change the "meaning" of a phoneme. Prosodic features also provide information about the emotional and physical characteristics of a talker. Prosodic features may be given more attention than the phonemes in such activities as reciting poetry or singing. This occurs when the embellishments (or the way in which something is said) are more important than the content (or what is said).

We listed prosodic features above in their subjective form. In order to carry out experiments and measurements relating to those features we must know what their objective counterparts are. (Keep in mind that there is not a simple one-to-one relationship between the subjective and objective quantities. Hence, the relationships that we describe are not without some ambiguity.) As we have done previously, we will assume that pitch is most significantly dependent on fundamental frequency. Stress is dependent on fundamental frequency, duration, sound level, and spectrum, in that relative order of importance. Rhythm is dependent on relative durations of syllables and pauses and the number of syllables produced per second.

Normal Spoken English

Consider first some examples showing the role of prosody in spoken English. The two sentences "The light housekeeper is gone" and "The lighthouse keeper is gone" both involve essentially the same phonemes; yet the two meanings are quite different. In Fig. 5.D-1 fundamental frequency and sound level as functions of time are traced out for each of the two sentences.

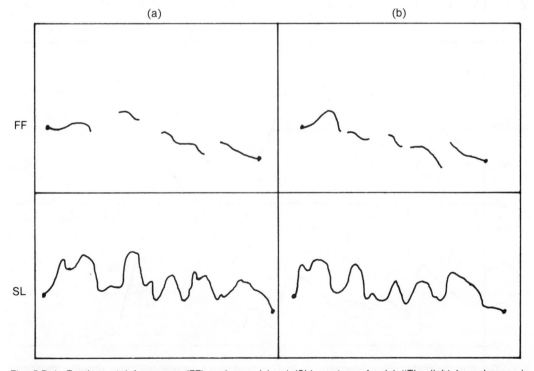

Fig. 5.D-1. Fundamental frequency (FF) and sound level (SL) contours for (a) "The light *house*keeper is gone" and (b) "The *light*house keeper is gone."

Note that in the "light housekeeper" version the emphasis or stress is on *house,* as shown by higher relative fundamental frequency, a longer duration, and higher relative sound level. In the "lighthouse keeper" version the emphasis is on *light.* Commonly (though by no means always), the fundamental frequency, duration, and sound level change together in a particular direction. For instance, when the intensity is increased the fundamental frequency has a natural tendency to increase unless other adjustments are made in the speech apparatus. It often happens that when something is made more intense its duration is increased, or vice versa. What we expect to observe in many cases is that fundamental frequency, intensity, and duration tend to increase or decrease together.

A further example of the influence of stress is provided in four versions of the utterance "John drove to the store." The four versions were spoken without stress, with stress on *John,* with stress on *drove,* and with stress on *store.* Fundamental frequency contours for these four versions are shown in Fig. 5.D-2. There is a very pro-

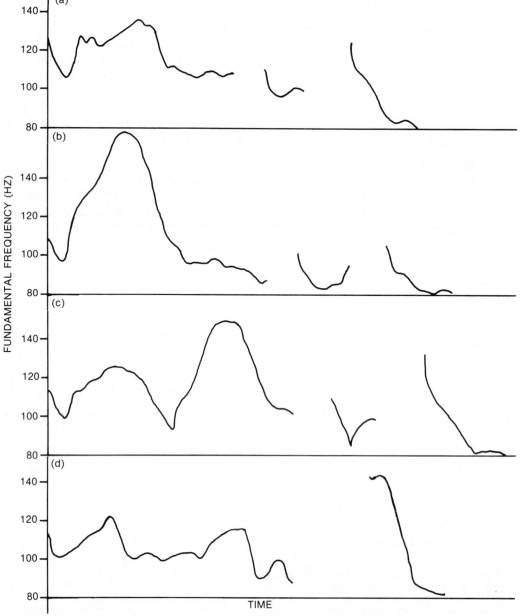

Fig. 5.D-2. Fundamental frequency contours for four versions of a sentence: (a) "John drove to the store." (b) "*John* drove to the store." (c) "John *drove* to the store." (d) "John drove to the *store.*" (Courtesy of A. Melby and R. Millett.)

184

nounced peak in the fundamental frequency contour of the stressed word in each case. There is also a noticeable increase in duration.

Another way in which prosodic features can be used to change meaning is in the contrast between a declarative sentence, such as "They are going home now," and its question counterpart, "They are going home now?" Note in Fig. 5.D-3 that fundamental frequency for the declarative sentence decreases toward the end of the sentence. In contrast, the same sentence as a question exhibits a fundamental frequency that is rising (or not falling) at the sentence end. (You might try uttering several declarative-question sentence contrasts to see how this works.)

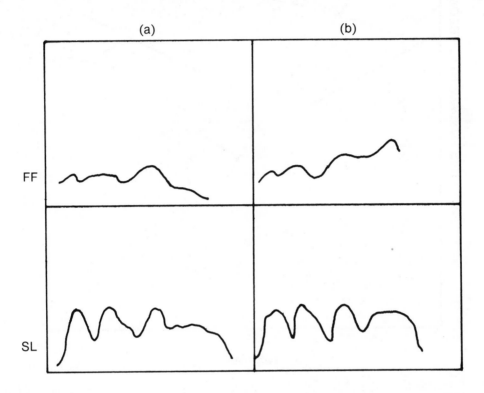

Fig. 5.D-3. Fundamental frequency (FF) and sound level (SL) contours for (a) "They are going home now" and (b) "They are going home now?"

Tone Languages

Tone languages are languages in which a particular phoneme can take on one of several different meanings depending on the tone (or fundamental frequency contour) with which it is spoken. Mandarin Chinese is an example of such a language. In addition to the specification of a particular phoneme, a tone (1, 2, 3, or 4) is also specified. The same phoneme can then take on one of four different meanings, depending on which tone is used with it. Typical fundamental frequency contours with their durations, for Mandarin, are shown in Fig. 5.D-4.

Perceived Personal State

Prosodic features can provide information about the physical and emotional characteristics of a talker. The prosodic features thus help to determine our perception of a talker's "personal state." We are often able to determine the sex of a talker on the basis of fundamental frequency and to a lesser extent on the basis of formant frequencies. The age of a person may be partly apparent on the basis of voice quality. We are often able to perceive the emotional state of a talker via speech. We form our perceptions of an individual's personality from hearing his/her speech, often in

185

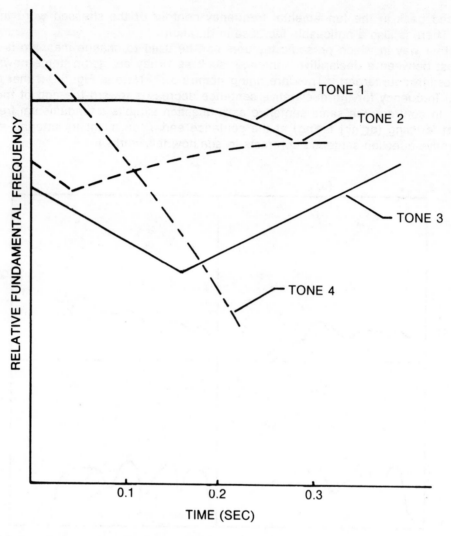

Fig. 5.D-4. Average fundamental frequency contours and durations for the tones of Mandarin. (After Ting, 1971.)

the absence of visual cues. (Some of the things we will discuss are not well defined at the present time and should be considered in that light.)

There is evidence that prosodic features are a medium through which the emotional state of a person can be observed. In studies of actual and simulated emotional conditions, it was found that generally when a person speaks in anger his durations were shortened and his fundamental frequency and sound level was increased relative to his normal speech. Durations were increased and fundamental frequency and sound level were decreased for a talker in a depressed state.

Studies have been done in which listeners were asked to rate different tape-recorded voices in terms of many pairs of opposites (e.g., kind-unkind). The results of these ratings were then combined and reduced and plotted along two perceptual dimensions labeled "benevolence" and "competence." The same tape-recorded voices were then modified in various ways by means of computers and other electronic devices. The modified voices were then played to listeners for their judgments, and the results were plotted in the two-dimensional perceptual space. Three factors were varied in the modified voices: rate of speaking, variation of pitch, and average pitch.

The study found that perceived personality was most strongly dependent on changes in speaking rate and less dependent on average pitch and variation of pitch. Perceived competence appears to increase and decrease with rate in an almost linear fashion; that is, increased rate of speaking gives rise to judgments of increased competence, etc. Perceived benevolence appears to show an inverted "U" relationship with rate; very high or very low rates produce decreased benevolence, whereas the middle range of speaking rate gives rise to the highest benevolence rating. As mentioned, the rate effects were very strong. Increased variation of pitch (i.e., making the speech less monotone) causes perceived competence and benevolence to increase, whereas decreased variation of pitch (i.e., more monotone) caused the perception of both to decrease. Changes in average pitch seem to have little effect on perceptions of benevolence and competence.

Exercises

1. Get one of your associates (perhaps a budding dramatist) to speak the sentence "You always get the same results," while speaking normally and then while simulating anger (or excitement) and depression. For each condition, list the nature of each of the following features: pitch, stress, and rhythm.

2. Get several different people to speak the sentence "You always get the same results," while speaking normally. For each person, list the nature of the following features: pitch, stress, and rhythm.

3. Choose a sentence similar to "John drove to the store." Determine different meanings that can be imposed on it by stressing different words. Draw some idealized fundamental-frequency contours that you might expect to obtain from the different versions.

4. Get a competent speaker of a tone language to illustrate the tones of the language. Note the pertinent features.

5. A talker with an "average" speaking rate has a recording of his speech speeded up. How will others perceive his benevolence to change? His competence?

6. Repeat exercise 5 with the speech slowed.

7. A person's speech is made more monotone. What effect does this have on perceived personality (i.e., competence and benevolence)?

* 8. If you have access to a sound spectrograph or a pitch tracker, produce sound spectrograms or pitch contours for selected sentences, as produced normally, in anger, and in depression. What do you observe for fundamental frequency and duration of the angry and depressed versions as compared with the normal?

* 9. Repeat exercise 8, but measure fundamental frequency for stressed words in sentences.

*10. Repeat exercise 8, but measure fundamental frequency for various tones of a tone language.

Further Reading

Denes and Pinson, chapter 8
Ladefoged, chapters 5, 10
Singh and Singh, chapter 8
Brown, B. L., W. J. Strong, and A. C. Rencher. 1974. "Fifty-four Voices from Two: The Effects of Simultaneous Manipulations of Rate, Mean Fundamental Frequency, and Variance of Fundamental Frequency on Ratings of Personality from Speech," J. Acoust. Soc. Am. *55,* 313-18.
Fry, D. B. 1958. "Experiments in the Perception of Stress," Language and Speech *1,* 126-52.
Lehiste, I. 1970. *Suprasegmentals* (M.I.T. Press).

Melby, A. M., W. J. Strong, E. G. Lytle, and R. Millett. 1977. "Pitch Contour Generation in Speech Synthesis: A Junction Grammar Approach," Am. J. Computational Linguistics, microfiche no. 60.

Ting, A. 1971. "Mandarin Tones in Selected Sentence Environments: An Acoustic Study," J. Acoust. Soc. Am. *51,* 102.

Williams, C. E., and K. N. Stevens. 1972. "Emotions and Speech: Some Acoustical Correlates," J. Acoust. Soc. Am. *52,* 1238-50.

Audiovisual

1. Tape illustrating prosodic features in speech
2. *The Sounds of Language* (32 min, B&W, 1962, IU)

Demonstrations

Use a person capable of a wide range of expression in speech to demonstrate. Perform fundamental frequency analyses if possible.

5.E. DEFECTS IN SPEECH PRODUCTION AND PERCEPTION

We will consider only three causes of speech defects in this section: cleft lip or palate, impaired vocal cords, and impaired hearing. Each of these defects varies in severity from one case to another, and each has a different effect on the use of spoken language. Impaired hearing, if it is substantial, is probably the most debilitating in terms of its effects on the whole process of spoken language communication.

Cleft Lip or Palate

Speech produced by a person with cleft palate or cleft lip has a quality that is perceived as nasal. Let us digress a bit at this point and consider what produces a nasal quality for a normal speaker. We have assumed that the vocal tract is a single tube of variable cross-section, as shown in Fig. 5.E-1a. This assumption is valid for most speech sounds, with the exception of the nasal sounds: /M/, /N/, and /NG/. (The vowel sounds can also be produced with a nasal quality. Nasalized vowels can be considered to be mixtures of nasal sounds and vowel sounds.) The soft palate serves as a valve in the vocal tract to control the amount of sound going into the nasal cavity. For nasal sounds the soft palate is lowered and the nasal cavity is connected to the rest of the vocal tract. For nasal sounds the mouth cavity typically is closed at some point, and we have a tube combination, as shown in Fig. 5.E-1b. The combination of two tubes gives rise to a more complicated set of resonances than those of a single tube. The resulting spectrum for a nasal sound gives the sound its particular nasal quality.

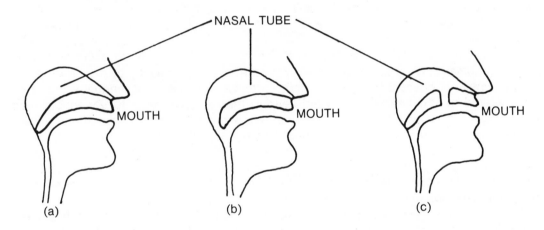

Fig. 5.E-1. Vocal tract configurations for different nasal tract couplings to the vocal tract: (a) nonnasal, (b) nasal, (c) cleft palate.

In a person having a cleft lip or palate (physically this amounts to a lip or palate with a hole in it) there is always an opening from the mouth cavity into the nasal cavity, as shown in Fig. 5.E-1c. Since the nasal cavity is always open to the vocal tract, the resulting speech sounds have a nasal quality. The usual correction for cleft lip or cleft palate is surgery to close the hole in the lip or palate, and it is generally successful.

Impaired Vocal Cords

The vocal cords may become inflamed because of illness or through abuse. The inflammation typically gives rise to a hoarse voice quality. If the inflammation is very severe, the ability to produce voiced sounds may be lost entirely for some period of time. Usually resting the cords and receiving proper medical treatment, if necessary, are adequate to overcome the problem of inflammation of the vocal cords.

Growths (malignant and otherwise) can occur on the vocal cords. Sometimes these growths obstruct the cords so that they cannot achieve a complete closure; in other instances they cause the cords to vibrate erratically. Either of these conditions can cause a harsh vocal quality. Sometimes surgery is necessary to remove growths from the vocal cords, and occasionally the entire cord must be removed if the growth is extensive and malignant.

If the vocal cords are missing or impaired to the point of being useless for voiced speech production, an alternate means of producing "voiced" speech, such as esophageal or buccal speech, may be developed. Another method of producing quasi-voiced speech is the use of an artificial larynx, a device for producing a buzzing sound that can be coupled into the vocal tract through openings in the tract or through some of the softer tissues of the tract in the vicinity of the larynx. The vibrating head of the artificial larynx is placed in contact with the external larynx structure and caused to vibrate while the user produces the vocal tract configurations that are appropriate for the intended speech sounds.

Impaired Hearing

Scientists at Bell Telephone Laboratories have coined the term "speech chain" to refer to the collection of different levels that are involved in the speech communication process. They suggest five levels in the speech chain: (1) the linguistic level of the talker (the concept or idea to be communicated), (2) the physiological level of the talker (neural and muscular activity of the speech production mechanism), (3) the acoustical level (speech sounds transmitted from talker to listener), (4) the physiological level of the hearer (mechanical and neural activity of the speech reception mechanism), and (5) the linguistic level of the hearer (the concept or idea received by the listener).

For the normally hearing child the development of speech seems to occur so "easily and automatically" that we are really unaware of the complexity of the process. It is only when we observe the great difficulties in the acquisition of language that are encountered by children with impaired hearing that we can begin to appreciate the intricacies of the language process. Spoken language is the most common and widely used form of language. Persons without relatively normal spoken language communication are placed at a great disadvantage in a society in which spoken language is the basic form of interpersonal communication. Indeed, it is not unreasonable to view the written language as being parasitic upon the spoken language in normal language development.

Substantial hearing impairments (particularly when they occur before a child has acquired facility in the use of language) can adversely affect a person in three major areas: reception of spoken language, production of spoken language, and acquisition of language. When the reception of spoken language is impaired, the normal development of language is hindered. It is possible to create substitutes for the spoken language, such as various manual languages, but these are often not as rich in the features of the language as is the spoken language. Furthermore, in societies in which the spoken language normally serves as the host for language development, substitutes for the spoken language are inconvenient and impractical in many instances even if they were adequate on other counts.

Some substitute systems for the spoken language have proven partially successful for the reception of language and for the acquisition of language. However, they have not produced significant successes in speech production for people with impaired hearing. One reason may be that substitute systems do not give the talker any feedback on his own speech production. Another possible reason is that the substitute systems are not rich enough in spoken language features to provide adequate feedback.

Consider the elements of a feedback system. The term *system* implies a collection of several components that function together to accomplish some desired end. A *feedback system* is a system in which a means has been provided for sensing the output of the system and then modifying the input of the system on the basis of the measured values of the output. A feedback system is illustrated in Fig. 5.E-2.

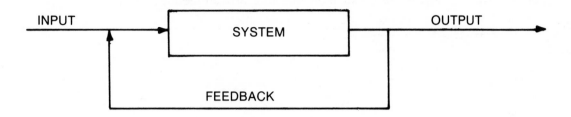

Fig. 5.E-2. Elements of a feedback system.

Many systems in our everyday world are feedback systems. The output of the speech system is sensed by the hearing system, and appropriate modifications are made in the control signals being sent to the speech production mechanism. When driving an auto on the highway, you are able to keep it appropriately placed on the road by means of visual feedback. (Imagine how absurd it would be for a blind person to attempt to drive an automobile.) Even such common tasks as walking and eating require feedback in order to succeed.

A hearing impairment tends to break the feedback loop of the speech chain and thus to impair speech production because it cannot be monitored. A severe impairment in a very young infant often causes the infant to cease to vocalize altogether, even though the infant may have begun babbling at the normal age. Hearing ability permits a speaker not only to monitor his own speech but also to monitor his speech in relation to the speech of others. A person with normal hearing can continually calibrate and recalibrate his own speech production against that of others, but a person with severely impaired hearing is unable to monitor his own speech or to receive the speech of others with facility.

The effects of a broken feedback loop on speech production can be seen quite clearly by comparing Fig. 5.E-3 with Fig. 5.E-4. Fig. 5.E-3 shows three formant frequencies, fundamental frequency, and sound level graphed as functions of time for

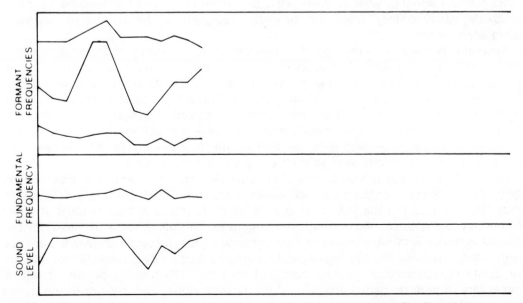

Fig. 5.E-3. Speech parameters for normal-hearing talker. The utterance was "The leaves will be. . . ."

the utterance "The leaves will be," spoken by a normal-hearing talker. Fig. 5.E-4 shows similar information for the utterance "The leaves will," spoken by a severely hearing-impaired talker. The time scales are the same in both figures. Note the overall difference in duration in the two cases and the difference in detail of the curves. The differences are presumably due to a broken feedback loop.

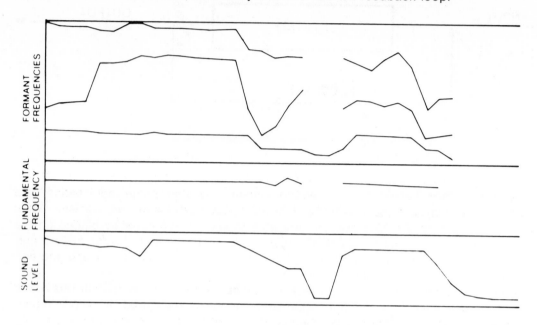

Fig. 5.E-4. Speech parameters for severely hearing-impaired talker. The utterance was "The leaves will . . ."

The most commonly used device for aiding the hearing impaired is the hearing aid, discussed in section 3.E. Acoustical amplification is probably the most useful sensory aid for the hearing impaired when there is adequate residual hearing. More sophisticated hearing aids make it possible to provide frequency selective amplification, frequency compression, and amplitude compression of the speech signal. Some hearing aids have been designed to take part of the high-frequency energy of speech and present it at low frequencies where the hearing-impaired person usually has the most residual hearing. Most of these latter aids have met with only marginal success, probably because they were not properly designed or because they were inadequately tested.

Several aids have been developed to perform partial analyses of the speech signal and then to present this information to the user, either visually or tactually. An attempt has been made in a number of these devices to create signals that would clarify the information that a person could gain from lipreading. Other devices have been designed to train a hearing-impaired person on selective speech production deficiencies. These include proper nasalization, adequate production of fricatives, proper pitch level, proper intonation patterns, and so on. In many cases the devices were not tested very extensively, and so their utility value is not known.

Let us close this section on a note of speculation and conjecture. We assume that, generally speaking, reception aids will prove more useful than training aids because they become a part of the individual and he learns to interpret their outputs in terms of his day-to-day encounter with spoken language. The utility value of these devices should increase with the passing of time because of the subject's greater experience with them. Because the hearing-impaired person's frequency bandwidth for speech reception is substantially reduced over that of a normally hearing person, it will be necessary to perform partial analyses of the speech signal and present to the user's ear a modified signal whose bandwidth requirements are much less than that of normal speech. For instance, such a device might perform a formant-frequency analysis of the speech signal and then present this information to the hearing-impaired person via his residual hearing or, if that is inadequate, via his tactile or visual senses.

Exercises

1. The speech produced by a person having a cleft lip or palate usually has a nasal quality associated with it. Why?

2. What is the best "cure" for a cleft lip or palate? What is done in this "cure"?

3. How does an artificial larynx function when used in the production of quasi-voiced speech? Is the vocal tract used at all when an artificial larynx is used? Get an artificial larynx and try using it to produce speech.

4. Describe the components, their functions, and their interrelationships for the talker's physiological level in the speech chain.

5. Describe the components, their functions, and their interrelationships for the listener's physiological level in the speech chain.

6. Describe and discuss some speech features that are available, to a listener, in the acoustical level of the speech chain that may not be available in some manual means of communication. How important are these features in the perception of speech? In the acquisition of language? In the production of speech?

7. Consider a warm air furnace as a feedback system. Describe the function of each of the following components in the furnace: thermostat, blower, cold air, warm air, gas, firebox. What happens if the thermostat is not working?

8. What are the components of the speech system? What is the feedback in the speech system? What is the input? What is the output? What senses (or measures) the output?

9. What is the role of feedback in singing? In performing on a musical instrument?

10. What musical instruments do you suppose a severely hearing-impaired person can perform on most successfully?

11. There are indications that development of speech in young children proceeds about as rapidly as the development of the speech and speech feedback mechanisms. What happens if the feedback mechanism fails to develop properly?

12. Suppose you have a task of developing a speech reception aid for the severely hearing-impaired. What speech parameters would you attempt to extract from the speech signal? How would you present them to the user? How much of a frequency bandwidth would be required?

13. Comparison of the speech parameters of normal and hearing-impaired talker's speech shown in Figs. 5.E-3 and 5.E-4
    a. Describe what you observe for relative movements of the formants for the two talkers.
    b. What are the differences in the fundamental-frequency contours? Which is more monotonic?
    c. What can you say about the relative rates of speaking for the two talkers?
    d. What can you say about the rhythm for the two talkers? Which has the more smoothly flowing speech?

*14. The role of normal feedback in speech can be demonstrated in the following way. If a talker hears his speech much later than he speaks it (delayed feedback) his speech often becomes somewhat confused and disoriented. This can be demonstrated by having a talker speak into a microphone and then listen to his speech over headphones that seal out the direct sound. A time delay is introduced between the microphone and the headphones.

If you have access to a delayed feedback apparatus, set it up and, with the headphones on, adjust the gain so that you hear only the delayed speech. Now read several sentences of some printed material and record the result with an auxiliary microphone. Play the auxiliary recording back several times while you listen and note any deviations from normal pitch, stress, and rhythm.

Further Reading

Denes and Pinson, chapters 6, 7, 8

Flanagan, chapter 7

Fletcher, chapter 19

Gerber, chapter 12

Stevens and Warshofsky, chapter 7

Calvert, D. R., and S. R. Silverman. 1976. *Speech and Deafness* (Alexander Graham Bell Association, Washington, D.C.).

Ling, D. 1976. *Speech and the Hearing Impaired Child: Theory and Practice* (Alexander Graham Bell Association, Washington, D.C.).

Pickett, J. M. and J. Martony, ed. 1978. *Proceedings of the Research Conference on Speech-Processing Aids for the Deaf* (Gallaudet College, Washington, D.C.)

Stark, R. E. 1977. "Speech Acquisition in Deaf Children," Volta Review *79,* 98-109.

Stark, R. E., ed. 1974. *Sensory Capabilities of Hearing Impaired Children* (University Park Press, Baltimore).

Strong, W. J. 1975. "Speech Aids for the Profoundly/Severely Hearing Impaired: Requirements, Overview, and Projections," Volta Review *77,* 536-56.

Audiovisual

1. Tape recordings of utterances produced by severely hearing-impaired talkers
2. *The Speech Chain* (19 min, color, 1963, BELL)
3. *Alaryngeal Speech* (21 min, B&W, 1966, UKANMC)
4. *Children With Cleft Palates* (29 min, color, 1957, UMICH)
5. *Listen* (30 min, color, 1973, USNAC)
6. *Silent World, Muffled World* (28 min, color, 1966, USNAC)
7. *The Function of the Pathologic Larynx* (24 min, color, ILAVD)

Demonstrations

1. Delayed-feedback speech to illustrate importance of feedback in production.
2. Use of artificial larynx

5.F. DEGRADED SPEECH

Consideration of degraded (or altered) speech is important for at least two reasons. One very practical reason is that almost all speech suffers from some kind of degradation in going from a sender to a receiver. The degradation is typically due to an added noise, limited band-width of the transmission system, or peak clipping. The second reason is that intentionally degraded speech can help us to gain a great understanding of the speech features that are important for preserving speech intelligibility or talker identity.

Intelligibility Testing

In practice, an *intelligibility test* consists of sending a set of speech stimuli (signals) through some kind of speech transmission system, such as a telephone line, a public address system, or even from one point in a room to another point. At the receiving end of the system, listeners are asked to identify the speech stimuli that were sent over the system. The intelligibility score is just the percentage of the speech signals that were correctly identified. The speech stimuli most often used in intelligibility tests are single words. This is probably true in part because in test situations single words are easier to deal with than sentences. The words used typically come from one of the following categories: spondees (two-syllable words), phonetically balanced words (single-syllable words that have occurrences of different phonemes in the same proportions as they occur in the natural language), and rhyming words (words that differ in a single phoneme, such as pat, bat, or mat). Intelligibility tests run with spondees and phonetically balanced lists require the listener to write what he heard and require considerable training of the listeners before the results are stable. Rhyming lists are typically used with untrained listeners, who are required only to mark the word that was received out of a set of two or more possible words. Rhyme tests can be administered with less effort on the part of the listeners, and they do not have to be specially trained.

Intelligibility tests using single words eliminate contextual information that is so very important in everyday speech communication. However, there is a fairly close relationship between the intelligibility score and the success with which a normal conversation might be expected to be carried on. For instance, intelligibility scores below about 50% would indicate that normal conversation would be difficult to carry on. However, scores from 80% to 100% would indicate that normal conversation should be very successful; i.e., with the additional information supplied from context virtually all of the conversation should be intelligible to the participants.

Additive Noise

Additive noise is probably the most common degradation of speech in communication systems. For instance, it is not uncommon when using the telephone for speech communication (particularly on long-distance calls) to get a "bad connection" or a "noisy line." Sometimes the talker is in a noisy environment and the rest of the transmission system simply transmits the noise along with the speech; or the listener may be in a noisy environment. In any one of these cases, noise (some unwanted signal) is competing with the speech signal. The reduction of the intelligibility scores in the presence of noise depends upon the energy in the speech signal in relation to the energy in the noise signal. Table 5.F-1 gives typical values for intelligibility, expressed in percent, for different values of signal-to-noise ratio, expressed in dB. A signal-to-noise ratio of 0 dB means that the speech and the noise are equal in sound level. A positive value of say 6 dB tells us that the speech sound level is 6 dB greater than the noise sound level, whereas -6 dB indicates the speech is lower by 6 dB than the noise. Note that when the signal-to-noise ratio is 6 dB or greater we might expect communication to be relatively successful.

Table 5.F-1. Intelligibility (percent) versus speech signal to noise ratio (dB).

Signal to noise ratio	−12	−6	0	6	12	18	24	30
Intelligibility		0	15	42	67	84	93	96	98

Bandwidth Limitations

Limited bandwidth is probably the second most common cause of speech degredation. Normal speech may have energy at frequencies ranging from 60 to 10,000 Hz. Most communication systems are not equipped to transmit this wide range of frequencies. (For instance, a telephone system transmits frequencies in a range of about 300-3000 Hz, and typical public address systems are probably poorer.)

It is of interest to know what frequencies must be transmitted in order to achieve good speech intelligibility. (We can presume that the range of frequencies transmitted by telephones has been based on these considerations.) This can be determined by taking speech that has an intelligibility score near 100%, passing it through various filter sets, and determining new intelligibility scores for the speech under the filtered conditions. The effect on the spectrum of passing white noise (a random signal having equal amounts of energy at all frequencies) through three different filter setups is illustrated in Fig. 5.F-1. Note that the low-pass filter leaves the low frequencies in the spectrum but eliminates, or at least reduces, the higher frequencies. The high-pass filter lets high frequencies get through and eliminates low frequencies. The band-pass filter lets a band of frequencies get through, but it eliminates frequencies that are either lower or higher than the pass band. Table 5.F-2 illustrates how low-pass and high-pass filtering affect the intelligibility of speech.

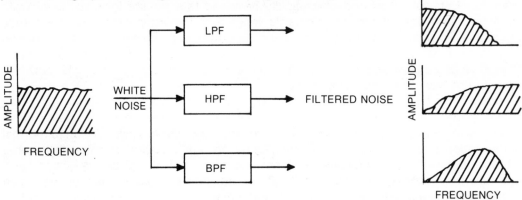

Fig. 5.F-1. Resulting spectra for filtered white noise.

Table 5.F-2. Intelligibility scores of speech, low-pass filtered or high-pass filtered, at different cutoff frequencies.

Cutoff frequency (Hz)	Intelligibility of low-passed speech (%)	Intelligibility of high-passed speech (%)
100	0	98
200	0	97
500	7	96
800	18	93
1000	26	89
1250	38	85
1500	50	80
1750	64	74
2000	70	65
3000	88	30
4000	93	13
5000	95	5

Waveform Clipping

Many speech transmission systems have limits in terms of the amplitudes of the speech signal they are able to transmit without producing appreciable distortion. When a signal amplitude exceeds the limits of a system, peak clipping often occurs. Peak clipping is illustrated in Fig. 5.F-2a. Note that the peaks of the signal that exceed the limits of the system are chopped off. Peaks of the signal that do not exceed the limits of the system pass through the system essentially undistorted.

Center clipping of a speech signal is of less practical interest because it does not occur in typical speech transmission systems. Fig. 5.F-2b illustrates the results of center clipping. Note that the portion of the signal that has the smallest amplitudes is chopped out.

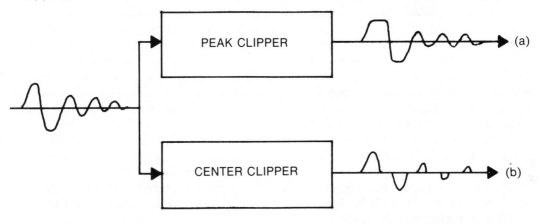

Fig. 5.F-2. Effects of waveform clipping: (a) peak clipping and (b) center clipping.

Peak clipping and center clipping both reduce the naturalness of speech. This is probably because the clipping process introduces additional amounts of high-frequency energy into the signal. (Any time a waveform is modified so that it changes in a more abrupt fashion from one instant to the next, additional high-frequency energy is introduced.) Although peak clipping reduces the naturalness of the speech signal, it leaves the intelligibility fairly intact. However, center clipping reduces the intelligibility of the speech signal as well as reducing its naturalness. This suggests that the operations of peak clipping and center clipping affect the speech signal in two different ways.

Modified Parameters in Speech Synthesis

Parameter modification is a form of speech degradation that is intentional. It is done so that the relative importance of different speech features can be assessed for speech intelligibility, speech recognition, speaker identification, and naturalness of speech. There are basically two ways in which parameter modification can proceed: (1) analysis of natural speech and then modification of the resulting parameters, or (2) synthesis of speech from artificially supplied parameters. Usually the latter procedure is guided, at least in part, by the former procedure; otherwise it is too inefficient.

It is possible with modern speech processing to take naturally produced speech and analyze it in terms of different sets of time-varying parameters. Some parameters that are being used are formant frequencies, energies in different frequency bands, or predictor coefficients. The above sets of parameters are used in some speech bandwidth compression systems termed vocoders (for voice coders). The parameter sets can be used in speech synthesizers that reconstruct a version of the speech signal. The synthesized speech is in many cases almost as intelligible as the original speech and in a more restricted number of cases almost as natural sounding as the original speech.

If speech synthesized from a parameter set is virtually indistinguishable from the original speech in terms of intelligibility and naturalness, then it is probably reasonable to assume that the parameter set adequately characterizes the speech. We can then use this parameter set to systematically study various speech features. Let us suppose that we wish to discover the primary differences between male and female speech. We would probably start by noting how the typical values of the parameters differ in the two cases. We would probably find that on the average the fundamental frequencies used by females are about an octave higher than those used by males. We would also probably find that the formant frequencies (if these are parameters accessible to us in the analysis) of female speech are about 15-20 percent higher on the average than for male speech. To test our observations, we might then take male speech and systematically increase the fundamental frequency and/or the formant frequencies and ask listeners to make judgments on whether the speech was produced by a male or a female. Similarly, the fundamental frequency and/or the formant frequencies of different samples of female speech could be adjusted to lower values and a similar listening test run. We would probably find that there is considerable overlap in our parameters in terms of speech that is perceived as spoken by females or males. However, we would probably find that at the extremes of the parameter ranges the judgments are more consistent.

We can test the effect of the fundamental frequency on speech intelligibility by taking the speech parameters and synthesizing speech after making modifications to the fundamental frequency. We can then run tests to assess what effect, if any, the fundamental frequency has on intelligibility. We might run comparable tests to assess the influence of fundamental frequency on naturalness. The kinds of tests that we might run are almost limitless.

Of practical significance are tests on what parts of the speech signal are most important to our perception of plosives, fricatives, vowels, etc. We might address ourselves to such questions as: What role do formant transitions play in the perception of stop consonants? How necessary is the burst of noise energy to the perception of unvoiced plosives? Do the relative durations of the burst and interval to onset of voicing have a significant influence on distinguishing between voiced and unvoiced plosives? What spectral characteristics are necessary to distinguish between /S/ and /SH/ for example? The answers to these and other questions might be acquired by analyzing natural speech and then systematically modifying various parameters, such as the elimination of formant transitions, elimination of noise bursts, change of the interval from burst to onset of voicing, and so forth.

An alternate procedure to that of analyzing speech, modifying parameters, and synthesizing speech, is to create stylized parameters for controlling a speech synthesizer. For instance, with a three-formant speech synthesizer (similar to the one discussed in section 5.C) we could investigate the role of formant transitions in the perception of various phonemes. The Pattern Playback machine developed by Haskins Laboratories has been used extensively to investigate some of the foregoing questions. However, more modern speech synthesis devices might still be used effectively to investigate these and other questions and to provide refinements to questions already studied.

Exercises

1. What do intelligibility tests measure?

2. What intelligibility score would you expect to obtain for a "good" telephone transmission if its bandwidth is limited to the range 300-3000 Hz?

3. What intelligibility score would you expect to obtain for a high-quality tape recording system if the overall system response is 60-15,000 Hz?

4. What is the most common form of speech degradation in the speech transmission systems in use today? Why?

5. What is probably the second most common form of speech degradation in speech transmission systems? Is this form of degradation a problem in face-to-face transmission?

6. What does a band-pass filter do?

7. If you had a band-pass filter having a pass band of 1000 Hz, but otherwise adjustable, where would you place the pass band to achieve maximum speech intelligibility?

8. Peak-clipped speech is quite intelligible, even though it does not sound natural. What is the difference between intelligibility and naturalness?

9. Center-clipped speech is neither natural sounding nor intelligible. If the claim that intelligibility of speech is largely dependent on an adequate formant structure is true, what does center clipping do to the formant structure of speech? What does the peak clipping do to the formant structure?

10. Suppose that with a speech analysis-synthesis system you are able to analyze speech, obtain the formant frequencies and amplitudes, and then modify the control parameters before synthesizing the speech. Suppose you synthesize speech using only one of the formants. You then run intelligibility tests on the synthetic speech. Which formant will you probably find as being most significant for speech intelligibility? Which formant will probably be least significant for speech intelligibility? Give some reasons to support your supposed results. (Hint: Use Tables 5.B-1 and 5.F-2.)

11. Suppose for some reason you want to analyze the speech of an adult male talker and then synthesize it so that it sounds more like that of an adult female talker. How would you modify the formant frequencies and the fundamental frequency?

12. A pulse train (a series of repetitive pulses) having a flat-line spectrum as shown is substituted for the white noise in Fig. 5.F-1. It passes through low-pass, high-pass, and band-pass filters. Draw stylized line spectra that are appropriate for each case.

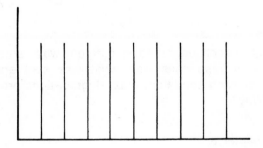

13. The waveform shown below is put through a transmission system that is limited to a range of −1 to 1 volts. Draw in colored pencil or pen the waveshape that will appear at the output of the transmission system.

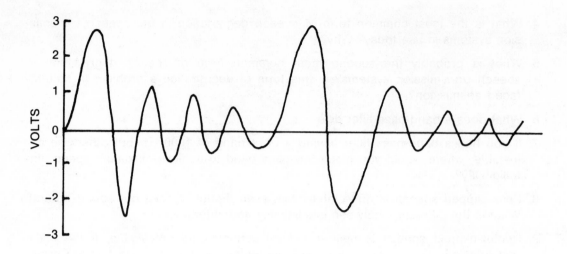

14. Four waveforms are shown in the diagram below. One of them came through a system in which there was very little degradation. The others passed through systems in which there was additive noise, peak clipping, or center clipping. Label each waveform to indicate which system it passed through.

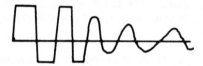

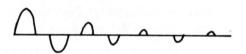

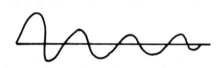

15. Listen to tapes of speech degraded in different ways and write down what you hear. Compare your responses with what was said and determine your intelligibility score for each condition. What degradations produce the greatest loss of intelligibility? Why?

Further Reading

Denes and Pinson, chapter 8

Flanagan, chapter 7

Fletcher, chapters 6, 15, 16, 17, 18

Gerber, chapter 11

Cooper, F. S., P. C. Delattre, A. M. Liberman, J. M. Borst, and L. J. Geerstman. 1952. "Some Experiments on the Perception of Speech Sounds," J. Acoust. Soc. Am. *24*, 597-606.

Keeler, L. O., G. L. Clement, W. J. Strong, and E. P. Palmer. 1976. "Two Preliminary Studies of Predictor-Coefficient and Formant-Coded Speech," IEEE Trans. Acoust., Speech and Signal Processing *ASSP-24*, 429-32.

Voiers, W. D. 1977. "Diagnostic Evaluation of Speech Intelligibility," to appear in *Benchmark Papers in Acoustic* series (Dowden, Hutchinson, and Ross, Stroudsburg, Penn.)

Audiovisual

1. Tapes of degraded speech: additive noise, peak clipped, center clipped, modified synthesis parameters
2. *Science Behind Speech* (8 min, color, 1964, BELL)

Demonstrations

1. Band-pass filter for illustrating effects of bandwidth limitation

5.G. MACHINE PROCESSING OF SPEECH

It is of interest to many scientists to try to understand natural processes to the extent that they can create substitute means for carrying out the processes. Much of what we will discuss about machine processing of speech falls into this category. Indeed, we often feel that we have a better understanding of natural processes if we are able to create substitutes for them. The four areas of machine processing of speech to be discussed in this section are (1) bandwidth compression of speech, (2) machine synthesis of speech, (3) machine recognition of speech, and (4) speaker verification and identification.

Bandwidth Compression

Speech bandwidth compression is of interest any time bandwidth limitations on a transmission channel are such that we are unable to transmit as many different conversations as we wish. This is often of primary concern when the transmission is over long distances, such as transcontinental or transoceanic cables. For example, suppose a particular undersea cable has a channel capacity of 1,000,000 Hz. If we transmit essentially full-bandwidth speech signals (10,000 Hz), we will be able to transmit only 100 different speech signals concurrently (actually fewer if we allow some "frequency space" as a buffer between contiguous signals). If we transmit telephone-quality speech (bandwidth of approximately 3000 Hz), we will be able to transmit 300 different speech signals concurrently over the same cable. If we could devise a means for further compressing the speech signal without degrading it too much, we might be able to transmit even more speech signals concurrently over the same cable. (If satellite communications make unlimited bandwidths available at low cost, then some of the following discussion will be mainly of academic interest.)

Before we discuss particular bandwidth compression systems, let us suggest a different way of talking about bandwidth or data rate. Suppose we use telephone quality speech as our standard. It has a bandwidth of about 3000 Hz. A roughly equivalent description is to consider telephone-quality speech digitized by means of an A/D converter. We sample the speech signal 6000 times per second. (According to the rules of the game the digitization rate must be equal to or greater than twice the bandwidth.) We find that by allowing 9 bits for each sample we get good quality. We think of bits as digits in a number system to the base 2 instead of to the base 10. For instance, to the base 10 we have ten digits from 0 to 9 and to the base 2 we have two digits, 0 and 1. Nine bits of information means that we can specify any one of 512 (2 to the ninth power) different values for any particular speech sample. We can then discuss our bandwidth or information rate in terms of bits per second. For the example we are considering, we get a rate of 54,000 bits/second (6000 samples/second times 9 bits/sample), which for convenience we round off to 50,000 bits /second as the information rate for telephone quality speech.

Considerable effort has been expended in trying to code speech so that its bandwidth is substantially reduced. Several different voice coder systems (vocoders) have been devised, including a channel vocoder, a formant vocoder, and a "phoneme" vocoder. A channel vocoder consists of an analysis part (illustrated in Fig. 5.G-1), a transmission channel, and a synthesis part. The analysis part is composed of a set of band-pass filters adjusted so that they cover the speech bandwidth. Each filter lets a signal pass through that consists of energy at the frequencies in its pass band. The signal coming out of the filter is rectified and then low-pass filtered to produce a signal varying rather slowly in time, which is a measure of the energy in the channel at a particular instant. This signal is sampled 40-100 times per second with 3-6 bits per channel. An additional channel is provided for the fundamental frequency of voicing. The signals resulting from the analysis are combined in some manner and then transmitted over some appropriate transmission line.

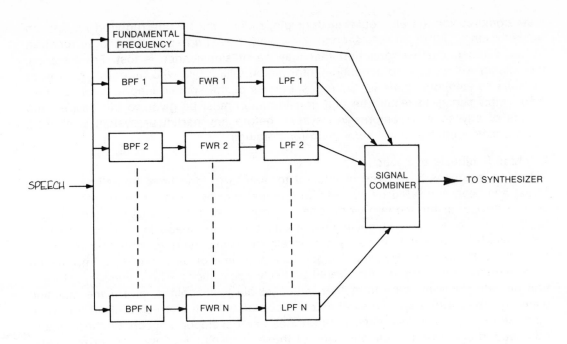

Fig. 5.G-1. Analysis part of channel vocoder.

The transmitted signals are received by the synthesis part of the system and used to reconstruct a facsimile of the speech signal. This is accomplished by using the fundamental-frequency signal to control the repetition rate of a pulse generator. The resulting pulse train is input to a set of band-pass filters similar to the ones used in the analysis part of the system. The individual channel signals are then used to control the amount of energy coming from each channel in the synthesizer that is to be used to synthesize the speech. The resulting signal is received by the listener.

For the sake of discussion of bandwidth reduction, consider a channel vocoder with 20 channels. Assume that each channel is sampled 100 times per second with 6 bits precision, and assume that the fundamental frequency channel is also sampled 100 times per second with 6 bits precision. The resulting bit-rate is then 21 channels times 6 bits per channel times 100 samples per second, or 12,600/second. This represents about one-fourth the bit-rate associated with telephone-quality speech. This particular vocoder could then compress speech so that we could transmit four conversations in the same channel capacity as we can transmit one normal telephone conversation. (Vocoders have actually been built that operate in the range of 1000 to 4000 bits per second and produce reasonable quality speech.)

A formant vocoder works on principles similar to those of the channel vocoder. However, the analysis is considerably more sophisticated because formant frequencies and amplitudes must be extracted. At the synthesis end of the vocoder, a pulse train (controlled in frequency by the fundamental frequency signal) is used to excite several formant filters whose frequencies are controlled by the formant amplitude signals. Their outputs are added together to form a facsimile speech signal that is received by the listener. The synthesis part of a formant vocoder was shown in Fig. 5.C-7.

A "phoneme" vocoder represents the ultimate in bandwidth compression. No such vocoder has been successfully implemented to date. In principle it would consist of an analysis part in which the incoming speech is processed in such a way that the phonemes are extracted. A phoneme code would then be transmitted. At the end of the system the message could be typed out to the "listener" or it could be used to control a speech synthesizer to produce a speech signal for the listener. This ultimate compression system could, in principle, reduce the bit-rate required for speech transmission to 20 or 30 bits/second. However, if a typed speech signal is received by the listener, much information that may be of significance is lost. For instance, the listener will receive no acoustical information telling him who the talker is or what the talker's emotional state is. Clearly, subtleties of the spoken language will be lost. This simply serves to remind us that consideration must be given to the desired end results of any speech compression system before any particular system is chosen. We are simply unable to get something for nothing.

Machine Synthesis of Speech

Talking machines have been of interest to man for a long time. Sometimes this interest has been motivated by curiosity and sometimes by a real or imagined need for such a device. In the modern world one can conceive of several useful applications. Most would be in trying to get a machine to produce speech in situations that involve much repetition and that are basically nonstimulating for the human talker. Speech synthesizers to give the temperature and time of day or to give information on departing planes, trains, etc., would be likely applications. The necessary information already exists in some form that is accessible to machines, and if the machine were equipped with a speech synthesizer it could "tell" an interested client the available information. Bell Telephone Laboratories has developed speech synthesis methods used in practical applications. One of these involves telephone answerback service in which the service provides standard information of some kind via a synthetic spoken message. The other of these involves recorded wiring lists that are used by technicians in the installation of telephone equipment. The wiring lists exist in "written" form in a computer. This information in the computer is used as the input to a speech synthesis system to create tape-recorded versions of the lists, which are then used by the technicians in the field.

Two exciting and potentially very useful applications of speech synthesis are in reading machines for the blind and in computer-aided instruction. Both are the object of considerable current research. A reading machine for the blind would consist of a page-scanning device coupled to a speech synthesis system. The page scanner would "read" the various written characters on the page, and the information would then be passed on to the synthesizer. The system would need to convert character strings into word, phrase, and sentence strings, with rules imposed to create the correct pitch contours, etc., that would serve as input to a speech synthesizer. Computer-aided instruction in its infancy depended primarily on typed messages and graphical messages for communicating with its user. Since audio communication is so important in the human communication process, it seems reasonable to suppose that a machine capable of spoken output for the user (and conceivably spoken input) would be better able to "communicate" with a user in a computer-aided instruction environment.

Tape-recorded messages have been in use for some time. When a user makes a request, the playback mechanism is actuated and the requester receives some kind of recorded message. Two speech synthesis methods that are being developed would make this procedure much more flexible than it now is. Speech synthesis permits a reduction in the speech data required to create any of many different utterances. If the storage medium is limited, this may represent an important savings in a manner quite analogous to the bandwidth savings of speech compression systems.

One method of speech synthesis is word concatenation, that is, linking words together. This method uses words spoken in isolation as its data base. When a sentence is desired, the words in the data base are concatenated, with simultaneous appropriate changes in the prosodic features (primarily in the pitch), to produce the desired utterance. Preliminary experiments found the concatenated words to be as satisfactory to listeners for comprehension as naturally spoken words when applied in a system dealing with telephone numbers. One disadvantage of word concatenation may be that it will require too much storage if a large vocabulary is to be used. Furthermore, telephone numbers would seem to be a natural application of word concatenation because they are spoken more or less a digit at a time; more elaborate phrases and sentences will probably make more severe demands on prosodic features.

Another method of speech synthesis uses phonemes as its data base. When words and sentences are desired, the phonemes in the data base are concatenated, and appropriate prosodic information is supplied to produce the desired utterance. In this method one must be concerned about how neighboring phonemes affect each other and about how the control parameters for synthesizing the various phonemes must be modified when the phonemes are concatenated. The prosodic information must also be available from the written input or supplied by the rules in some standard fashion. This method gets down to the basics of speech synthesis. It offers the potential advantage of greater economy for large vocabulary applications and greater flexibility for many applications.

Machine Recognition of Speech

Machine recognition of spoken language is a most intriguing challenge. It is probably the most difficult of the tasks that we are discussing in the general category of machine processing. When machine recognition is sufficiently developed, many applications will likely be forthcoming. One such application might be in the area of aids for the hearing impaired, in which case a machine recognizer of speech would give the user a written version of the message to work with. Other applications might be in the general area of man-machine communication, such as information services or computer-aided instruction. A viable machine recognizer of speech would make it possible for the machine to "listen" and "understand" as well as to speak to the user.

Machine recognition of speech is more difficult than machine synthesis of speech because in recognition the machine must carry a greater burden. In machine synthesis of speech the human listener can compensate for many of the machine's deficiencies and can probably successfully receive the message with a high degree of intelligibility even when the quality is not that of natural speech. The human as a speech processor is very flexible in his/her ability to adapt the processing procedure to accommodate a wide range of speech input; hence the processor's ability to deal with machine-produced speech, along with many different talkers, many different dialects, etc. However, machine recognizers to date have not begun to achieve this human flexibility. A machine geared to accept a rather limited vocabulary for one person may have a difficult time if it tries to accept the same vocabulary from a different talker. Many of the problems that we tend to ignore in everyday speech communication with one another must be dealt with in one way or another in the machine recognizer.

One finds that it is not possible to perform an acoustical analysis of running spontaneous speech and extract a phoneme string that corresponds clearly to the "intended phoneme string." When we speak spontaneously we tend to produce a speech signal that carries no more information than is necessary for the task at hand. If we are conversing with a friend on a well-known subject, we will probably tend to be somewhat more careless in our phoneme production than if we are conversing with a stranger on a less well-known subject. We are able to make the necessary adjustments in our reception and perception of speech almost automatically. We capitalize on our knowledge of the subject and the vocabulary to provide contextual clues and supply missing or distorted phonemes. Indeed, in conversations of this nature we can often anticipate what our friend is going to say before he/she has completed saying it. The acoustical signal then serves to confirm what we had anticipated, and a few missing or distorted phonemes or even a few distorted words do not impair the effectiveness of our conversation. However, if we suddenly change the subject of conversation, even with a friend, the success of the communication is very often reduced until the friend has had time to become reoriented. (You might try this on a friend or make notice of the same effect happening to yourself.)

A problem arises in machine recognition of speech: the phoneme string we are able to extract from an acoustical speech signal does not correspond on a one-to-one basis with the idealized phoneme string that the talker intended. It is almost as if we perform the recognition and then say what phonemes are present even though we are unable to detect the presence of some of the phonemes before the recognition. We might postulate that any machine recognizer of speech will need to have in its rules access to additional information about language structure. Any phoneme string that is extracted from speech must be convertible into a word string that is acceptable to the vocabulary limits imposed by the task at hand. The word strings must generally satisfy some grammar of the language that is adequate for the task at hand. The sentences must generally satisfy certain semantic constraints that have been imposed on the task.

One current approach to machine recognition assumes the form of a task-oriented problem; i.e., the task is well defined in terms of the vocabulary and the grammar and the kinds of sentences that have meaning. Systems that are being developed for continuous speech typically involve vocabularies under 1000 words, use some simplified grammar, and have tasks that involve limited numbers of meaningful sentences (such as Computer Chess). Other systems that are being developed are essentially single-word recognizers. An example is a digit recognizer that might be used in voice dialing of telephones.

As an ultimate application we might think of a machine interpreter which would serve to accept one spoken language and produce another spoken language. This machine interpreter might consist of a machine recognizer (that would convert spoken language into written language), a machine translator (that would convert one written language into a different written language), and a machine synthesizer (that would convert written language into spoken language). A sufficiently advanced version of a machine interpreter would then permit two talkers, whose native languages are different, to carry on a conversation with each talker speaking and hearing only his own language. Although this is very speculative at this point, it is worth noting that we rarely accomplish something unless we have first dreamed about it.

Speaker Verification and Identification

We know as a practical matter that we can identify several different people quite reliably from their voices. Sometimes we accomplish this identification by noting the particular words and sentences that a person says. However, even when two persons have said nominally the same thing on the telephone we are still able to differentiate between them. Presumably we use the person's prosodic features for identification. In other words, it is not so much a matter of *what* the person says as *how* he/she says it that is important in identification.

With the advent of powerful speech-processing machines the automatic verification of a talker's identity by means of his spoken utterances becomes of interest. Assume the following situation: A selected group of 30 persons has access to a company's confidential files. When any one of the authorized persons wishes to gain access to the files he must properly identify himself. One way of doing this is by typing in a code. However, to increase the security of the files an additional verification of the person's identity is desired. One way that this can be accomplished is to require him to repeat some standard sentence, then analyze the sentence and compare the analysis results to a previously accumulated set of results. If the analysis results compare within a sufficiently low tolerance, the person is accepted as the same person whose code was typed. If not, he is rejected as an imposter.

Most of the automatic schemes that have been attempted work with one or more of the prosodic features of the talkers' speech. Some schemes work by comparing average spectra over long times, with similar long-time averages. Some schemes work by comparing the fundamental frequency contour of an unknown talker's utterance with a known sentence having previously computed and stored contours. All of these schemes use sophisticated analysis techniques. However, as an idealized example, consider the three stylized fundamental-frequency contours shown in the diagram of Fig. 5.G-2. The first and second contours are the standards for two talkers. The final contour is from an unknown talker, but is presumed to be the contour for one of the two talkers. The dashed lines represent the superposition of the third contour onto the first and second contours. You can see that the unknown contour fits the second contour better than it does the first contour. Hence, if the fit is sufficiently close, a judgment might be made that the unknown is in fact the second talker.

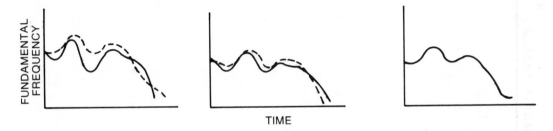

Fig. 5.G-2. Stylized fundamental frequency contours for three talkers.

There has been considerable discussion about using speech spectrograms for speaker identification. The discussion becomes somewhat heated when considered in the area of forensics, where the issue of establishing guilt is involved. At this point in time it appears that spectrographic evidence is not sufficiently reliable for legal purposes because of the great variation that can occur in the acoustical signal.

Exercises

1. Data rates of natural and compressed speech
   a. Use the data rate of 50,000 bits/second suggested in the text for the data rate of natural speech.
   b. Assume that you have a channel vocoding system similar to the one described in the text. Suppose your system has 16 channels, each of which is sampled with 4 bits accuracy 100 times each second. Suppose the fundamental frequency is sampled 100 times each second with 6 bits accuracy. What is the total bit-rate of your vocoder in bits/second? What percentage of the bit-rate of natural speech is this?
   c. Suppose that you sample each of the signals in your vocoder only 40 times each second instead of 100 times. What would be the bit-rate? What is this as a percentage of the bit-rate of natural speech?
   d. Assume that you have a formant vocoding system similar to the one described in the text. Suppose your system transmits information on the frequencies and amplitudes of only three formants and on the fundamental frequency. Suppose that each of the 7 signals is sampled 100 times each second, with an accuracy of 6 bits. What is the bit-rate of this formant vocoder? What is this as a percentage of the bit-rate of normal speech?
   e. Assume that you have a phoneme vocoder similar to the one described in the text. The system involves the use of 32 phonemes. How many bits are required to represent this set of 32 phonemes? Suppose the phonemes are transmitted at an average rate of 10 phonemes/second. What is the bit-rate of the system? What is this as a percentage of the bit-rate of normal speech? What has been lost in achieving this great bandwidth reduction?

2. Speech synthesis by rule

You are given the task of creating a reading machine for the blind. The optical scanning mechanism is already available. You are to develop a speech synthesizer that will accept codes for written characters at its input and produce synthetic speech at its output. You decide that the best kind of speech synthesizer for the project is a three-formant speech synthesizer.

a. In the graph below draw stylized control signals for the synthesizer (three-formant frequencies and sound level) for the sentence "These controls produce speech."

b. How natural would you expect the resulting speech to sound? How intelligible?

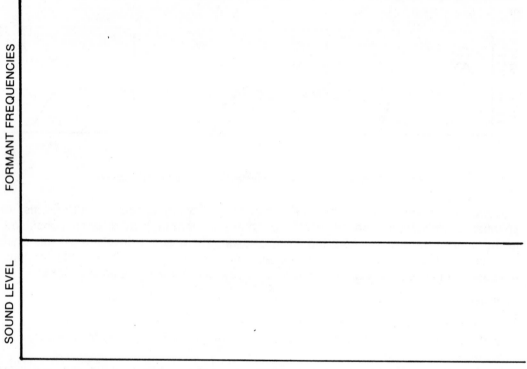

TIME

3. Digit recognition

A digit recognizer consists of an acoustical analyzer that extracts the three most intense formant frequencies from the speech signal. When the signal is voiced, the formant is represented with a solid line; when unvoiced, with a dashed line. The 20 sketches below represent stylized outputs of an acoustical analyzer for each of the spoken digits (zero, one, two, three, four, five, six, seven, eight, nine) as spoken by each of two talkers.

You are to be the rest of the "recognition machine." Based on your knowledge of formant patterns for various phonemes and voiced-unvoiced information of various phonemes, label each sketch with an appropriate digit label.

4. Speaker identity via prosodic features

The first three tracings below show fundamental frequency as a function of time. The fourth tracing on the page shows a fundamental frequency curve for an unknown talker who may or may not be one of the previous three talkers.

a. In your judgment, is the fourth talker one of the first three? Give reasons for the judgment that you make.

b. If you judged the fourth talker to be one of the first three talkers, which one was he? Why?

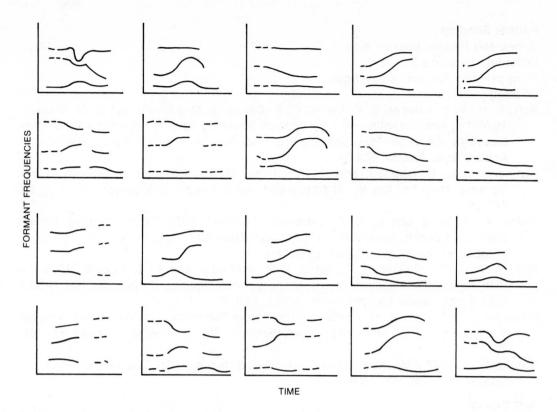

FORMANT FREQUENCIES

TIME

Exercise 3

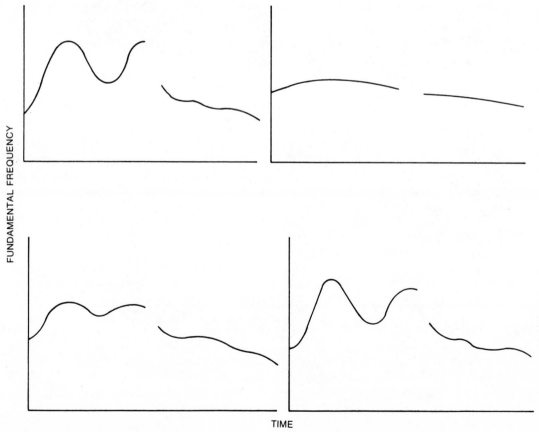

FUNDAMENTAL FREQUENCY

TIME

Exercise 4

209

Further Reading

Denes and Pinson, chapter 8

Flanagan, chapters 5, 6, 8

Flanagan and Rabiner, entire book

Ladefoged, chapter 8

Bolt, R. H., F. S. Cooper, E. E. David, P. B. Denes, J. M. Pickett, and K. N. Stevens. 1970. "Speaker Identification by Speech Spectrograms: A Scientist's View of Its Reliability for Legal Purposes," J. Acoust. Soc. Am. *47*, 597-612. See also J. Acoust. Soc. Am. *54*, 531-37.

Flanagan, J. L., K. Ishizaka, and K. L. Shipley. 1975. "Synthesis of Speech from a Dynamic Model of the Vocal Cords and Vocal Tract," Bell System Tech. J. *54*, 485-506.

Melby, A. M., W. J. Strong, E. G. Lytle, and R. Millet. 1977. "Pitch Contour Generation in Speech Synthesis: A Junction Grammar Approach," Am. J. Computational Linguistics, microfiche no. 60.

Nye, P. W., J. D. Hankins, T. C. Rand, I. G. Mattingly, and F. S. Cooper. 1973. "A Plan for the Field Evaluation of an Automated Reading System for the Blind," IEEE Trans. Audio Electroacoust. *AU-21*, 265-68.

Rosenberg, A. E., and M. R. Sambur. 1975. "New Techniques for Automatic Speaker Verification," IEEE Trans. Acoustics, Speech, Signal Processing *ASSP-23*, 169-76.

Strong, W. J. 1967. "Machine-Aided Formant Determination for Speech Synthesis," J. Acoust. Soc. Am. *41*, 1434-1442.

Audiovisual

1. *The Human Voice . . . and the Computer* (An IEEE Soundings Tape #70-S-04)
2. *The Speech Chain* (19 min, color, 1963, BELL)

Chapter 6

Musical Acoustics

6.A. PRODUCTION AND PERCEPTION OF MUSICAL TONES

Many of the physical properties of an instrument that affect its tone production are peculiar to the particular class of instruments (i.e., strings or winds) and even to the particular instruments within a class (i.e., violin, viola, cello, double bass). Some general features of tone production and perception, however, apply over a broad range of instruments, and we will now discuss some of them under the headings of transients in musical tones, dynamics of performance, families of musical instruments, and uncertainty in musical tones.

Transients in Musical Tones

A certain amount of time is required to "start" musical tones and to "stop" them once they are started. The idealized waveform shown in Fig. 6.A-1 illustrates these features. The part of the waveform labeled "attack" is that part associated with the starting up, or onset, of the tone. The part labeled "decay" is that part of the waveform associated with the shutting down, or ending, of the tone. The "steady state" is the part of the waveform between the attack and decay and may be fairly steady or it may be quite unsteady.

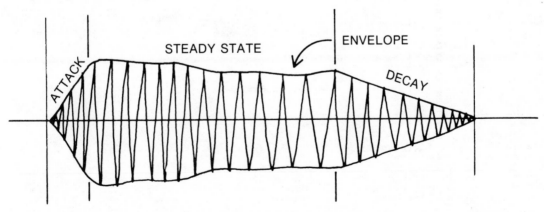

Fig. 6.A-1. Idealized waveform illustrating transients.

Percussive tones would not be expected to exhibit a steady state since they are excited with an impulse, and once the excitation has ceased the tone begins to decay. Nonpercussive tones, on the other hand, would be expected to exhibit all three features with a steady state that is determined by how long the excitation (blowing, bowing, etc.) is applied.

From a physical point of view it might be expected that the large instruments of a particular family would have longer attack times than the smaller instruments of the same family—e.g., the double bass longer than the violin, the bassoon longer than the oboe. This should be so because in a larger instrument the disturbances produced by the exciting force on the vibrator must travel longer distances before standing waves can be set up—e.g., the longer strings of the double bass vs. the shorter strings of the violin, the longer tube of the trombone vs. the shorter tube of a trumpet. By the same argument, it might be expected that the high-frequency tones on some instruments would have shorter attack times than the lower-frequency tones because the instrument is made shorter by fingering.

In addition to the size of an instrument, its relative energy losses should influence the attack and decay times, with the more lossy instruments, such as double reeds, exhibiting shorter times than less lossy ones, such as the clarinet. This should be so because musical instruments store energy and it takes time for the energy to be stored when the instrument is being excited and time for the store of energy to be exhausted when the excitation is shut off. A low-loss system takes a long time to store and exhaust an energy supply, whereas a high-loss system stores or exhausts an energy supply more quickly (see sections 1.C, 1.D, and 4.B).

Attack and decay times can also depend on the style of playing and the particular performer. However, we cannot give a very firm description of this dependence. There is some problem in defining the duration of an attack. It might be defined as the time from the beginning of the signal until some percentage of the final amplitude has been reached. It could also be defined in terms of the slope of the amplitude envelope over some portion of the beginning of the signal. Luce and Clark

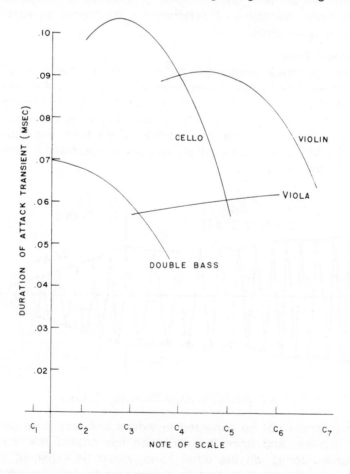

Fig. 6.A-2. Duration of the attack transients of stringed instruments. (From Luce and Clark, 1965.)

212

(1965) have measured attack times using the criterion of 50 percent of final amplitude. Their results are reproduced in Figs. 6.A-2 through 6.A-4. The figures basically exhibit the features we have talked about: decreasing attack times for higher-frequency tones and longer attack times for large instruments in a family. (The strings seem to violate this convention, perhaps because of a stronger influence of playing style on attack time.)

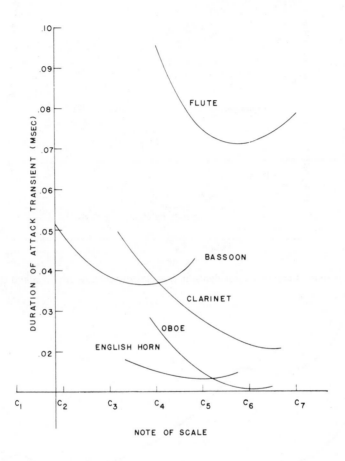

Fig. 6.A-3. Duration of the attack transients of woodwinds. (From Luce and Clark, 1965.)

Attack transients of non-percussive tones have been shown to be of considerable importance in the perception of musical tones (Clark, 1963). Even short segments of a tone that include the attack transient enable a listener to identify a musical instrument much more accurately than a substantially longer segment of a tone from which the attack has been deleted. One reason for the aural significance of the attack is no doubt the great spectral evolution that takes place during the attack. The spectrum at the beginning of the attack is usually characterized mostly by low-frequency partials, but by the end of the attack high frequencies have been added, as seen in Fig. 6.A-5. The decay of nonpercussive tones is probably of subordinate importance, whereas the decay of percussive tones is likely to be almost all important (Clark, 1959).

Dynamics and Performance

We now consider some acoustical features that result from asking a performer to play at different dynamic markings (i.e., pp, mf, ff). In addition to the obvious in-

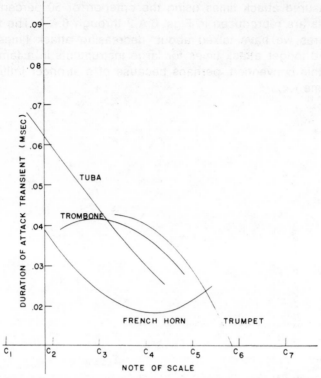

Fig. 6.A-4. Duration of the attack transients of brass instruments. (From Luce and Clark, 1965.)

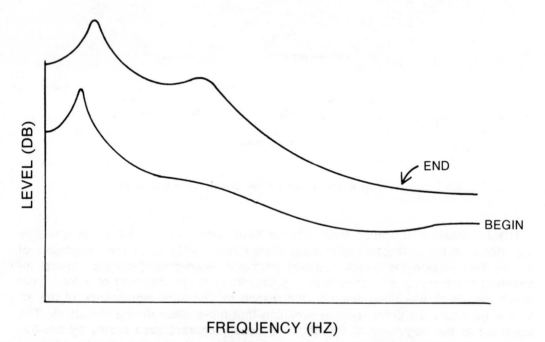

Fig. 6.A-5. Spectra measured at the beginning and ending of an attack transient. (Courtesy of J. D. Dudley.)

crease in loudness produced when an instrument is excited more strongly, there is reason to expect that the timbre may change. It is a fairly general rule that any time a vibrating system is driven more vigorously the high-frequency modes gain in energy relatively more than the low-frequency modes. This should be true in part because when vibrators are weakly excited their motion tends to be smoother and more sinusoidal, which should then produce strong low-frequency partials and rather weak high-frequency partials. On the other hand, a strongly driven vibrator may exhibit abrupt changes in its motion (e.g., as when a clarinet reed beats against the mouthpiece), which will result in a spectrum much richer in high-frequency energy. Spectra actually observed for loud tones are typically much richer in high-frequency energy than those of soft tones. (See, for example, Backus, 1963; Fletcher, 1975; Luce and Clark, 1967.)

Perceptually, the timbre of nonpercussive musical instruments is a weak function of the intensity with which they are played, even though the relative amounts of energy at low and high frequencies may be quite different. This has been demonstrated in experiments in which the loudness of tones produced at different dynamic markings was equalized before presenting them to musically competent listeners (Clark and Milner, 1964). The subjects were able to differentiate the soft tones from the loud tones on the basis of timbre rather than volume.

The dynamic intensity range of instruments as a function of different dynamic markings is also of interest. For nonpercussive musical instruments it is found to average about 11 dB between the dynamic markings of pp and ff (Clark and Luce, 1965). The woodwinds exhibit dynamic ranges smaller than the average, while the strings and brasses exhibit a greater range, as shown in Fig. 6.A-6 (Clark and Luce, 1965). Dynamic range and dynamic level are also dependent to some extent on whether the instrument is performing in its low range or high range, with the differences being larger for double bass, flute, and French horn than for the other instruments. Details of this can be seen in Figs. 6.A-7 to 6.A-10. (Clark and Luce, 1965.)

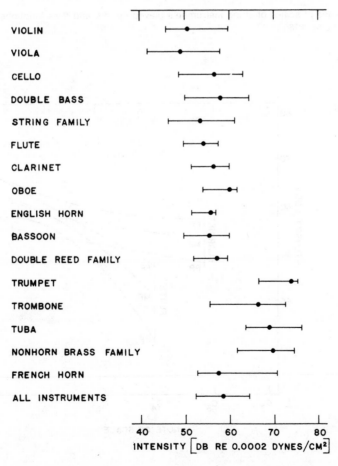

Fig. 6.A-6. Average levels of scales played at the dynamic markings of *pp, mf,* and *ff* for various instruments. The left-most, short, vertical bar indicates the average level of a scale played *pp;* the dot the average of a scale played *mf;* the right bar the average of a scale played *ff.* (From Clark and Luce, 1965.)

215

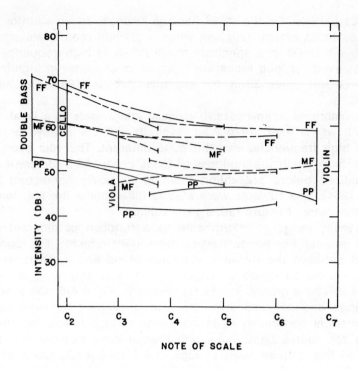

Fig. 6.A-7. Intensities of scales of string instruments played *pp, mf,* and *ff* as functions of the note sounded. (From Clark and Luce, 1965.)

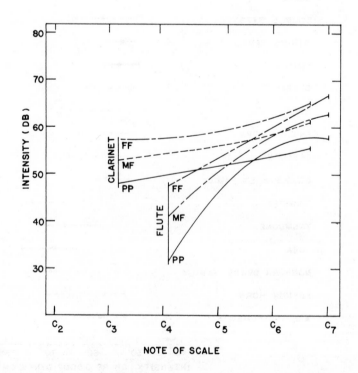

Fig. 6.A-8. Intensities of scales of the clarinet and the flute played *pp, mf,* and *ff* as functions of the note sounded. (From Clark and Luce, 1965.)

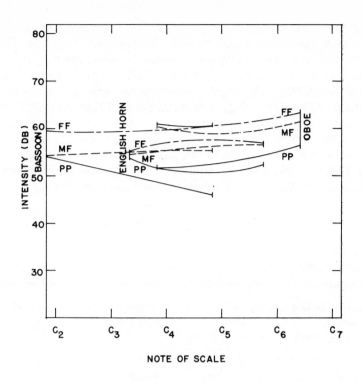

Fig. 6.A-9. Intensities of scales of double reed instruments played *pp, mf,* and *ff* as functions of the note sounded. (From Clark and Luce, 1965.)

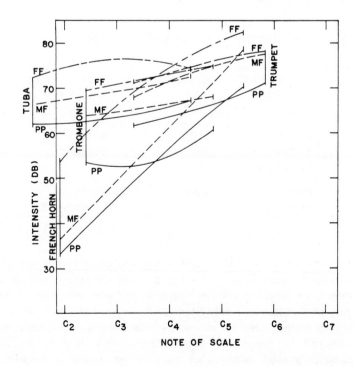

Fig. 6.A-10. Intensities of scales of brass instruments played *pp, mf,* and *ff* as functions of the note sounded. (From Clark and Luce, 1965.)

Families of Musical Instruments

There may be several bases on which musical instruments could be categorized into families. The notion of a family suggests similar features for the members of a family, with perhaps size or scaling being a major difference among instruments within a family. Physical similarities that might be considered would include the method of excitation (plucking, striking, bowing), characteristics of the resonator (string, cylindrical tube, conical tube), radiation characteristics (as from violin body, trumpet bell, clarinet fingerholes, drum head), and nature of the excitation source (bow, single or double mechanical reed, lip reed, air reed). Table 6.A-1 shows one way in which instruments might be assigned to families on a physical basis. (There are, of course, other ways in which the instruments might be placed in families.)

Table 6.A-1. Families of musical instruments

Family	Nature of Vibrator and Excitation	Resonator	Radiation	Examples
Mechanical Reed	blown single reed or double reed	cylindrical tube conical tube	fingerhole openings	clarinet oboe
Lip Reed	blown lips or vocal cords	conical tube cylindrical tube variable shaped tube	bell mouth	trumpet voice
Air Reed	air blown against sharp edge	conical tube cylindrical tube	fingerholes end of tube	flute organ
Bowed String	bowed string	string and body	primarily from body	violin
Percussive String	struck string or plucked string	string and body	from body or sound board	piano guitar
Percussives	struck membrane or bar or plate	same as vibrator	same as vibrator	tympanum xylophone cymbals
Electronic	electronic oscillators	filters	from loudspeaker	synthesizer

If, in fact, members of an instrument family are scaled versions of one another (i.e., larger and smaller realizations of the same basic model), it is reasonable to suppose that the tones they produce will be frequency-scaled versions of each other—that is, low-frequency characteristics exhibited by a large instrument in a family will be observed at higher frequencies for the small instruments of a family. In other words, the spectra measured for large and small instruments in a family should be approximately frequency-scaled versions of each other. There is evidence to support this idea, particularly for some instrument families. The example for average spectra of trumpet, trombone, and tuba shown in Fig. 6.A-11 demonstrates this quite convincingly (Luce and Clark, 1967).

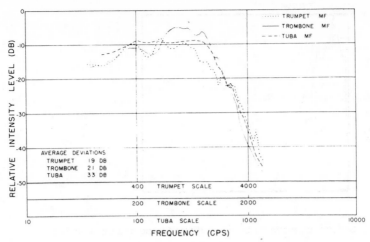

Fig. 6.A-11. Trumpet, trombone, and tuba spectra; envelopes scaled in frequency. (From Luce and Clark, 1967.)

One interesting way of describing an instrument's spectral characteristics is in terms of *formants* which are pronounced peaks in the spectrum. The idea of formants has been used extensively for many years to describe the vowel spectra of speech. More recent evidence from analyses of musical tones indicates that the idea is applicable to many nonpercussive musical instruments. (See, for example, Beauchamp, 1974; Fletcher, 1975; Luce, 1963; Strong and Clark, 1967.) The formants may be determined for musical instruments by plotting the spectra for each of many different tones in a chromatic sequence superimposed on each other and then by drawing a curved line that "represents" the average spectrum over all of the tones, as shown in Fig. 6.A-12 for violin tones. Examples of spectral envelopes for bassoon and flute are shown in Fig. 6.A-13 and 6.A-14. The formants of an average spectrum, then, represent the most intense frequencies for the instrument on the average, although the partials of individual tones may deviate substantially from the average. The formants might be expected to exhibit similar shapes for instruments of a particular family, although having different relative frequencies, as seen for the brasses in Fig. 6.A-11. The formants may also shift in frequency as a result of loud versus soft playing. This has been noted for flute tones (Fletcher, 1975) in which the spectral peaks shifted to higher frequencies for the more loudly blown tones.

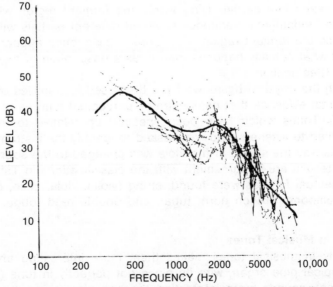

Fig. 6.A-12. Superimposed spectra for several *mf* violin tones with "average spectrum" line drawn in. (After Beauchamp, 1974.)

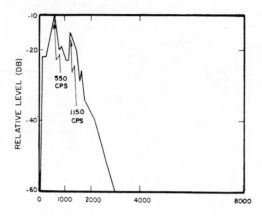

Fig. 6.A-13. Spectral envelope for bassoon. (From Strong and Clark, 1967.)

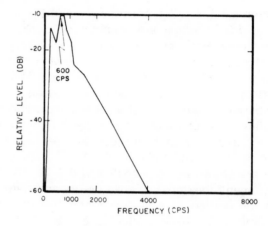

Fig. 6.A-14. Spectral envelope for flute. (From Strong and Clark, 1967.)

Insofar as the notion of formants is applicable, the average frequency characteristic of an instrument can be specified in terms of fixed-frequency bands of emphasis, which means that partials lying within the formant region will be the strongest. Thus, as the fundamental frequency is varied different partials will be emphasized as they lie within the formant region. The formant description appears to be more representative of what actually happens than does a description of fixed relative harmonic amplitudes. (See section 3.B.)

Along with the physical bases we have discussed for families of instruments, there is experimental evidence that demonstrates perceptual families of instruments (Clark, et al, 1964). Tones which have been speeded up relative to the recording speed were presented to listeners who were asked to identify the instruments producing the tones. In this way the sound of the viola was changed to the sound of the violin, and so on. The results are in agreement with the classifications in Table 6.A-1 as the following perceptual families were found: string (violin, viola, cello, double bass), brass (trumpet, trombone, French horn, tuba), and double reed (oboe, English horn, bassoon).

Uncertainty in Musical Tones

Uncertainty in musical tones makes them more interesting and exciting. For instance, a typical pipe organ, with its pipes not perfectly in tune (in an exact mathematical sense), sounds more interesting than an electronic organ in which the oscillators are mathematically in tune relative to each other. Uncertainty may be

220

inherent in the tone production process or it may be introduced by the player as a desirable ornament. We will consider the following tonal uncertainties: vibrato, tremolo, and choral effects.

Vibrato is an ornamental frequency modulation usually produced intentionally by the performer. It is used most extensively in singing and in performing on the bowed string instruments. Vibrato is accomplished on the bowed strings by alternately shortening and lengthening the string with a low-frequency rolling motion of the finger stopping the string. The partials produced are higher in frequency when the string is short and lower when the string is long. Associated with the frequency modulation is a spectral modulation that can be quite pronounced. Fig. 6.A-15 illustrates typical spectral differences between the "low" and the "high" parts of the vibrato (Fletcher and Sanders, 1967). An amplitude modulation usually accompanies vibrato, but it is probably not too significant perceptually. The amount of frequency changes from the average frequency (0.6% to 9%) and the rate of the frequency changes (about 5-8 complete changes per second) are typically dependent on the particular performer and on the musical situation (Seashore, 1967).

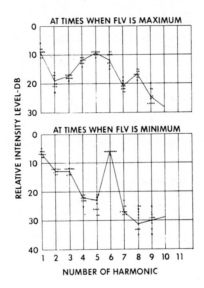

Fig. 6.A-15. Harmonic structure curves for E on the A string at times when frequency level is maximum and minimum. (From Fletcher and Sanders, 1967.)

Tremolo is an ornamental amplitude modulation usually produced intentionally by a performer. It is the ornament used most often with the flute (which is an air reed instrument), where it can be produced by modulating the blowing pressure. Tremolo rate (5 to 8 changes per second) is similar to that for vibrato, but the amount of amplitude modulation can be as much as 80% of the amplitude. Again, as with vibrato, tremolo is accompanied by a spectral modulation and, to a lesser extent, by a frequency modulation (Fletcher, 1975).

When two or more instruments perform the same musical notes together, several features appear that are not present when a single instrument is performing. *Choral tone* features probably arise unintentionally for the most part, although performers may invoke them or exaggerate them for particular musical reasons. Suppose, for example, that two performers are asked to perform a note in perfect time synchronism and with perfect intonation. Even if they could accomplish this impossible task, the result would be different from that of increasing the sound level of a single instrument; there would be two spatially separate sound sources which give rise to a different perception. The performers would be unable to play in perfect time synchrony, which would result in a slight staggering of the attacks of the two instruments, thus further modifying the sound perception of the instruments playing together as opposed to the sound of a single instrument amplified. Staggered attacks can also occur on an organ when the performer depresses the keys nonsynchronously, or when different ranks of pipes have different attack times. Two performers will be unable to perform perfectly in tune (in a mathematical sense) and this will give rise to beating between neighboring partials in the combined tones, further coloring the percept. Organ pipes that are not perfectly in tune (mathematically) exhibit these same features. It has been found that electronically synthesized organ tones can be made to resemble the tones of a pipe organ by detuning them slightly (about 0.3-1.2%) and adding the detuned signal to the original (Fletcher, et al., 1963).

Exercises

1. Which instruments (percussive or nonpercussive) are likely to exhibit the longer attack times? Why?

2. What tones (percussive or nonpercussive) have little or no steady state?

3. What determines the amount of time required for the attack? For the decay?

4. What determines the length of the steady state?

5. Put the following instruments in order of increasing attack time and give typical attack times (when possible): violin, guitar, piano, clarinet, oboe, saxophone.

6. Determine the attack time, the steady state time, and the decay time for the idealized waveform envelope shown in Fig. 6.A-1.

7. Which of the instruments shown in Fig. 6.A-6 has the greatest dynamic range? The smallest?

8. Which instruments have dynamic ranges that increase with increasing frequency? That decrease?

9. Which instruments exhibit the greatest dynamic level (i.e., are the most powerful)? Which the least? Do these results agree with your experience?

10. In what ways can the musical instruments be organized into families other than the way shown in Table 6.A-1?

11. The spectra in Fig. 6.A-5 can also be viewed as showing the average spectra of an instrument blown softly and loudly. Which curve corresponds to each condition?

12. The curve in Fig. 6.A-14 is the spectrum envelope for a flute. What might be a reasonable estimate for the spectrum envelope of a "bass flute" if it performs an octave lower?

13. What might be a reasonable spectrum envelope for an oboe if it performs an octave higher than a bassoon? (See Fig. 6.A-13.)

14. A trumpet has a formant centered at 1000 Hz. Which partial (by number) will be the strongest when the trumpet is sounded with a fundamental frequency of 250 Hz? 300 Hz? 500 Hz? 1000 Hz?

15. A vibrato produces frequency variations of 1% above and 1% below the average frequency. What are the highest and lowest frequency values for each of the first five partials if the fundamental frequency is 200 Hz?

16. Similar organ pipes each have spectra exhibiting 15 partials. If the nominal fundamental frequency of each pipe is 200 Hz, what beat frequencies are produced by the adjacent partials when the two pipes are sounded together if they are 0.5% out of tune?

Further Reading

Roederer, chapters 1-5

Winckel, chapter 3, 6, 7, 10

Backus, J. 1973. "Acoustical Investigation of the Clarinet," Sound 2, 22-25.

Beauchamp, J. W. 1975. "Analysis and Synthesis of Cornet Tones Using Nonlinear Interharmonic Relationships," J. Audio Eng. Soc. 23, 778-95.

Beauchamp, J. W. 1974. "Time-Variant Spectra of Violin Tones," J. Acoust. Soc. Am. 56, 995-1104.

Clark, M. 1959. "Several Problems in Musical Acoustics," J. Audio Eng. Soc. 7, 2-4.

Clark, M., D. Luce, R. Abram, H. Schlossberg, and J. Rome. 1963. "Preliminary Experiments on the Aural Significance of Parts of Tones of Orchestral Instruments and on Choral Tones," J. Audio Eng. Soc. 11, 45-54.

Clark, M., and P. Milner. 1964. "Dependence of Timbre on the Tonal Loudness Produced by Musical Instruments," J. Audio Eng. Soc. 12, 28-31.

Clark, M., P. Robertson, and D. Luce. 1964. "A Preliminary Experiment on the Perceptual Basis for Musical Instrument Families," J. Audio Eng. Soc. 12, 199-203.

Clark, M., and D. Luce. 1965. "Intensities of Orchestral Instrument Scales Played at Prescribed Dynamic Markings," J. Audio Eng. Soc. 13, 151-57.

Fletcher, H., E. D. Blackham, and D. A. Christensen. 1963. "Quality of Organ Tones," J. Acoust. Soc. Am. 35, 314-25.

Fletcher, H., and L. D. Sanders. 1967. "Quality of Violin Vibrato Tones," J. Acoust. Soc. Am. 41, 1534-44.

Fletcher, N. H. 1975. "Acoustical Correlates of Flute Performance Technique," J. Acoust. Soc. Am. 57, 233-37.

Jansson, E. V., and J. Sundberg. 1975. "Long-Time-Average Spectra Applied to Analysis of Music. Part I: Method and General Applications," Acustica 34, 15-19. See also Acustica 34, 269-80.

Luce, D., and M. Clark. 1965. "Durations of Attack Transients of Nonpercussive Orchestral Instruments," J. Audio Eng. Soc. 13, 194-99.

Luce, D., and M. Clark. 1967. "Physical Correlates of Brass-Instrument Tones," J. Acoust. Soc. Am. 41, 1232-34.

Patterson, B. 1974. "Musical Dynamics," Scientific Am. 231 (5), 78-95.

Strong, W., and M. Clark. 1967. "Synthesis of Wind-Instrument Tones," J. Acoust. Soc. Am. 41, 39-52.

Strong, W., and M. Clark. 1967. "Perturbations of Synthetic Orchestral Wind Instrument Tones," J. Acoust. Soc. Am. 41, 277-85.

Audiovisual

1. Tape of speeded and slowed brass tones
2. Tape of synthetic organ tones

6.B. THE SINGING VOICE

As noted earlier in section 5.A, the vocal cords produce periodic puffs of air which are filtered and modified as they pass through the vocal tract. The air flow through the glottis is periodic, but not sinusoidal, and consists of many harmonic components which are generated because of one or more of the following conditions: (1) air flow does not vary directly with glottal area; (2) pressure across the glottis does not remain constant; (3) the glottal area does not vary sinusoidally. The frequency of oscillation of the vocal cords is determined primarily by their effective mass and tension and is approximately independent of the vocal tract shape. The formant frequencies (resonances) of the vocal tract in relation to the fundamental voicing frequency determine which voicing harmonics are emphasized.

Several phenomena occur in singing that are not usually present in speech production. Pitch vibrato (modulation of vocal cord fundamental frequency) is one of these and has been discussed in section 6.A. Timbre and loudness modulations accompany the pitch modulation and all affect the quality of the resulting tone. There are some shifts in vowel quality in singing that result from vocal tract configurations differing from those in normal speech production. A lowering of the larynx results in a lengthening of the vocal tract and a lowering of the formant frequencies, which in turn cause the "darker" vowel quality in "covered singing." Other significant features peculiar to singing are vocal registers, singing formant, and formant tuning, which will now be discussed.

Vocal Registers

In the discussion of speech production no specific mention was made of vocal register because speech is normally produced within a single register. However, the situation is different for singing, there being some two or more registers commonly used in singing. (The reader is referred to the references at the end of this section for discussions on the multiplicity of registers.)

A *vocal register* is characterized by one of the modes or combinations of modes in which the vocal cords vibrate. The mass on a spring vocal cord model discussed in section 3.A (or even more elaborate, two masses on springs models) are too simplified. They work reasonably well for speech modeling, although even there they fail to capture significant features of the vocal cord motion because they do not treat vocal cord motion exhibiting wave phenomena. For instance, they represent the glottal area (opening between the vocal cords) as rectangular, even though it is more nearly oval in shape. They also fail to account for motion of the vocal cords in the direction of air flow through the larynx. They are inherently incapable of representing some of the more string-like behavior of the vocal cords observed in high-pitched singing. A many-mass model of the vocal cords more accurately represents some of these features because it can show either wavelike or single- or double-mass behavior. We will discuss briefly the two registers that lie at the extremes of the male voice—the modal and the falsetto. There are roughly comparable registers for the female voice; the descriptions given for the male voice can be thought of as applicable to female voice by assuming appropriate reductions in size.

The modal (or chest) register is used for low fundamental frequency production and is the register commonly used in speech production. A combination of modes typical of this register is shown in the first row of Fig. 6.B-1. The top view (a) shows the characteristic oval opening produced when the two ends of the cord are fixed and the middle portion moves. This is essentially a half-wavelength mode. The end view (b) shows that the upper and lower portions of the cord tend to vibrate out of phase with each other. This is essentially a half-wavelength "thickness mode," with the upper and lower surfaces of the cord free to move. The edge view (c) shows the cord moved upward in the direction of airflow. Again, this is a half-wavelength mode in the vertical direction, with the ends fixed. Keep in mind that all of these modes are vibrating together so that the motion of the vocal cords is rather complex. In addition, this register is characterized by the following features.

1. Vocal cords are comparatively short and thick, with a relatively large effective mass.
2. Vocal cord tension is comparatively low.
3. Vibration amplitude of the cords is large.
4. The vocal cords tend to close completely over part of the vibratory cycle.
5. The signal produced by the vocal cords tends to be rich in harmonics.
6. The energy supplied by the lungs is converted rather efficiently into sound energy.

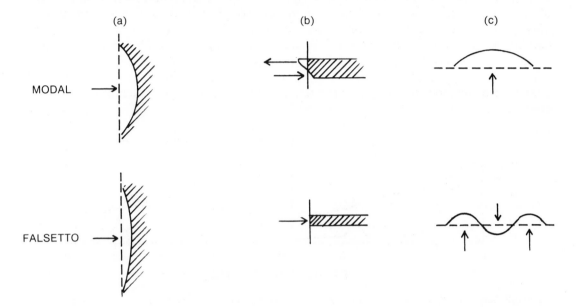

Fig. 6.B-1. Vocal cord modes typical of modal and falsetto registers: (a) superior view (looking from above). (b) coronal view (looking along the length of the cord from one end). (c) sagittal view (looking at the edge of the cord). (After Titze and Strong, 1975.)

The falsetto (or head) register is used for high fundamental frequency production. A combination of modes typical of this register is shown in the second row of Fig. 6.B-1. The top view (a) shows a characteristic oval opening. The vibration amplitude is smaller than in the modal register. The mode shown is a half-wavelength, although sometimes higher modes involving several half-wavelengths may also be involved. The end view (b) shows upper and lower portions of the cord moving together as a unit, which is different from the modal register. Note that the cord is much thinner. The edge view (c) shows a three half-wavelength mode in the vertical direction with the vocal cord ends fixed. The single half-wavelength mode shown in (c) for the modal register may also be involved in the falsetto register. Again, keep in mind that these modes are vibrating together. In addition to the modes, the following features are characteristic of the falsetto register.

1. Vocal cords are longer (up to 30%) and thinner, with a smaller effective mass.
2. Longitudinal vocal cord tension is comparatively high.
3. Vibration amplitude of the cords is small.
4. Vocal cords tend to lack complete closure during any part of the vibratory cycle. This results in increased airflow and a more breathy quality.
5. The signal produced by the cords tends to be poorer in harmonics.
6. The conversion of "lung energy" into sound energy is less efficient than in the modal register.

Some of the above described features have been studied in computer simulation with a many-mass vocal cord model. The simulation verifies some of these features and demonstrates that the modal register (large mass and low tension) exhibits natural frequencies that are lower than those of the falsetto register.

Singing Formant

A first-class opera singer can be heard above relatively high levels of sound from an orchestra and in relatively large halls. This is apparently due to the "singing formant," a strong peak in the spectrum at about 3.0 kHz for sung vowels. The spectra of a spoken and sung /OO/ (as in cool) are shown in Fig. 6.B-2. The dashed lines in the figure represent the spectrum of a typical orchestra averaged over long periods of time. If a singer produced a spectrum comparable to that of ordinary speech (Fig. 6B-2a), his song would be masked by the orchestra and would not be very perceptible as an entity in its own right. However, the spectrum produced by a trained singer (Fig. 6.B-2b) differs from that of normal speech primarily because of the singing formant. The singing formant is in a frequency region where sounds of the orchestra are reduced in energy, which enables the singer to be heard above the orchestra.

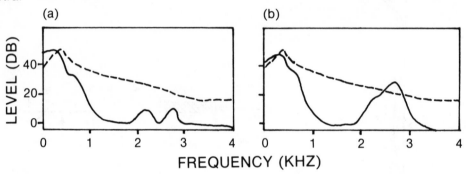

Fig. 6.B-2. Solid line: spectra of the vowel /OO/; (a) spoken and (b) sung. Dashed line: average energy of orchestra. (After Sundberg, 1974.)

The origin of the singing formant appears to be an extra formant in the singer's vowel spectrum. We noted in section 5.B that frequencies in the vicinity of formants are emphasized. Furthermore, when two formants lie close together in frequency the ehancement is even more marked for frequencies in the vicinity of either of the two formants. This can be seen in the vowel spectra in Fig. 5.B-7, where frequencies around 2.5 kHz are emphasized because of the proximity of the second and third formants in the vowel /EE/ (as in heat) and where frequencies around 1.0 kHz are emphasized because of the proximity of the first and second formants in the vowel /O/ (as in hot).

The introduction of an additional formant in the vicinity of the third and fourth formants would result in significant enhancement of the energy at about 3.0 kHz. This phenomenon has been demonstrated in more detailed models of the larynx-pharynx system than we have considered. We have considered the larynx-pharynx system to consist of a single tube of smoothly varying cross-section terminated at the vocal cords (see Fig. 5.B-2). Anatomical evidence indicates that this is an oversimplified model, especially for singing when the larynx is lowered. A more realistic model is one in which the vocal cords connect to the pharynx via a small cavity (the laryngeal ventricle, or sinus Morgagni) and a narrow tube, as seen in Fig. 6.B-3. The width of the pharynx is much larger than the narrow tube, which results in an acoustical mismatch where they join. The acoustical mismatch means that the small cavity and narrow tube can be considered a resonator in its own right. Its resonant frequency furnishes the extra formant at about 3.0 kHz when the larynx has been lowered and the pharynx widened as noted.

226

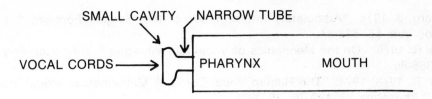

SMALL CAVITY NARROW TUBE

VOCAL CORDS⟶ PHARYNX MOUTH

Fig. 6.B-3. "Singer's" vocal tract. (After Sundberg, 1974.)

Formant Tuning

The first formant ranges from about 300 to 900 Hz for female vocal production, depending on the vowel being produced. When the singing voice is producing very high fundamental frequencies (as would be the case with sopranos), it is possible that the fundamental frequency will not lie close to a formant frequency. This can result in substantial loss of amplitude (and hence loudness), since the further a harmonic is from a formant the greater its attenuation. Furthermore, the loudness of the tones produced would fluctuate depending on the relation of the fundamental and first formant frequencies.

The singer's (primarily soprano) solution is to move the first formant frequency so as to match it more closely to the fundamental frequency being produced, especially when the fundamental lies above the first formant. This is accomplished by varying the vocal tract shape via jaw opening or by shortening the tract via lip control. This minimizes the loudness variation for different vowels sung at different pitches. This "formant tuning" to the fundamental might be expected to result in the loss of vowel intelligibility. It probably does, but in singing, tonal quality is usually of more importance, with intelligibility often of lesser concern.

Exercises

*1. Get a good tenor to illustrate the difference between his spoken and sung vowels. Do a spectrum analysis of them if possible and look for the "singing formant." Compare your results to Fig. 6.B-2.

*2. Get a good soprano to illustrate the effects of formant tuning for high pitches. Note any differences in loudness and vowel quality. Do spectrum analyses of the different productions if possible.

3. You might produce something akin to the "darker" vowels of "covered" singing by singing a vowel at a given pitch under normal circumstances and then with a short length of tubing surrounding the lips. This lengthens the vocal "tract" and lowers the formant frequencies. You can use your cupped hands to provide the short length of tube, although this will result in a fairly large increase to the vocal tract length. (In actual covered singing the length of increase of the tract is at the other end of the tract, in the vicinity of the vocal cords.)

*4. Get a good singer to produce a tone at a given pitch in both modal and falsetto registers. Perform spectrum analyses on the tones if possible.

Further Reading

Backus, chapter 11

Benade, chapter 19

Flanagan, chapters 2, 3

Seashore, chapter 4, 20

Large, J., ed. 1973. *Vocal Registers in Singing* (Mouton and Co., The Hague).

Sundberg, J. 1976. "The Acoustics of the Singing Voice," Scientific American *236* (3), 82-91.

Sundberg, J. 1975. "Formant Technique in a Professional Female Singer," Acustica *32,* 89-96.

Sundberg, J. 1974. "Articulatory Interpretation of the 'Singing Formant'," J. Acoust. Soc. Am. *55,* 838-44.

Titze, I. R. 1976. "On the Mechanics of Vocal-fold Vibration," J. Acoust. Soc. Am. *60,* 1366-80.

Titze, I. R. 1973, 1974. "The Human Vocal Cords: A Mathematical Model, Parts I and II," Phonetica *28,* 129-70; *29,* 1-21.

Titze, I. R., and W. J. Strong. 1975. "Normal Modes in Vocal Cord Tissues," J. Acoust. Soc. Am. *57,* 736-44.

Vennard, W. 1967. *Singing: The Mechanism and Technic.* (Carl Fischer, Inc., New Jersey).

Audiovisual

1. *Function of the Normal Larynx* (21 min, color, 1956, ILV)

Demonstrations

See exercises.

6.C. LIP REED INSTRUMENTS

The lip reed, or brass, instruments include the trumpet, cornet, trombone, tuba, and French horn, as well as some less well-known instruments like the fluegelhorn and ophicleide. The common physical features of brass instruments are (1) a cup-shaped mouthpiece, (2) sections of cylindrical and conical tubing, (3) valves or slides to change the length of the tubing, and (4) a flared output end called a bell. These features can be observed in Fig. 6.C-1. The air column contained within the tubing is excited by the vibrating lips of the player. The resulting tone quality of a brass instrument depends on the shape of the mouthpiece, the shape and size of the tube, and the flare of the bell. In this section we will first consider the general properties of lip-reed instruments; then we will consider these instruments separately in more detail.

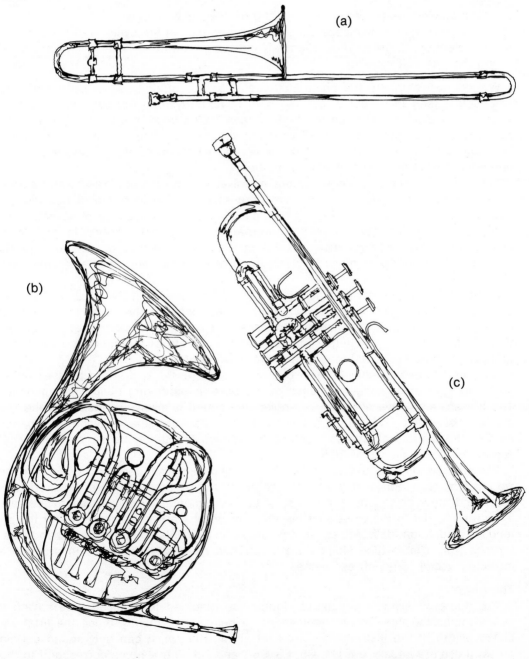

Fig. 6.C-1. Brass instruments: (a) trombone, (b) French horn, (c) trumpet. (Not drawn to scale.)

General Characteristics

The following observations can be made about the brass family of instruments.

1. The lip reed acts as a valve to allow alternately a greater and smaller air flow. This gives rise to alternating condensations and rarefactions traveling in the instrument.
2. The trumpet, trombone, and French horn have bores which are mostly cylindrical but partly conical, with an abruptly flaring bell at the end.
3. The tuba and the fluegelhorn have bores which are predominately conical (most of the tubing increases in diameter from mouthpiece to bell).
4. In general, the narrower the tube for a given length, the more brilliant the tone quality produced.
5. A large bell produces a more mellow tone; a smaller bell causes a more brilliant tone.
6. All harmonics are present in the spectra of these instruments.
7. The lip reed generally closes completely during part of the cycle.
8. The spectrum contains proportionately more high-frequency energy for hard blowing than for soft blowing.

We will now compare the brass instruments to two other types of instrument: the mechanical reed instruments and the human vocal system. Although the brass instruments use a bore like the mechanical reed instruments, the vibrating lips of the player are more similar to a set of vibrating vocal cords than to a mechanical reed. Furthermore, the mechanical reed has a very small mass; so its frequency of vibration is controlled more completely by the bore.

Since the lips are considerably more massive than mechanical reeds, and accessible to greater conscious control, the bore of the brass instruments has less influence on the player's lips. The lip reed can, in effect, choose one of several available bore resonance frequency combinations to excite. There is usually a considerable frequency variation possible in the vicinity of a resonance as a player can "lip" the note up or down by a semitone or so by tensing or relaxing the muscles controlling the lips.

In the vocal mechanism the vocal tract has little influence on the vibration frequency of the comparatively massive vocal cords. Also, because the vocal tract does not usually have harmonic resonance frequencies, it does not make "strong" suggestions to the vocal cords; hence the vocal cords vibrate at a frequency determined primarily by their mass and tension and the blowing pressure.

Another difference between the woodwinds and brasses is that the brasses utilize many more of the resonance modes of the bore in producing their different notes. The brasses also have no open side-holes; the sound is always emitted from the bell of the instrument. To obtain notes which are between the available resonance modes, the brass instruments increase the overall length of the air column by inserting additional pieces of tubing.

The woodwinds and brasses utilize combinations of cylinders and cones. In the woodwinds, resonances of the bores are perturbed by the addition of covered tone holes and other deviations in the bore of the instrument. The shifted resonances are a source of grief to the woodwind players, who must learn to compensate for the attendant intonation deficiencies. In the brass instruments, on the other hand, distortions from simple tube shapes are introduced deliberately in order to produce a musically useful series of resonances.

The Bugle

The bugle consists of a cupped mouthpiece attached to a coiled tube having a slowly increasing flare which terminates in a bell-shaped mouth. Since the total uncoiled length of the instrument is fixed at about 160 cm, it can only sound certain notes: these are usually C4, G4, C5, E5, B5^b, and C6. These notes correspond to the natural frequencies of this horn. An idealized plot of the input impedance showing the resonances for a bugle is given in Fig. 6.C-2. The bugle player can select which of these modes he wants to "play on" by tightening or relaxing his lips. The modes used most often are the third through sixth, corresponding to the notes G4 through G5. The second mode (C4) is used infrequently, and the first mode is never used.

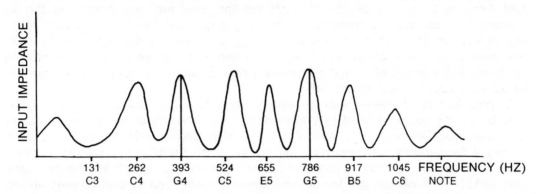

Fig. 6.C-2. Idealized input impedance curve for bugle showing resonance peaks for different modes.

How do the lips cause the steady air stream from the lungs to set up an oscillatory motion in the bore of the instrument? While there is still some uncertainty in how the air inside the instrument is set into vibration, we can describe the action of the lips, which act as a throttling valve to the air stream. Assume that we start with the lips stretched taut but closed (as shown in Fig. 6.C-3a). The pressure of the air stream exerts a force against the lips, which causes them to blow open, as shown in Fig. 6.C-3b. When the lips reach their maximum opening (Fig. 6.C-3c) the air flow through the opening is quite large. Because the lips still form a constriction, the speed of the air is increased, thus causing a pressure reduction by Bernoulli's Law (see section 1.B). Because of the reduced pressure, and because of the increased tension on the lips due to the large opening, the lips begin to close, as shown in Fig. 6.C-3d. The lips continue to close until they resume the position of Fig.6.C-3a, where the entire process begins anew. The frequency at which this process takes place will be one of the resonant frequencies of the bugle. In just this manner, a steady stream of air is converted into a pulsating flow which is very rich in harmonics. Many of these harmonics will be reinforced by the natural modes of the tube. This same basic process applies to lip vibration for all the brass instruments.

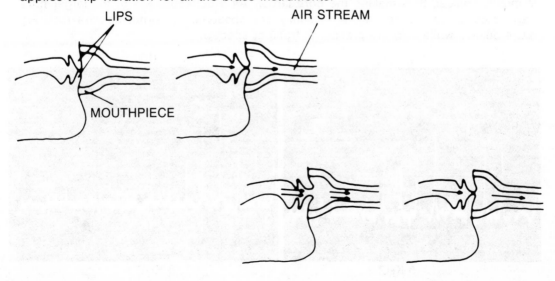

Fig. 6.C-3. Lip position and airflow from player's mouth at different times in a lip vibration cycle. (After Olson.)

231

Some interesting points in regard to the production of a pulsating air stream can be made. The area of the opening between the lips varies in an almost sinusoidal fashion. The air pulses would be almost sinusoidal and have little harmonic content if they varied as the lip opening area. However, there are other means (discussed in section 5.A) by which the air pulses are enriched with a good deal of harmonic content. For one thing, the air flow through the lips does not vary directly as the lip opening because the resistance of the opening to air flow does not vary with opening. In addition, the standing waves in the bugle give rise to a fluctuating pressure in the mouthpiece which affects the flow. These two features are primarily responsible for producing a series of air pulses having a rich harmonic content even though the lip opening varies almost sinusoidally.

Another feature of interest with lip reed instruments is the "privileged notes" made possible by the cooperation of several modes of the tube. We consider for illustration the bugle input impedance curve in Fig. 6.C-2. Suppose we wish to produce a tone having a fundamental frequency of 131 Hz at which there is no resonance peak. However, there are resonance peaks at 262 Hz, 393 Hz, etc., which are integer multiples of 131 Hz. If we can get the second harmonic produced by the vibrating lips to coincide with the 262 Hz mode, the third harmonic with the 393 Hz mode, etc., we will end up with a fundamental frequency of 131 Hz even though there is no mode present to encourage it. A "privileged note" in this sense is a note occurring at a frequency with which no natural modes coincide. It is made possible because higher harmonics of the vibrator coincide with other tube modes; these other tube modes need to have a nearly integer relationship with the desired fundamental frequency if they are to cooperate effectively in its production. The "privileged frequencies" associated with a collection of modes can be determined by seeing what frequency differences between pairs of modes occur. In the bugle case, if we consider modes having frequencies of 262, 393, and 524 Hz, we get frequency differences of 131 Hz and 262 Hz, which would both be "privileged frequencies."

The Trumpet and Cornet

The coiled tube of a trumpet is cylindrical for about the first one-third of its 140 cm (uncoiled) length. The rest is conical, with a very slow rate of flare (except for the last 25 cm, which flares rapidly to form the bell). Three piston valves are provided to change the length of the air column, as will be described later. The cornet is constructed to be more compact than the trumpet, even though its uncoiled length is only slightly shorter. Although the cornet resembles the trumpet very closely, its bore is mostly conical, thus making the bore larger. As a consequence of the larger bore, harmonics above the seventh are usually not observed in cornet spectra (see Fig. 6.C-4 below), while they are present in trumpet spectra.

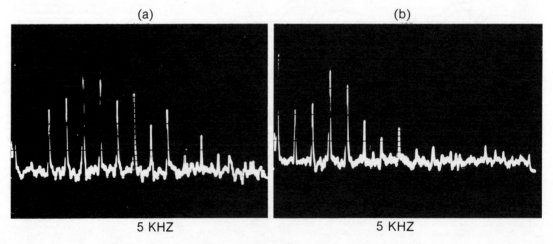

(a) (b)

5 KHZ 5 KHZ

Fig. 6.C-4. Brass spectra: (a) trumpet and (b) cornet.

Since the lips of the player essentially close off the end of the instrument, and since the trumpet tube is mostly cylindrical, we would expect to find only odd harmonics present. But brass players all know that they can play all harmonics (odd and even). We can explore how this is possible by considering the evolution of a trumpet from a cylindrical piece of electrical conduit pipe having a length just slightly shorter than a trumpet. By buzzing our lips in the conduit pipe we find a set of resonances as given in the first column of Table 6.C-1. Note that these frequencies are those we would expect for a cylindrical closed-open tube. The second column gives the musical notes that are nearest to the frequencies shown in column 1.

Table 6.C-1. Resonance frequencies of various idealized approximations to a trumpet.

Conduit		"Trumduit"		Trumpet		Harmonic Series	
Frequency	Note	Frequency	Note	Frequency	Note	Frequency	Note
65	C_2	65	C_2	85	E_2	116	B_2-flat
195	G_3	195	G_3	232	B_3-flat	232	B_3-flat
325	E_4	323	E_4	345	F_4	348	F_4
455	B_4-flat	446	B_4-flat	450	B_4-flat	464	B_4-flat
585	D_5	570	D_5	557	D_5	580	D_5
715	F_5-sharp	695	F_5-sharp	682	F_5	696	F_5

As a first modification to the conduit, in order to make it musically more useful, we attach a trumpet mouthpiece. This new instrument, which we will name a "trumduit" has resonances, in which the fundamental frequency is essentially unchanged, but the upper resonances are all lowered. This lowering occurs because the mouthpiece is an expanded tube, as shown in Fig. 6.C-5.

Fig. 6.C-5. Cross section of trumpet mouthpiece.

Now we add a trumpet bell to our trumduit; if at the same time the length of the tube is changed slightly, the higher resonances will be left unchanged, but the lower resonant frequencies will all be raised. This is indicated in the third column of Table 6.C-1. Note that we now have a musically useful series of resonance frequencies, with the exception of the lowest, or fundamental, which is badly out of tune with the others. This is of limited consequence, however, because the actual lowest resonance (E_2) is not used in practice. The remainder of the resonances form an approximately harmonic series based on a fundamental which is actually not present (B_2-flat). If the note B_2-flat has a frequency of 116 Hz, the harmonic series formed on this note is given in the last two columns of Table 6.C-1.)

When we compare the harmonic series based on B_2-flat with the resonances of the trumpet we find that, with the exception of the fundamental, the trumpet resonances are very close to the harmonic series. The playable frequencies on our trumpet are then all approximately integer multiples of a fundamental (B_2-flat) which is *not* an actual trumpet resonance, and which might be termed a nonexistent fundamental. The so-called *pedal tone* in the trumpet is the octave below the first useful mode (B_3-flat) and hence is identical to the nonexistent fundamental. The pedal tone is produced, even though there is no actual resonance present, by buzzing the lips at a frequency of 116 Hz. The upper harmonics of the lips are integer multiples of 116 Hz (232, 348, etc.), which very nearly coincide with the resonances of the trumpet, thus reinforcing the vibration. In other words, the presence of harmonics 2, 3, 4, etc., in the tone will cause the lip and air column vibrations to repeat at the fundamental frequency of 116 Hz. Actual resonance curves for a trumpet are shown in Fig. 6.C-6.

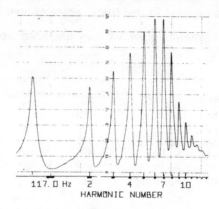

Fig. 6.C-6. Input impedance curves for a trumpet (open valves) showing resonance peaks. (From Backus, 1976).

The addition of the trumpet bell has several other effects. First, it helps the trumpet produce a louder and clearer tone by increasing the sound radiation. Secondly, it not only changes the resonance frequencies of the tube, but it reduces the number of resonances which are present. Beyond a certain cut-off frequency, the pressure waves in the instrument are not reflcted by the bell to produce standing waves, but leak out through the bell, thus reducing not only the number of resonance peaks, but their overall amplitude as well. This is dramatically illustrated in Fig. 6.C-7 for a simple cylindrical tube with and without a bell.

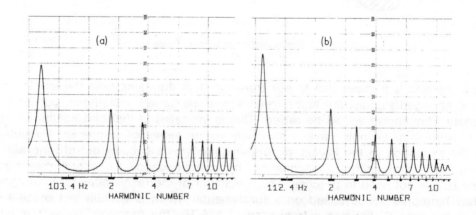

Fig. 6.C-7. Input impedance curves for (1) cylindrical tube (b) tube with bell. (From Backus, 1976).

The trumpet we have designed so far is actually a type of bugle. It can play notes corresponding to the harmonics of the instrument, but that is all. In order to make the instrument more useful musically, we would like to be able to play all the notes within the range of the instrument. This is accomplished for the trumpet by adding a

234

set of valves and extra tubing which can be used to increase the length of the tube. Every time the length of the instrument is changed, a new set of resonances, based on a different fundamental frequency, is produced.

The Trombone and the Tuba

The trombone bore consists of a cylindrical U-shaped tube which is doubled back on itself and connected to a tapered section which flares into a bell. The total length of this tube is about 270 cm. The telescopic section makes it possible to lengthen the tube, thereby altering the resonant frequencies. Since the tube can be set at any length, it is possible to vary the frequency continuously. The set of resonances which a performer can choose from is determined by the position of the slide, but the player picks the desired pitch by adjusting his lip tension and blowing pressure to play on one of these natural frequencies. The natural frequency peaks for a trombone are given in Fig. 6.C-8. You may note the similarities between this curve and Fig. 6.C-6 for a trumpet.

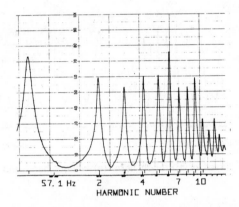

Fig. 6.C-8. Input impedance curve for trombone with slide in. (From Backus, 1976.)

The coiled bore of the tuba is a slowly flaring cone which terminates in a large bell, the entire (uncoiled) length being about 720 cm. The tube usually has four valves, with their associated plumbing, in order to give the player greater flexibility and ease of performance.

Another instrument of the brass family, which is now obsolete, is the ophicleide. Although this instrument was blown with the lips like the brass instruments, it had tone holes along its tube like a woodwind. For this instrument the bell would have served a useful purpose only for the two or three lowest notes in the instrument's range.

The French Horn

The French horn consists of a coiled tube about 360 cm in length, with a slow conical flare terminated in a very large bell. The instrument is equipped with three valves. The French horn differs from the other brass instruments in one important aspect. The trumpet and trombone use the first eight or nine resonance modes for playing, whereas the horn uses the series up to an octave higher, as far as the sixteenth resonance. This requires that the higher resonances be pronounced and distinct, which is accomplished by having considerably more of the French horn tube to be conical and less of it cylindrical. The resonance curves for a French horn are shown in Fig. 6.C-9a. Note that they are quite pronounced out to the twenty-second or so. Also note that, since a more nearly conical tube was used, the fundamental frequency coincides more nearly with the nonexistent fundamental.

Since the horn typically plays in a frequency region where the modes are very close together, the player has considerably more flexibility than do players of other brass instruments. It was discovered empirically, however, that by placing one's hand in the bell of the instrument, one could lower the pitch by an amount that depends on how far the hand is inserted. This technique was developed to a point where a complete chromatic scale was possible over the upper part of the instrument's range. When a hand is inserted in the bell, the area is constricted, which increases the acoustic mass at the end of the instrument, thus lowering the resonant frequencies. In addition to lowering the frequencies, the number of harmonics is usually increased as the hand is inserted, thus making it possible to play in resonance modes which were not sufficiently well developed before the hand was inserted. A comparison of Fig. 6.C-9b with Fig. 6.C-9a shows the effect of inserting the hand into the bell of a French horn.

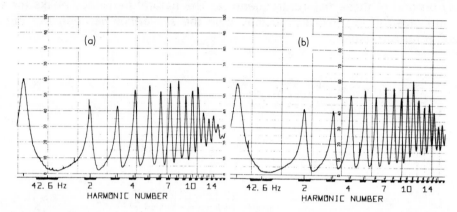

Fig. 6.C-9. Input impedance curves of a French horn: (a) without hand and (b) with hand. (From Backus, 1976.)

When the hand is thrust as far as possible into the bell, a tone of quite different quality, called the *stopped* tone, is produced. The stopped tone has a pitch which is a semitone higher than the normal horn tone. The apparent rise in pitch is caused by the large increase in acoustic mass as the hand is inserted. The mass actually increases so much that each mode is pulled down in frequency until it is within a semitone of the original frequency of the one below it. A given note is then played in the next higher mode with the horn stopped, the mode previously used now having gone much too flat. This can be demonstrated by following a note down in pitch as the hand is inserted. The note can be followed down to its new low pitch, but if the player tries to hold his pitch the note jumps to the next higher mode, which is a semitone sharper than the unstopped mode.

Exercises

1. How can a bugle, with no valves (and hence a single length), play tones at more than one fundamental frequency?

2. Determine several "privileged note" frequencies for the bugle discussed in the text.

3. Refer to the bugle input impedance curves in Fig. 6.C-2. The solid lines indicate the first two partials in the spectrum for a tone of 393 Hz (G_4). Draw similar lines indicating the partials for a tone having a frequency of 262 Hz. For 524 Hz. How do the spectra of these tones compare?

4. The trombone is similar to the trumpet except that it is much longer. What effect does this have on normal mode frequencies?

5. The trombone and French horn have similar overall lengths. Are their spectra significantly different? How does this influence their tone quality?

236

6. What effect do different shaped mouthpieces have on brass tones?

7. If the trumpet has only three valves (giving eight different combinations) how can it play tones at 30 different frequencies?

8. How is it possible to produce the pedal note of a brass instrument in which the lowest normal mode frequency lies several semitones below the pedal note frequency?

9. What is the effect of a leaky valve in a brass instrument?

10. Why is it possible to sing all notes (within a certain range) without changing the length of the resonator (your vocal tract), while musical instruments only produce frequencies close to a normal mode frequency?

11. Use the following table to explain why mechanical reed instruments are dominated by the bore of the instrument, the vocal mechanism is dominated by the vibrator, and the brass are a mixture of the above two.

    Mass of clarinet reed $\simeq$ 0.001 gram
    Mass of vocal cords $\simeq$ 0.24 gram
    Mass of lips $\simeq$ 0.10 gram

12. Answer the following questions by referring to both spectra of Fig. 6.C-4.
    a. How many harmonics are present in each spectrum?
    b. What are the approximate frequencies of the harmonics?
    c. What is the fundamental frequency and to what musical note does it correspond?

Further Reading

Backus, chapter 12

Benade, chapter 20

Culver, chapters 12, 13

Kent, part II

Backus, J. 1976. "Input Impedance Curves for the Brass Instruments," J. Acoust. Soc. Am. *60,* 470-80.

Backus, J., and T. C. Hundley. 1971. "Harmonic Generation in the Trumpet," J. Acoust. Soc. Am. *49,* 509-19.

Bate, P. 1966. *The Trumpet and Trombone* (W. W. Norton, New York).

Beauchamp, J. W. 1975. "Analysis and Synthesis of Cornet Tones Using Nonlinear Interharmonic Relationships," J. Audio Eng. Soc. *23,* 778-95.

Benade, A. H. 1973. "The Physics of Brasses," Scientific Am. *228,* (1), 24-35.

Benade, A. H., and D. J. Gans. 1968. "Sound Production in Wind Instruments," Ann. N.Y. Acad. Sci. *155,* 247-63.

Cardwell, W. T. 1966. "Working Theory of Air-Column Design," J. Acoust. Soc. Am. *40,* 1252.

Igarashi, J., and M. Koyasu. 1953. "Acoustical Properties of Trumpets," J. Acoust. Soc. Am. *25,* 122-28.

Long, T. H. 1947. "The Performance of Cup-Mouthpiece Instruments," J. Acoust. Soc. Am. *19,* 892-901.

Martin, D. W. 1942. "Lip Vibrations in a Cornet Mouthpiece," J. Acoust. Soc. Am. *13,* 305-8.

Martin, D. W. 1942. "Directivity and Acoustic Spectra of Brass Wind Instruments," J. Acoust. Soc. Am. *13,* 309-13.

Morley-Pegge, R. 1960. *The French Horn* (Ernest Benn, London).

Pyle, R. W. 1975. "Effective Length of Horns," J. Acoust. Soc. Am. *57,* 1309-17.

Stauffer, D. 1954. *Intonation Deficiencies of Wind Instruments* (Catholic Univ. of Am. Press, Washington, D.C.).

Audiovisual

1. *Descriptive Acoustics Demonstrations—Part 6.A.* (Cassette, 1977, GRP).
2. *Sounds in Your Ears—Brass Instruments* (3⅓ ips, 2 track, 30 min, UMAVEC).

Demonstrations

1. Cylindrical tube with clarinet and with brass mouthpieces

6.D. MECHANICAL REED INSTRUMENTS

The mechanical reed instruments are one subclass of the woodwind group. (The other subclass will be considered in section 6.E.) Like the brasses, the mechanical reed instruments have a bore consisting of cylindrical and/or conical sections and an air-throttling device known as the reed. As with the brasses, the mechanical reed alters the steady air stream from the player's lungs and causes the air column enclosed by the bore to begin oscillating. Unlike the brass instruments, however, the vibration frequency of the mechanical reeds is controlled almost completely by the bore and very little by the reed. By opening a tone hole along the body of the instrument the effective length is shortened, thus causing the reed to vibrate at a higher frequency.

The mechanical reed instruments can be classified by the shape of the bore and by the type of reed used. The two types of tube are cylindrical and conical; the two types of reeds are single reed and double reed. We will consider the three instruments shown in Fig. 6.D-1 as representative of their respective families: clarinet

(a)

(b)

(c)

Fig. 6.D-1. Mechanical reed instruments: (a) clarinet, (b) oboe, (c) saxophone. (Not drawn to scale).

239

(single reed and cylindrical tube), oboe (double reed and conical tube), and saxophone (single reed and conical tube).

The following general observations can be made about the mechanical reed instruments.

1. The reed end of the instrument is approximately closed.
2. The reed acts as a valve to allow alternately a greater and smaller air flow.
3. The clarinet has a bore which is essentially cylindrical, while the other mechanical reed instruments have bores which are essentially conical.
4. The clarinet spectrum has predominately odd harmonics; other reed instrument spectra have all harmonics.
5. The reed does not close completely for soft blowing; the reed may be closed for half or more of the cycle for hard blowing.
6. The spectrum contains proportionately more high-frequency energy for hard blowing than for soft blowing.

In this section we consider the different reed and bore types in some detail. Then we investigate the physical properties of the clarinet, the oboe, and the saxophone. We then describe some recent research on the double-reed instruments which may have ramifications on the design of these instruments in the future.

Reed and Bore Types

Instruments utilizing a single reed are members of the clarinet and saxophone families, and instruments which use a double reed are oboe, English horn, and bassoon. Figure 6.D-2a presents a cross-sectional view of a clarinet mouthpiece and reed. The player places his upper lip on the mouthpiece and his lower lip on the reed, partly closing the opening between the reed and mouthpiece. When he blows with sufficient pressure the instrument sounds. The air pressure in the mouthpiece acts on the reed in such a way as to alternately push it farther away and pull it closer to the mouthpiece. The rate of vibration of the reed is controlled by the length of the bore (the tube through which the sound travels) of the instrument up to the first open tone hole. As a player moves his fingers, this distance changes, and thus the rate of vibration changes. The oboe reed, shown in Fig. 6.D-2b opens in a similar manner, except there are two reeds which flex such as to open and close the space between them.

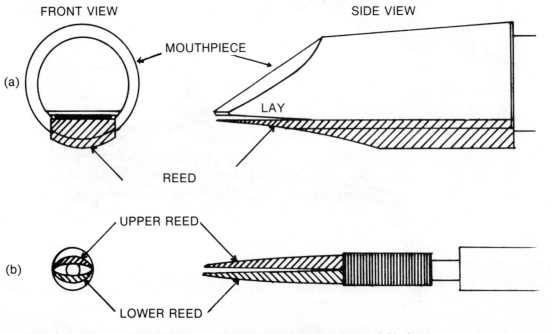

Fig. 6.D-2. Reed and mouthpiece for (a) clarinet and (b) oboe.

There are four basic shapes of Bessel horns that might be considered for the bore of woodwind instruments, illustrated in Fig. 6.D-3.

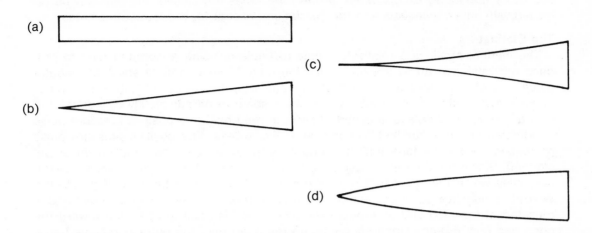

Fig. 6.D-3. Types of Bessel horns: (a) cylindrical, (b) conical, (c) increasing flare, (d) decreasing flare. (After Benade, 1960.)

Woodwinds are designed with cylindrical or conical bores because these bores have resonant frequencies which are harmonically related to the fundamental. The woodwinds are designed with a series of tone holes which effectively shorten the length of the vibrating tube as they are opened. The rate of vibration of the reed is controlled by the length of the bore of the instrument up to about the first open hole. Thus, to play an ascending scale, the player merely opens the appropriate tone-holes, continually shortening the tube, while blowing on his reed. To play three octaves in this manner, however, would require the presence of an inordinate number of tone-holes, as well as being impractical in terms of the length of the "shortened bore." So, after the lowest octave is played on an oboe, the player pushes a register key, thus opening a small hole near the center of the instrument, which shifts the frequency of the lowest mode substantially. This results in the instrument sounding in the second mode, which is an octave higher for the oboe. The same tone-holes can now be reused for the second octave on the instrument. Register keys work well for the cylindrical and for the conical bores, because the resonances present remain unchanged when different registers are employed, thus making it possible for the same tone-holes to be used in two different registers. For tubes shaped like (c) and (d) in Fig. 6.D-3 this is not possible. The upper registers would be out of tune with the lower, so that the same tone-holes could not be utilized.

When a reed instrument is played, a standing wave, having a pressure antinode at the reed end and a pressure node at the first open hole, is generated inside the instrument. The player maintains the standing wave by the power supplied through blowing. Sound is radiated through the open holes, primarily through the first open hole below the mouthpiece. The mechanical reed instruments are extremely inefficient; they radiate only a fraction of a percent of the energy put into them. The input energy is given up in the following ways in the approximate amounts shown (Benade and Gans, 1968).

Thermal and viscous losses to walls of tube (80%)
Reed losses (10%)
Losses to mouth and vocal tract of player (8%)
Turbulent losses (2%)
Radiated as acoustical energy (0.2-3%)

The sound one hears from a woodwind, of course, is not the internal standing wave in the tube, but the radiated sound. The bell of the instrument is not the radiating source of the sound except when all the tone holes are closed; the bell can be removed with little consequence to the instrument, except for the lowest few notes.

The Clarinet

The clarinet provides a relatively simple example of using a vibrating reed to produce oscillations in an air column. Since the reed opening is quite small, the mouthpiece end of the instrument is a pressure antinode. Let us follow a pressure pulse down a hypothetical clarinet tube which we assume to be completely cylindrical, but with a clarinet mouthpiece attached. When the reed opens, a positive pressure pulse is admitted into this "tubinet," as shown in Fig. 6.D-4a. The positive pressure pulse propagates down the tube until it encounters the open end (bell end on an actual clarinet). Since the open end must be a pressure node, the reflected wave is a negative pulse which then propagates back down the tube toward the reed (Fig. 6.D-4b). When this negative pulse encounters the reed, it tends to pull the reed closed and is reflected at the reed end as a negative pulse which is again propagated toward the open end (Fig. 6.D-4c). Upon encountering the open end, the pulse is reflected as a positive pulse which propagates back toward the reed. When this pulse encounters the reed, it now helps to open the reed, thus admitting another positive pressure pulse, and the motion repeats.

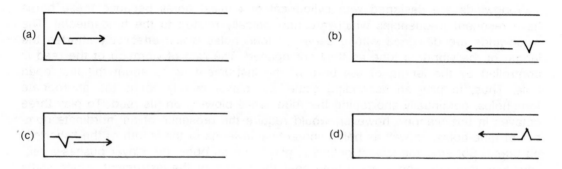

Fig. 6.D-4. Pressure pulse propagation in a "tubinet."

Since it takes the pulse four trips along the tube before one cycle is completed, the wavelength of this vibration must be four times the length of the tube, a situation we have come to expect for closed-open tubes. Furthermore, since only odd harmonics are allowed in closed-open tubes, the frequencies present will be odd multiples of the fundamental. If we measure the input impedance of our tubinet using the method described in section 2.D, we obtain a result similar to the solid curve shown in Fig. 6.D-5.

The resonant frequencies (peaks) of the curve are those frequencies where oscillations are favored. (Note the odd multiples of the fundamental frequency, f_1.) Likewise, the places where weak oscillations of the air column take place are seen to be the valleys, which occur at even multiples of f_1. When the shape of our perfect cylinder is altered in any way, such as by adding the bell, the resonant peaks are shifted in frequency. A slight enlargement at the open end will move the peaks upward in frequency, but the lowest peak will be moved the most and each consecutive peak will be shifted less, as shown by the dashed curve in Fig. 6.D-5. Consversely, a flaring at the mouthpiece end will displace the peaks downward, but the higher-frequency peaks are now affected more than the lower ones, as seen in the dotted curve in Fig. 6.D-5. In general terms, any alteration at the reed end has the opposite effect from the same alteration at the open end.

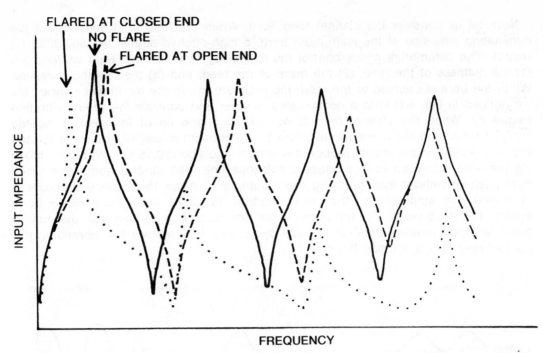

FLARED AT CLOSED END
NO FLARE
FLARED AT OPEN END

INPUT IMPEDANCE

FREQUENCY

Fig. 6.D-5. Input impedances of "tubinet" with and without flare showing shift of natural mode frequencies.

Now consider the resonance peaks of an actual clarinet for the lowest note (E3). Naturally the effect of the bell will be appreciable here, as you can see by inspecting Fig. 6.D-6. The effect of the bell is to raise the fundamental frequency to 153 Hz, thus making the other harmonics multiples of the fundamental. Also, note that the resonant frequencies depart more and more from the odd harmonics as we go to higher frequencies. The eighth harmonic, for instance, coincides more nearly with a resonance than either the seventh or ninth. Consequently, the eighth harmonic is quite predominant in this tone. In general, actual clarinet tones only show the odd-harmonic structure up to around 2000 Hz; beyond this frequency even and odd harmonics are present to about the same extent.

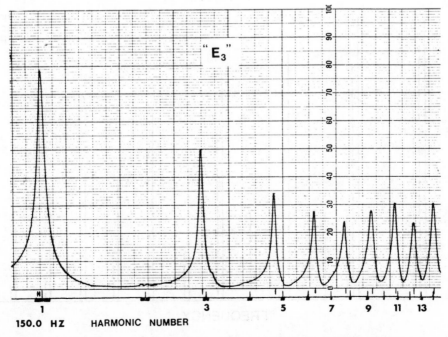

"E₃"

150.0 HZ   HARMONIC NUMBER

Fig. 6.D-6. Input impedance for "E3" of clarinet. (From Backus, 1974.)

243

Now, let us consider the clarinet reed itself. When the reed is played without the dominating influence of the instrument bore, a high-pitched squeal of 1500-3000 Hz results. The parameters which control the frequency at which this reed vibrates are (1) the stiffness of the reed, (2) the mass of the reed, and (3) the blowing pressure. When the bore is coupled to the reed, the pressure inside the mouthpiece varies (as we noticed in Fig. 6.D-4) in a complicated manner and controls the reed's vibration frequency. When the clarinet reed is not vibrating, the tip of the reed is located about 0.1 cm from the mouthpiece. When the instrument is played, the lower lip and the air pressure in the mouth reduce this distance to about 0.05 cm. When the blowing pressure is low, so as to produce a soft tone, the reed vibrates about its equilibrium position without ever touching the mouthpiece. When the blowing pressure is increased, the amplitude of the reed vibration increases until it begins to bang against the mouthpiece. For fortissimo tones, the reed is in contact with the mouthpiece for approximately one-half of each cycle. Fig. 6.D-7 shows the vibrations of a clarinet reed for pp, mf, and ff tones.

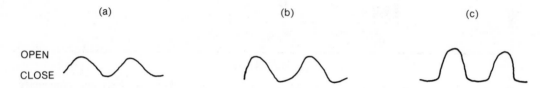

Fig. 6.D-7. Clarinet reed openings as a function of time for (a) pianissimo, (b) mezzo forte and (c) fortissimo dynamic markings. (After Backus, 1963)

Note that the vibration for pp tones is essentially sinusoidal. As the blowing pressure increases, more and more of the top of the sine wave is clipped off (by the reed banging into the mouthpiece). Clipping a sine wave in this manner adds more and more high-frequency energy; in the extreme case, the sine wave is clipped to such a great extent that it almost becomes a square wave. (You may recall from chapter two that square waves have a large number of odd harmonic frequencies present.) Fig. 6.D-8 shows how the higher frequency harmonics increase relative to the low fre-

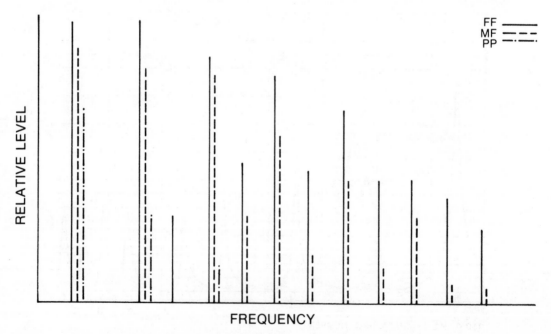

Fig. 6.D-8. Clarinet spectra for pp, mf, and ff markings. (Data from Backus, 1963.)

quencies as the clarinet is played at higher dynamic levels. Note that in going from a dynamic level of pp all the way to ff the change in amplitude of the low-frequency components is small. Rather, one is increasing loudness by increasing the number and amplitudes of the high-frequency components.

The Oboe and Bassoon

The oboe and the bassoon have bores which, to a first degree of approximation, may be represented as cones. (The cones, of course, must be truncated slightly so that the reed may be put into place, as shown in Fig. 6.D-9.)

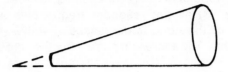

Fig. 6.D-9. Truncated conical approximation to double reed instrument bore.

A cylinder closed at one end has resonances which are odd-integer multiples of the fundamental. In the previous section, we saw that enlarging the open end or contracting the closed end of a cylinder moves the resonances upward in frequency. Suppose we take a cylinder and convert it into a cone by expanding the open end and at the same time contracting the closed end. The result is that all the resonances move upward in frequency by a constant amount. The fundamental resonance f_1 moves upward to the position of the former valley, $2f_1$. Likewise $3f_1$ moves upward to $4f_1$, etc., as shown in Fig. 6.D-10.

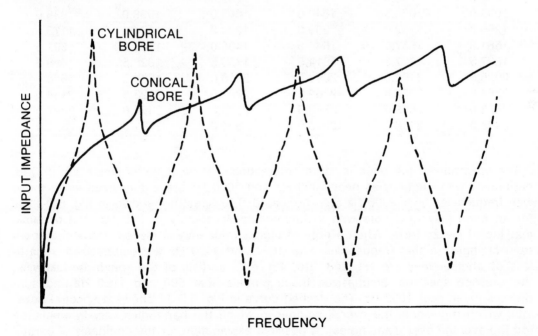

Fig. 6.D-10. Shift of resonance frequencies when a cylindrical tube is converted to a conical tube.

Note that two things have happened. A cone having a length equal to the cylinder has its fundamental resonance an octave higher (at $2f_1$ rather than at f_1). Secondly, the resonances of the cone are seen to be multiples of the new fundamental frequency ($2f_1$). If we call $2f_1 = f'$, then the resonances of the cone are given by f', $2f'$, $3f'$, etc. Since they are the same resonant frequencies as for a cylinder *open* at both ends, we can say that in some respects a cone acts like an open-open cylindrical

245

tube. Since the oboe and bassoon are basically conical instruments, they will have resonances which are multiples of the fundamental. The actual instruments, however, can only be approximately represented as a cone. In actuality the simple cone is modified in the following ways: (1) the cone is truncated so that the reed can be attached; (2) the bore consists of several truncated cones, each conical to a different degree, joined end to end; (3) there is a noticeable discontinuity between the end of the bore and the reed staple; (4) there are closed tone holes; and (5) the reed cavity has a larger diameter than the bore. The results of these sequential changes are summarized in Table 6.D-1. They consist of the following: (1) the resonances are slightly stretched; (2) the fundamental resonance is lowered, but the upper resonances are raised slightly; (3) all the resonant frequencies are raised slightly (because the area at the top of the cone is decreased slightly); (4) all the resonant frequencies are lowered slightly (by increasing the effective cross-sectional area); and (5) the high-frequency resonances are lowered much more than the lower ones, which somewhat compensates for the stretching of the higher modes due to the truncation.

Table 6.D-1. Natural mode frequencies (Hz) of different approximations to an oboe. (From Plitnik, 1972)

One joint	Five joints (continuous)	Discontinuity at staple	Closed holes	Reed cavity	Losses
256.0	241.0	246.5	234.0	230.0	230.0
513.0	517.0	523.0	504.0	481.0	480.5
772.0	793.0	788.0	762.0	708.0	707.5
1033.0	1058.5	1049.0	1003.0	938.0	938.0
1296.0	1302.0	1289.0	1244.0	1174.5	1175.5
1561.0	1557.0	1543.0	1496.0	1388.5	1391.5
1827.5	1827.5	1814.5	1750.5	1620.5	1626.5
2095.0	2102.0	2093.0	2018.0	1860.0	1868.0
2364.0	2372.5	2369.5	2281.0	2123.5	2134.5
2633.0	2638.0	2640.5	2535.0	2384.0	2397.0
2903.0	2906.0	2915.0	2800.0	2631.0	2645.0

The spectrum of the oboe is rather interesting. For most notes which are played, harmonics with frequencies near 1000 Hz and 3000 Hz are emphasized while those with frequencies near 2000 Hz are attenuated. The spectrum shown in Fig. 6.D-11 is for an oboe playing a note with a fundamental frequency of 250 Hz at a dynamic marking of mezzo forte. When a note of higher frequency is played, the entire spectrum changes so that frequencies near 1000 and 3000 Hz are emphasized, regardless of whether they are 1st, 2nd, 3rd, 4th, etc., partials of the spectrum. Likewise, the bassoon spectrum emphasizes those partials near 500 and 1150 Hz and suppresses those near 1000 Hz. The dashed curve in Fig. 6.D-11 shows a spectral envelope which represents the overall manner in which the harmonics vary in amplitude (on the average over many notes). A possible explanation for the emphasis of certain frequency regions in both oboe and bassoon spectra is the manner in which stretched resonances of the truncated cone interact with the harmonics of the reed. When the harmonics of the reed, which are controlled primarily by the first few resonances of the bore, differ substantially from the upper resonances of the bore—a situation which occurs because the resonances are stretched—the harmonics are attenuated. Apparently this takes place (for most notes) at a frequency in the vicinity of 2000 Hz.

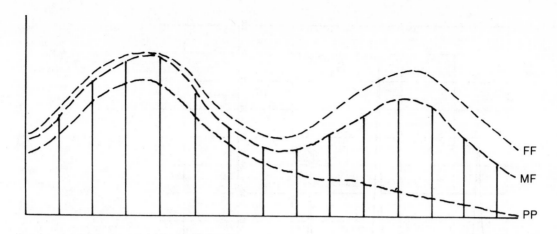

Fig. 6.D-11. Oboe spectrum for a 250 Hz tone played at a dynamic marking of mf. The dashed curves show approximate spectral envelopes for tones played at dynamic markings of pianissimo, mezzo forte, and fortissimo. (After Plitnik, 1972)

Now consider the effect of playing an oboe at dynamic markings of pianissimo, mezzo forte, and fortissimo. As with the clarinet, the oboe reed action is more abrupt for higher levels, thus adding more high-frequency energy to the spectrum. The dashed curves in Fig. 6.D-11 show how the spectral envelopes of an oboe vary with dynamic level. Note that the high-frequency components increase substantially as the dynamic playing level increases, while the lower frequencies change very little.

The Saxophone

A member of the saxophone family consists of a single mechanical reed fitted onto a conical tube which ends in a slight flare. Although the saxophone is a conical bore instrument, its action is most similar to that of the clarinet. In one detail, however, the saxophone is different from the other mechanical reed instruments; the diameter of the bore at the reed end is relatively large. This means that the input acoustic impedance will be fairly small; so the reed will be less tightly coupled to the bore than is the case for the other mechanical reed instruments. With a looser coupling, the saxophone reed, although still controlled by the bore, will have a somewhat faster attack than its cousins. (It does not take as long for the vibrating reed to reach its steady state.)

Impedance curves for two notes of an alto saxophone appear in Fig. 6.D-12. We can see that there are fewer resonances than for the other conical instruments (only two for C_6). Also, the two lowest resonances are, over most of the instrument's range, exactly one octave apart. This instrument will thus be observed to have a fairly uniform timbre throughout its range.

History and Recent Developments

Although the history of reed instruments is a long one, the common woodwind instruments of the mechanical reed variety are of more recent origin than is often realized. The oboe was invented during the latter part of the seventeenth century, and although it possessed only three simple keys, it was much closer to the present instrument than to earlier instruments known by the same name. The bassoon, which developed at about the same time as the oboe, was conceived as a bass to the oboe (sounding a twelfth lower in pitch) but with a tonal character all its own. Because of its large size, the bassoon is bent back on itself and the fingerholes are drilled at angles to avoid impossible side finger spacings. The forerunner of the English horn was the tenor oboe, but the English horn soon evolved its own unique tone color and can no longer be considered a tenor oboe. The globular bell on the English

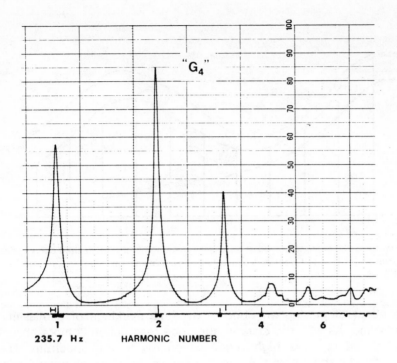

235.7 Hz   HARMONIC NUMBER

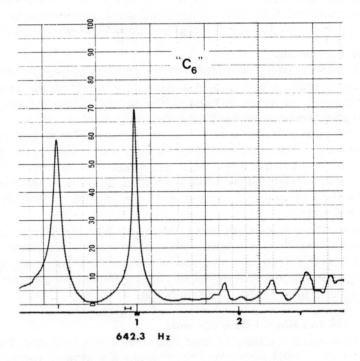

642.3 Hz

Fig. 6.D-12. Impedance curves for a saxophone. (From Backus, 1974)

horn has a profound visual effect on the instrument but contributes a changed tone color only to the lowest notes.

The clarinet was invented at the beginning of the eighteenth century, some forty or fifty years after the invention of the oboe. The precursors to the clarinet, although using a cylindrical tube, had a very limited range—the clarinet was invented when it was realized that a register key could be added which would extend the scale into the second register (the twelfth).

The next phase in the development of the mechanical reed instruments occurred during the nineteenth century when more keys were added and the key mechanism itself underwent considerable change and improvement. The oboe also underwent some minor changes to its bore, but the bassoon was completely reworked because of the unevenness and instability of some notes of the scale. During the mid-nineteenth century, the saxophone was invented as a reed instrument loud enough to provide a middle section to the military band. By the end of the nineteenth century the instruments had become more-or-less standardized into the familiar forms of today.

The resonances of an actual oboe are more complex than the descriptions earlier in this section might indicate. However, by means of good measuring techniques (Backus, 1974; Benade, chapter 9) and computer calculations (Plitnik, 1972) it has become possible to study mechanical reed instruments in some detail.

The lower plots of Fig. 6.D-13a-d show the results of using a digital computer to calculate the input impedance curve of an oboe from the actual physical dimensions of the instrument for four different notes. The upper plots of Fig. 6.D-13 are the actual measured input impedance for the same four notes. You will notice that there is a remarkably good agreement between the calculated and the measured impedance curves. The computer is able to predict almost all of the fine detail of these curves. By examining the impedance curves of an instrument we can tell a great deal about the tone quality and playing behavior of the instrument. Notice, for instance, that the low B (Fig. 6.D-13a) had many impedance peaks at approximate multiples of the frequency of the first peak. Since this plot resembles the impedance plot of a brass instrument, we expect a coarse and brassy tone, which is in fact the case for the lower notes of an oboe.

For a good, clear oboe tone one needs three impedance peaks which are at harmonics of the playing frequency. The E of Fig. 6.D-13b has its third peak slightly flat, and this turns out to be a rather bad note on this particular oboe. You may also note that above 1200 Hz waves in the oboe propagate down the entire length of the in-

(a)

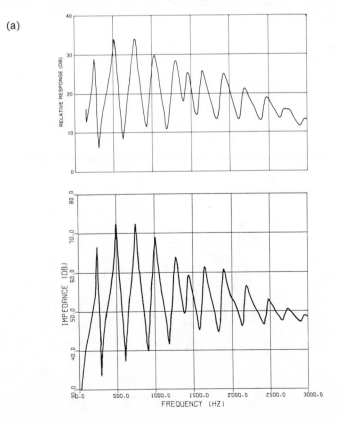

(Cont.)

(Cont.)

(b)

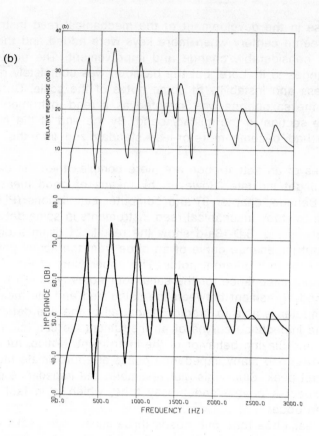

(c)

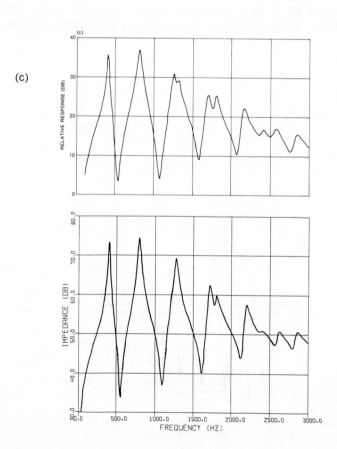

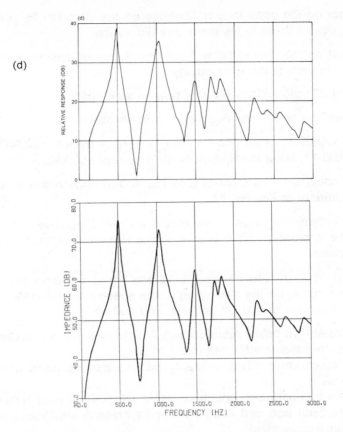

(d)

Fig. 6.D-13. Input impedance curves for different notes on an oboe: (a) B3, (b) E4; (c) G4, (d) B4. The upper curve in each case is an experimentally measured curve; the lower curve in each case is calculated numerically. (The experimental curves are courtesy of A. H. Benade.)

strument instead of being radiated. This additional active air column leads to a considerable irregularity in the height, shape, and placement of the peaks. Fig. 6.D-13c shows a fairly good note, G. Note that the cut-off frequency in this case occurs around 1600 Hz. The final curve, Fig. 6.D-13d for B, also shows a typical pattern for a fairly bad note. The cutoff frequency occurs around 1400 Hz, which results in the third peak being all but obliterated. We thus expect a fairly weak tone, which, as every oboist knows, is the case for B4.

Being able to compute the input impedance for each note, one could presumably improve a mediocre oboe by predicting what changes should be made to the bore and the tone holes, thus saving experimentation on the actual instrument. Since the input impedance can now be predicted with reasonably good accuracy, it may be possible to optimize the placement and size of each tone hole, as well as the size and shape of the bore, so as to produce an instrument having excellent tone quality throughout its entire range. The answer to this question must await the results of future research.

Exercises

1. Is the spectrum of a reed instrument primarily due to reed type (single or double) or bore type (cylindrical or conical)?

2. Refer to Fig. 6.D-4 to answer the following questions. Will the reed tend to open or close at c? At a?

3. Suppose, in the above exercise, that the tube is lengthened. How will this affect the number of reed vibrations in each second? If the tube is shortened?

4. What effect do the open-tone holes have on the vibration frequency of the reed? Does the size of these holes make any difference?

5. What effect do the closed holes have on the overall spectrum of the instrument? How does this affect the tone quality?

6. What is the function of the register hole in a clarinet?

7. What is the function of the register hole in an oboe?

8. Draw the spectrum of an oboe playing a note of frequency 500 Hz and compare it to Fig. 6.D-11. Draw the spectrum for a note of 1000 Hz.

9. Draw the spectrum of a clarinet (see Fig. 6.D-8) playing a note of frequency 250 Hz and compare to Fig. 6.D-11.

10. From the information given in the text, construct (approximately) the spectral envelopes for a bassoon playing at dynamic markings of pianissimo, mezzo forte, and fortissimo.

11. Sketch the approximate spectral envelope for a bassoon playing mezzo forte tones. Then draw, on the same graph, the spectrum for a bassoon playing notes having frequencies of 100 Hz, 250 Hz, and 500 Hz.

*12. The spectra shown below resulted from (a) clarinet and (b) saxophone tones.
   a. What is the fundamental frequency of each?
   b. Which spectrum exhibits primarily odd harmonics, particularly at lower frequencies?
   c. Does the odd-harmonic spectrum occur because of reed type or bore type?
   d. Why are both odd and even harmonics present with about equal amplitudes at higher frequencies?

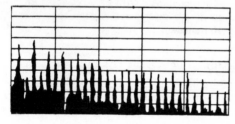

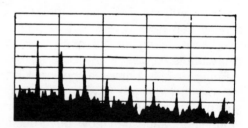

*13. a. Decide on the gross features of a wind instrument that you wish to construct.
   b. By consulting appropriate sources determine the physical properties of your instrument (such as length, diameter, fingerhole positions, bore shape, etc.) so that it will produce the desired frequencies.
   c. Construct the instrument.
   d. Take the instrument into the laboratory and make measurements to determine how well you have achieved your design goals.

Further Reading
Backus, chapter 11
Benade, chapters 21, 22
Culver, chapters 12, 13
Kent, part III

Backus, J. 1961. "Vibrations of the Reed and the Air Column in the Clarinet," J. Acoust. Soc. Am. *33,* 806-9.

Backus, J. 1963. "Acoustical Investigation of the Clarinet," Sound *2,* 22-25.

Backus, J. 1974. "Input Impedance Curves for the Reed Woodwind Instruments," J. Acoust. Soc. Am. *56,* 1266-79.

Bate, P. 1956. *The Oboe: An Outline of its History, Development and Construction* (Ernest Benn Ltd., London).

Benade, A. H. 1960. "On the Mathematical Theory of Woodwind Finger Holes," J. Acoust. Soc. Am. *32,* 1591-1608.

Benade. A. H. 1960. "The Physics of Woodwinds," Sci. Am. *203* (4), 145-54.

Benade, A. H., and D. J. Gans. 1968. "Sound Production in Wind Instruments," Ann. N.Y. Acad. Sci. *155,* 247-63.

Benade, A. H. 1959. "On Woodwind Instrument Bores," J. Acoust. Soc. Am. *31,* 137-46.

Langwill, L. G. 1965. *The Bassoon and Contrabassoon* (W. W. Norton, N.Y.).

Plitnik, G. R. 1972. "The Calculation of Input Impedance for Double-reed Wind Instruments, and the Time Variant Analysis and Synthesis of their Tones using Digital Computer Techniques" (Ph.D. dissertation, Brigham Young University).

Rendall, F. G. 1971. *The Clarinet: Some Notes Upon Its History and Construction,* 3rd ed. (W. W. Norton, N.Y.).

Strong, W., and M. Clark. 1967. "Synthesis of Wind-Instrument Tones," J. Acoust. Soc. Am. *41,* 39-52.

Strong, W., and M. Clark. 1967. "Perturbations of Synthetic Orchestral Wind Instrument Tones," J. Acoust. Soc. Am. *41,* 277-85.

Audiovisual

1. *Woodwind Instruments* (3 3/4 ips, 2 track, 30 min, UMAVEC)
2. *Clarinet* (10 min, color, 1968, SF)
3. *The Clarinet, Part 1* (20 min, color, 1965, OPRINT)
4. *The Clarinet, Part 2* (17 min, color, 1965, OPRINT)

Demonstrations

1. Slide clarinet consisting of one cylindrical tube to slide over a smaller cylindrical tube equipped with a clarinet mouthpiece
2. Clarinet mouthpiece on cylindrical, conical, decreasing-flare, and increasing-flare tubes

6.E. AIR-REED INSTRUMENTS

Lip reed and mechanical reed instruments discussed in the preceding sections depend on the alternate opening and closing of a physical reed to interrupt air flow. Air-reed instruments depend on the action of an air stream as it moves toward a sharp edge attached to a pipe to provide an oscillating air flow. Representative air-reed instruments are the flute (see Fig. 6.E-1), the recorder, and flue organ pipes. The following observations can be made about these instruments:

1. The excitation mechanism is an air stream moving past a sharp edge that oscillates into and out of the pipe to which the edge is attached.
2. The pipe characteristics largely determine possible vibration frequencies; parameters of the air stream determine which of the vibration frequencies is produced.
3. Increasing the blowing pressure may cause a pipe to sound in a higher mode.
4. All harmonics are present for the flute and open organ pipes; predominantly odd harmonics are present for closed organ pipes.

Fig. 6.E-1. Sketch of a flute

Air Stream and Pipe Combination

When a stream of air is directed through a narrow slit toward a sharp edge attached to a pipe, the air stream does not simply divide into two parts, but oscillates from flow into the pipe to flow out of the pipe, as shown in Fig. 6.E-2. The air stream causes the air in the pipe to vibrate, and the pipe in turn determines possible vibration rates of the air stream, resulting in the sequence labeled a, b, c, d in Fig. 6.E-2. In part (a) the air stream is shown entering the pipe, where it produces a positive pressure which moves toward the other end of the pipe. The positive pressure, however, tends to force the air stream out of the pipe, as seen in parts (b) and (c). The positive pressure meanwhile travels to the open end of the tube, where it is reflected as a negative pressure. As the negative pressure arrives back at the blowing end it lets the air stream enter the pipe again, as shown in parts (d) and (a), and the motion repeats. When a pipe is made to vibrate in this manner, the vibrator frequency is essentially the fundamental frequency of the pipe. As the pipe is blown harder, thus increasing the speed of the air stream, the frequency does not change, because the resonance of the pipe constrains the vibration frequency of the air stream to remain essentially the same. However, when the air speed is increased sufficiently the vibration frequency will suddenly jump to the next natural vibration frequency of the pipe. For an open-open pipe the next mode of vibration is twice the fundamental frequency; so the instrument now sounds the octave. The production of higher resonance modes in this manner is known as overblowing.

Possible pipe modes are determined by whether it is open or closed. Pipes associated with air-reed instruments may be regarded as open at the blowing end, which means that pressure nodes and air flow antinodes exist here. Pipes open at the other end are termed open pipes while pipes closed at the other end are called closed pipes. Because pressure nodes exist at the blowing end of air-reed instruments—rather than pressure antinodes, as with the lip and mechanical reed instruments—the

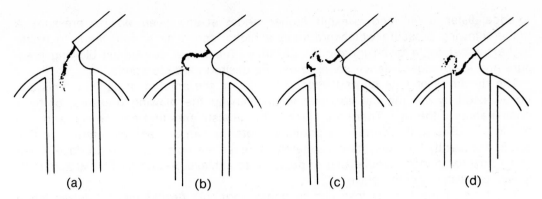

Fig. 6.E-2. Interaction of air stream and pipe open at both ends. (After Coltman, 1968.)

natural frequencies occur when the input impedance at the blowing end is a minimum. For example, if a clarinet reed is attached to a cylindrical tube, the predominant modes will have relative frequencies of 1, 3, 5, etc. corresponding to the peaks in the solid curve of Fig.6.D-5. If a flute embouchure is attached to the same cylindrical tube, the predominant modes will have relative frequencies of 2, 4, 6, etc., corresponding to the valleys of the solid curve of Fig. 6.D-5. The same tube used as a clarinet or flute will sound an octave higher as a flute with all harmonics present.

The travel time of the air stream from the slit to the sharp edge determines which of the natural pipe modes is excited. The travel time is determined by the distance between the slit and edge and the speed of the air stream. The speed of the air stream depends mainly on blowing pressure. Flute and flue organ pipes differ in terms of the flexibility of controlling the travel time from slit to edge. The flute player can vary this time appreciably by changing his blowing pressure or by changing his lip position relative to the edge. A flue pipe operates with a fixed pressure and a fixed slit-to-edge distance and thus a fixed travel time.

The Flute

Flutelike instruments date back to antiquity. Sometime in the early history of this instrument it was noticed that small finger holes along the bore of the instrument allowed one to change the pitch by uncovering the holes, thus effectively shortening the length of the tube. The primitive instruments utilized six holes; by successively uncovering these holes one could play a diatonic scale. By overblowing, a second octave may be added. The modern flute is a further development of this primitive instrument. It consists of a metal or wood tube about 66 cm in length. The inside bore is about 1.9 cm in diameter with an oval embouchure (mouth hole) near one end of the instrument. The flute has about 16 openings which are controlled by a series of levers connected to pads which cover the holes; these serve to change the effective length of the instrument. The instrument behaves very much like a flue organ pipe for the production of individual notes, except for a much greater flexibility.

The natural frequencies of a cylindrical flute are approximately determined by the distance between the embouchure and the first open tone hole and the speed of sound in air. The sound speed is affected by the air temperature and the relative proportions of carbon dioxide present. When a flute is excited externally by a loudspeaker and the resonances are measured it is found that there is some degree of stretching of the resonances, i.e., the resonance frequencies are more widely spaced than a harmonic series. When the flute is being blown, however, the stretching is less pronounced, partly because of the air flow and the way the player positions his/her lips.

As mentioned previously, the particular mode which is excited as the fundamental depends on the air stream travel time from slit to edge. The travel time should be equal to about one-half period of the mode to be excited. For the flute, a shorter lip-to-edge distance will tend to excite higher modes at any given blowing pressure. A higher blowing pressure which produces a higher-speed air stream tends to excite higher modes for a given lip-to-edge distance. The player can adjust blowing pressure and lip-to-edge distance so as to get the desired mode to sound.

The dynamics of flute playing depend largely on the amount of air flow. This can be adjusted by varying the lip opening, by varying the blowing pressure, or by a combination of the two. There are limits to the adjustments that can be made for dynamic purposes while sounding a particular note, as can be seen by referring to Fig. 6.A-8. Apparently flute players consistently increase the lip opening to produce greater air flow for louder outputs. Use of pressure to achieve greater air flow is apparently less consistent among players.

The tone of the flute is less rich in upper harmonic development than are oboe and trumpet tones. Fig. 6.D-3 shows spectra for three different flute tones, produced by four players. Note that the spectra are characterized by a formant region around 550 Hz for *forte* tones and a lower value for *piano* tones. Note also that the *forte* tones show a significantly greater upper partial development than the *piano* tones.

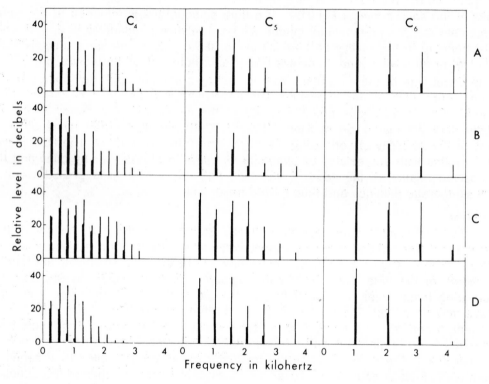

Fig. 6.E-3. Harmonic analysis of *forte* and *piano* notes produced by A, B, C, and D. (From Fletcher, 1975)

Although we have considered the flute to be a simple tube open at both ends, it is considerably more complicated than this. At the mouth end the hole is partially covered by the player's lip, and at the fingered end the opening (for all but the lowest note) is an open tone hole with a key pad directly above it. Furthermore, the small volumes added by the closed tone holes and the portion of the tube beyond the open tone holes all have an effect on the flute resonances. It is important to have several open tone holes beyond one determining the fundamental frequency so that the tube is effectively terminated; otherwise an appreciable standing wave can exist beyond the first open tone hole. This can be seen by comparing Figs. 6.E-4a and 6.E-4b, where the curves are proportional to the air flow of the mode being sounded. Note that there is a pronounced standing wave beyond the open hole when a single fingerhole is open (Fig. 6.E-4b), but not beyond the first open hole when several are open (Fig. 6.E-4a).

256

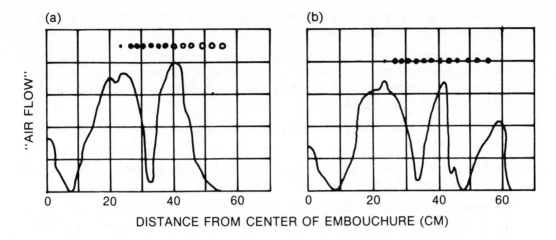

Fig. 6.E-4. "Air flow" at different positions in a flute bore with (a) a series of closed fingerholes followed by a series of open fingerholes to produce a 688 Hz tone and (b) a single open fingerhole to produce a 675 Hz tone. (After Young and Loughridge, 1936.)

Additional features are necessary in the design of good-quality flutes. Most of them are concerned with achieving better intonation over the range of the instrument. These include a contracted head joint to get more nearly harmonic frequency relations for the blown flute. Also, a proper cork to embouchure distance is required to help obtain more nearly harmonic frequencies. All of these features must be considered in terms of the instrument bore and player practice.

Organ Flue Pipes

An organ flue pipe functions in a manner similar to that shown in Fig. 6.E-2. The slit which produces the narrow stream of air required is formed by the passage way between the lower lip and the languid (see Fig. 6.F-4a). A sound is produced when this slit of air impinges upon the upper lip, as described previously. The final tone of a pipe is greatly affected by the process of voicing, whereby the height of the mouth and the width of the slit are adjusted. The pipes are tuned by adjusting the length of the pipe at the upper end.

The material from which the pipes are constructed has long been considered to exert considerable influence on the tone produced. For hundreds of years organ builders have constructed bright-toned pipes from tin, while lead was used to produce a duller sound. Most of the pipes were constructed from a mixture of lead and tin, but it was considered that the higher the percentage of tin, the brighter the tone. Recent scientific experiments, however, have laid this myth to rest (Boner and Newman, 1940; Backus and Hundley, 1966). Not only does the wall material not affect the tone quality, but the vibrations of the material are too small even to be measured. The flute pipes of an organ may either be open or stopped at the top and both wood and metal are used in constructing these pipes. As you recall, an open flute will have all harmonics present, while a stopped flute will contain primarily odd harmonics.

In the next section we will present considerably more detail on the pipe organ as an instrument and the tonal qualities of various organ pipes.

Exercises

1. A flute can be considered an open-open cylindrical pipe. What harmonics are present? Explain.

2. Sketch diagrams similar to Fig. 6.E-2 showing the behavior of an air stream in a closed tube. What is the fundamental frequency of the tube if its length is 200 cm? How does this frequency compare with that of an open tube of the same length?

257

3. Are flue pipes in the organ open-open or open-closed? What harmonics do they produce? Why?

4. When an open-open flue pipe sounding a frequency of 220 Hz is overblown, what frequency do you hear? When an open-closed flue pipe having a frequency of 220 Hz is overblown, what frequency do you hear? Explain why the result is different in these two cases.

5. Does the playing frequency of a flute increase or decrease with an increase of temperature? Explain.

6. If a flutist inhaled helium gas and then played a note on the flute, would it sound at its normal pitch? If not, how would it differ? Why?

7. For organ flue pipes, the slit-to-edge distance is kept as short as possible (for a given blowing pressure) without overblowing, in order to produce a brighter sound. Why does a shorter slit-to-edge distance produce a brighter sound?

8. If an organ flue pipe is voiced on a particular wind pressure, will the pipe speak properly on half that pressure? Why or why not?

9. If the material from which organ flue pipes are made does not affect their sound, why do wooden and metal pipes sound quite different? Could they be constructed to sound identical? Explain.

10. The spectrum shown is for a flutephone tone of about 800 Hz. Does it exhibit characteristics of an open pipe or a closed pipe? How can you tell?

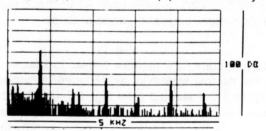

Further Reading

Backus, chapter 11
Benade, chapter 22
Culver, chapters 12, 13
Kent, part III
Olson, chapters 5, 6
Wood, chapter 8

Backus, J., and T. C. Hundley. 1966. "Wall Vibrations in Flue Organ Pipes and Their Effect on Tone," J. Acoust. Soc. Am. *39* 936-45.
Boner, C. P., and R. B. Newman. 1940. "Effect of Wall Material on the Steady State Acoustic Spectra of Flue Pipes," J. Acoust. Soc. Am. *12*, 83-89.
Coltman, J. W. 1968. "Acoustics of the Flute," Physics Today *21*, no. 11, 25-32.
Fletcher, N. H. 1975. "Acoustical Correlates of Flute Performance Technique," J. Acoust. Soc. Am. *57*, 233-37.
Young, R. W., and D. H. Loughridge. 1936. "Standing Sound Waves in the Boehm Flute Measured by the Hot Wire Probe," J. Acoust. Soc. Am. *7*, 178-89.

Audiovisual

1. *The Flute* (22 min, color, OPRINT)

Demonstrations

1. Clarinet tube blown with flute embouchure or clarinet mouthpiece.
2. Comparison of open and stopped organ flue pipes
3. Metal organ pipes (set of four) with adjustable jet, Sargent Welch #3280

6.F. THE PIPE ORGAN

If not the king of instruments, the pipe organ may at least be considered the dinosaur of the musical instrument world, since it has become (after the baroque era) an unwieldy giant which may have outlived the climatic conditions in which it prospered. This instrument is an assemblage of hundreds of resonant tubes of various shapes and sizes (see Fig. 6.F-1), each pipe producing only one pitch and one tone quality.

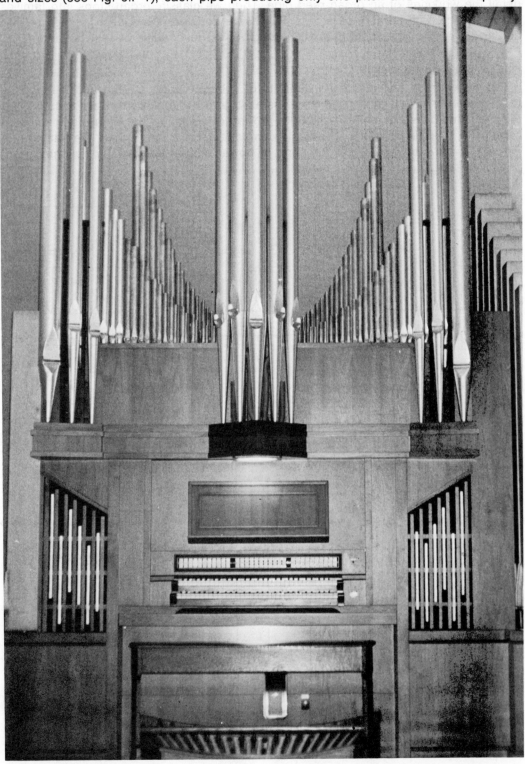

Fig. 6.F-1. The organ at Christ Lutheran Church, LaVale, Maryland. (Designed by G. R. Plitnik.)

The majority of the pipes found in organs are of the flue pipe variety, although about 20 percent of the pipes are usually of the mechanical reed variety. The pipes sit on a wind chest containing a set of valves, each of which admits air into the appropriate pipe when the valve is energized from the console. Each desired tone quality consists of a set of pipes, graduated in size (and hence pitch) through the entire manual or pedal division. In this section we consider several types of organ action; then we detail the workings of an organ reed pipe. Then we classify organ stops into families and relate the families to pipe scales. Finally we consider the speech of organ pipes and an analysis of the spectral qualities of pipes representative of each tonal family.

Construction and Action

The typical modern pipe organ consists of a large variety of flue and mechanical reed pipes controlled by two or more manuals and a pedalboard. Each manual and the pedalboard usually control separate groups of pipes; so the organ is in reality a composite of several organs. The basic components of a pipe organ, diagrammed in Fig. 6.F-2, are a blower to provide wind, a power supply (for electric actions), regu-

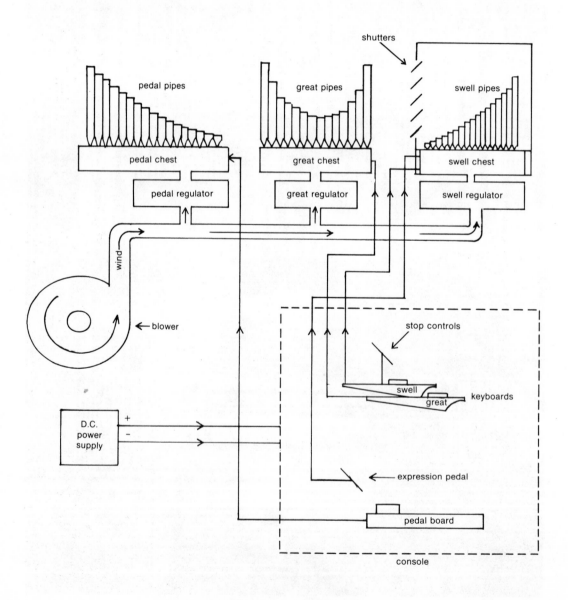

Fig. 6.F-2. Schematic diagram of a typical pipe organ.

lators to maintain constant pressure, wind chests (which hold the pipes), the pipes, shutters to control the volume (swell division), and a console to control it all. The upper manual, called the Swell, controls the pipes which are enclosed behind a set of movable shutters. When the shutters are open the sound can escape; by closing the shutters (by means of the swell pedal) the organist can reduce the volume to a mere whisper. The lower manual controls the pipes of the Great organ; these pipes are not usually enclosed.

The wind chest contains a set of valves which when opened will admit air into the appropriate pipe. There are three types of windchest actions in use today. The oldest type, used (in some form) for nearly a thousand years, is the tracker (or mechanical) action. No power supply is needed with this action because the keyboards are connected directly to the chests by a complicated set of mechanical linkages (called trackers). The second type of action, developed in the early decades of this century, is called electropneumatic because small electromagnets are used to exhaust pneumatic bellows. The final type of action is the direct electric action; in this system an electromagnet opens a pipe valve directly. Fig. 6.F-3 is a representation of each type

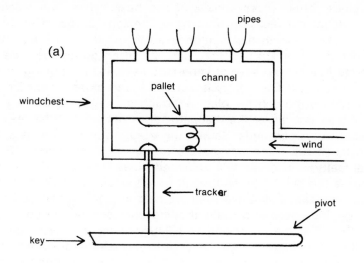

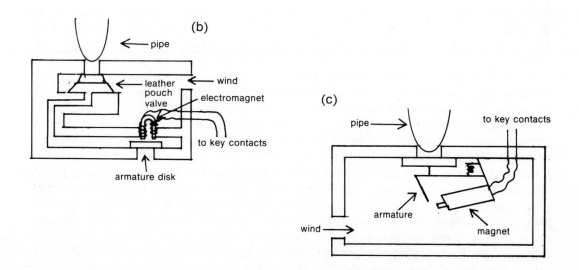

Fig. 6.F-3. Windchest actions: (a) simplified tracker; (b) electropneumatic; (c) direct electric.

of action. A careful inspection of the figure shows that for the tracker action the air is admitted into the pipes when a valve (called the pallet) is pulled down by pressing the key. For the type of electropneumatic action shown, air is admitted to the pipe when the leather pouch valve is exhausted via lifting the armature disc by the energized electromagnet. The direct electric valve opens whenever the armature is attracted by the energized magnet.

On considering the different types of action we have not considered the effect of the stop keys. The stops are small tablets, usually located above the top manual. Each stop controls a set of pipes called a *rank*. The pipes are graduated in pitch so that there is one pipe for each manual (or pedal) key, but all the pipes of one rank possess the same quality. When one key is depressed, one pipe will sound for every stop which is on. Thus, the keyboard controls various sets of pipes which can be selected according to the tone quality or volume desired.

Organ Reed Pipes

Having examined the basic construction of the organ, let us now direct our attention to the pipes. Organ pipes consist of two basic types: flue pipes (air reed) and reed pipes (mechanical reed). In both types the tone is the result of a vibrating air column; the difference is in the method used to excite the air column. We have already considered (in section 6.E) the operation of an organ flue pipe. The reed pipe differs from the flue pipe in several important ways: (1) a vibrating brass reed modulates the air stream, (2) the energy is applied at the closed end of the pipe, and (3) all of the air passes through the pipe. Although the organ reed is a mechanical reed instrument, the air column of the resonator does not control the reed as is the case for mechanical reed woodwinds. Rather, the vibrating organ reed is influenced by the resonator in much the same way as are the lip-reed instruments. Fig. 6.F-4 shows the essential features of organ flue and reed pipes. The sound-producing section of the reed pipe is the curved brass tongue (reed) which beats against the shallot at a certain frequency. The shallot and reed are held in place in the lead block by a small wooden wedge, the entire arrangement being enclosed by the boot. The shallot is a brass tube, closed at the lower end and flattened on the reed side. A long narrow hole in the flat side leads through the boot to the resonator, which is the only route by which air can escape from the boot. A sliding wire spring presses the tongue firmly against the shallot, which changes the vibrating length of the tongue and is thus used for tuning. When air enters the bottom of the boot it begins to escape through the opening under the curved reed tongue. But, by Bernoulli's Law, the reduced pressure of the rushing air pulls the tongue against the shallot, thus closing the opening and interrupting the escape of air. Since the tongue is elastic, however, and since a positive pressure pulse soon returns from the resonator, the reed again pulls away from the shallot, allowing more air to escape and thus repeating the entire process. The puffs of air are admitted to the resonator, thus setting up a standing wave. The lower end of the resonator will be a pressure antinode. If the resonator is cylindrical, the odd harmonics will be favored, as with the orchestral clarinet. A conical resonator will, as discussed in section 6.D, reinforce all harmonics, and for this reason most reed pipe resonators are conical.

Although the natural frequencies of the brass tongue and the resonator are generally about the same, there is a certain degree of latitude. When the tongue is lengthened it not only vibrates at a lower frequency but, being more flexible, also vibrates with greater amplitude. The sound of the reed becomes louder and richer through production of more upper harmonics. Since the pitch can be brought back to its original value by changing the length of the resonator, we have a method to regulate the tone of a reed pipe. If we want to change the loudness without any alteration of tone quality we have to modify the curve of the tongue (a job which requires a considerable amount of skill).

262

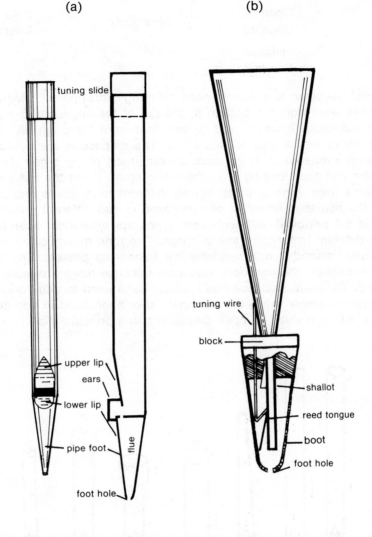

(a)   (b)

tuning slide

upper lip
ears
lower lip

pipe foot

flue

foot hole

tuning wire

block

shallot

reed tongue

boot

foot hole

Fig. 6.F-4. Parts of organ pipes: (a) flue; (b) reed.

Reed pipes are extremely sensitive. The curve of the tongue must be so perfectly graduated that it will roll smoothly down the shallot. Nonuniformities of the tongue or dust on the shallot will cause a most unpleasant sound when the reed is played.

Classification of Organ Stops

The separation of organ stops into flue pipes and reed pipes forms the basic classification of the instrument's tonal resources. Within each of these main divisions there are several important tonal families, as summarized below in Table 6.F-1, which we will consider in more detail.

Table 6.F-1. Classification of Organ Stops.

Flue Tone		Reed Tone	
Diapason	Open Principal	Classic	Chorus Regal
Flute	Open Stopped	Romantic	Solo Effect
String	Imitative Non-Imitative		

The word *diapason* is a Greek word meaning "through all." Many centuries ago not all ranks went from the bottom to the top of the keyboard, but were incomplete sets. The diapason, however, was a rank that went "through all" the notes of the keyboard—there was a pipe for each key. The diapasons are cylindrical open-open pipes having a tone which is the most characteristic of the organ. This tone, which is quite unlike that produced by any other instrument, forms the basis of the organ ensemble (rank upon rank of pipes having different tone colors and pitches). We can consider the diapason family to be composed of two different groups: the open diapason and the principal. Although both types are cylindrical open-open pipes, they are quite different in function and purpose. The tone quality of an open diapason is "round" and "smooth," with relatively few harmonics present. The tone quality of a principal is brighter than an open diapason because there are more partials present in the tone. The open diapason has typically been used to add "gravity" and volume to the organ ensemble, while the principals have been used to add clarity and brightness. Figs. 6.F-5a-b show an open diapason and a principal pipe.

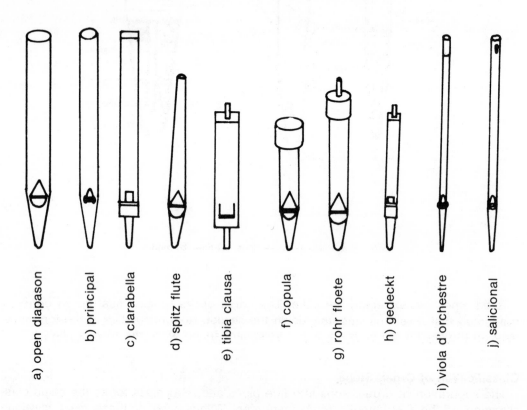

a) open diapason b) principal c) clarabella d) spitz flute e) tibia clausa f) copula g) rohr floete h) gedeckt i) viola d'orchestre j) salicional

Fig. 6.F-5. Flue pipes: (a-b) diapason; (c-d) open flute; (e-h) stopped flute; (i) imitative string; (j) non-imitative string.

The *flute* stops are those whose tone quality resembles somewhat the orchestral flute, although this is not their purpose. Organ flutes come in three different shapes (cylindrical, conical, and rectangular), which occur in each of the two tonal groups (open and stopped). The open flutes (see Figs. 6.F-5c-d) have a quality which is most similar to an orchestral flute, as would be expected from the fact that both instruments produce a tone by the same method and both have an open-open tube resonator. Since the open flute is an open-open resonator, all harmonics are present in the spectrum; typically, however, only the first three or four harmonics are prevalent. Sometimes the open flute pipes are made "harmonic." In this case, a small hole is drilled in the center of the pipe (approximately mid-way between the mouth and the tuning slide), which produces a pressure node at the center of the pipe. The pipe is thus caused to speak an octave higher—the length of the pipe is the same as the wavelength of the fundamental.

The stopped flutes, with their emphasis on odd partials, have relatively few harmonics in the spectrum. Typically, the stopped flutes are of rectangular cross-section, fitted at the end with an adjustable stopper (for tuning purposes). An illustration of wooden stopped flutes, the tibia clausa and the gedeckt, are shown in Fig. 6.F-5e and h. A capped metal flute, the copula, is pictured in Fig. 6.F-5f. The metal cap is adjustable so that the pipe can be tuned. One type of stopped flute, the rohr flute, consists of a cylindrical pipe fitted with an adjustable cap, but with an open-open tube of small diameter placed in the center of the cap. Although the large stopped pipe emphasizes the odd partials, the addition of the small "chimney" increases the amplitude of one of the upper partials. The length of the chimney is usually chosen so that the fifth or sixth partial is emphasized. A rohr flute is pictured in Fig. 6.F-5g.

The *string* family of organ stops slightly resembles the tone produced by the string section of the orchestra. Their tone quality can best be described as soft but rich in harmonic development. In many examples the upper harmonics have a greater amplitude than the fundamental. The imitative strings actually attempted to imitate their orchestral counterparts, insofar as this was possible with flue pipes, by producing a strident, "buzz-saw"-like tone. Such names as *'cello* and *viole d'orchestre* are common for the imitative string stops, such as the viole d'orchestre pictured in Fig. 6.F-5i. On the other hand, the nonimitative strings make no attempt to imitate orchestral instruments. Their tone tends to be placid and bland compared to the imitative strings because of less harmonic development. The salicional, pictured in Fig. 6.F-5j, is the best known example of this type of string pipe.

The reed pipes can be divided into two tonal families: the classic and the romantic. Classic reeds are the typical reeds found on most pipe organs today. The *chorus reeds* are used in ensemble combinations. When several ranks of chorus reeds (usually of different pitches) are present on the same manual, a reed chorus of unbelievable brilliance and beauty is created. In general, chorus reeds on the organ serve the same purpose as the brass instruments of an orchestra. Several typical chorus reeds are shown in Figs. 6.F-6a-d. Notice that each of these chorus reeds utilizes either a single or a double cone as a resonator.

The regal reeds, typically found only in larger pipe organs as a supplement to the chorus reeds, are characterized by being softer than the chorus reeds and often having very little fundamental energy but many harmonics. The attenuated fundamental occurs because fractional length resonators are often used. (In some cases the resonator may be only one-eighth or one-sixteenth of the fundamental wavelength.) Illustrations of regals are given in Figs. 6.F-6e-f.

The romantic reeds occur in two varieties: the solo reeds and the special effect reeds. The *solo reeds* are not useful in ensembles, but are chiefly valuable for solo passages. The solo reeds are analogous to the orchestral woodwinds, as can be seen from their names: clarinet, orchestral oboe, English horn, etc. Two examples of solo reeds are pictured in Figs. 6.F-6g-h. The special effects reeds are to the roman-

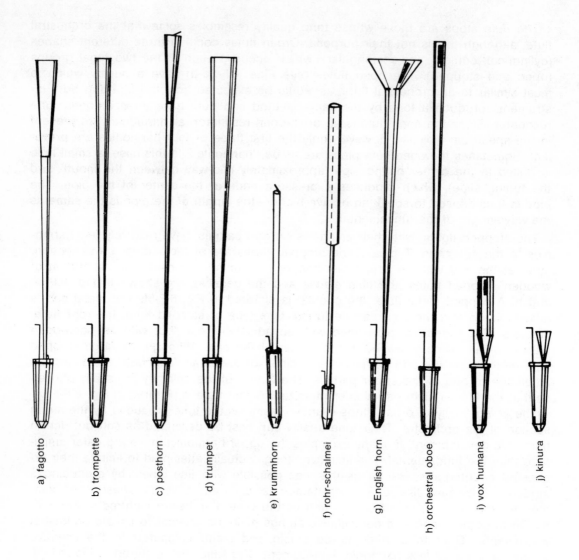

Fig. 6.F-6. Reed pipes: (a-d) chorus reeds: (e-f) regals; (g-h) solo reeds; (i-j) special effect.

tic reeds group as the regal reeds are to the classic reed group. Just as the regals can be used to color a baroque flue chorus, so can the special effect reeds be used to color romantically voiced flue-stops. The special effect reeds do not imitate any known sound; they produce an entirely unique, though not very useful, tone color. Like their distant cousins, the regals, the special effect reeds use fractional length resonators to produce a wealth of harmonics with an attenuated fundamental. The two best known special effect reeds, pictured in Figs. 6.F-6i-j, are the vox humana (the sound of which resembles that made by a cow dying in agony) and the kinura (which sounds remarkably similar to an angry bee trapped in a bottle).

The preceding classification scheme is entirely independent of pitch; on an actual pipe organ, ranks of pipes are present at many different pitch levels. An 8' manual stop is one whose lowest pipe has a pitch the same as an open-open pipe approximately 8' long. The third C from the bottom of the keyboard is then middle C. A 4' stop plays one octave higher while a 16' stop plays an octave lower. A 2' manual stop plays a pitch which is two octaves above middle C. The higher-pitched stops were found to add brightness to the ensemble by artificially adding upper partials. All the pitch levels so far mentioned can be grouped under one heading: foundation.

266

Foundation stops change the octave of a note without changing the note itself; in other words, "C" pipes, although at different pitches, always play from middle C on the keyboard. *Mutation* stops, on the other hand, do change the pitch characteristic—the pitch sounded has a different name from the key being depressed. The mutation stops are necessary to fill in the gaps in the harmonic series created by foundation stops of different pitches. As an example, suppose we depress middle C on the keyboard. When a 8' stop is flipped on, a pitch of middle C (C4) sounds. By adding a 4' stop a pitch one octave higher (C5) is added; this corresponds to the pitch ot the second harmonic of middle C. By adding a mutation stop (2 2/3') the pitch G5 is added to our ensemble: this is also the pitch of the third harmonic of C4. Next we add a 2' stop, which adds C6, the pitch of the fourth harmonic of C4. For the higher partials of the harmonic series a mixture stop (also a mutation) is usually provided. The mixture stop immediately adds from 2 to 6 ranks of pipes, each rank corresponding to some higher partial of C4. Mixture pipes are generally of diapason quality as they, along with their lower-pitched relatives, help form part of the diapason chorus.

Each group of pipes shown in Table 6.F-1 can be found at 16', 8', and 4' pitch levels. Flutes and diapasons are also typically present at higher pitch levels, while flutes, diapasons, and chorus reeds are sometimes also found at the 32' pitch level.

Pipe Scales and End Correction

The *scale* of a rank of pipes refers to the ratio of diameter to length for the lowest-pitched pipe of the set. Large-scale pipes (with a large cross-section) have a dominating fundamental frequency and few upper harmonics. Small-scale pipes (small cross-section in relation to length) have considerably more harmonic development. If we arrange the families of open-open flue organ pipes according to scale (from large to small), we get the following arrangement: wide scale flute, open diapason, principal, medium scale flute, nonimitative string, imitative string. This list then also corresponds to arranging the families in order of increasing harmonic development.

When the air within an open-open flue pipe is set into vibration, the maximum motion of the air does not occur right at the end of the pipe. Rather, because of an impedance mismatch between the tube and the outside air, the node of the enclosed pressure wave occurs a short distance above the pipe end. The same effect is much greater at the mouth of the pipe. The distance from the end (or mouth) of the pipe to the node of the fundamental pressure wave is termed the *end correction*. For an open-open cylindrical pipe, the end correction is approximately 0.6 times the radius at the open and 2.7 times the radius at the mouth (Ingersler and Frobenius, 1947). For the higher partials the end correction is proportionately less. Let us now consider how the end correction explains why small-scale pipes encourage, and large-scale pipes inhibit, upper harmonic development. Consider a tone with all harmonics of the fundamental tone having approximately equal amplitude. The harmonics actually present in the tone of a pipe will depend on how many of them are encouraged by the pipe resonator. For a large-scale pipe, only the first few harmonics are reinforced by the natural frequencies of the pipe. For the small-scale strings, the end correction is much smaller. Therefore, the discrepancy between the natural frequencies of the pipes and the harmonics is much less, and the upper partials are more strongly developed.

Speech Characteristics

In the previous subsection, we classified organ stops in several different tonal families: the classification is based primarily on the timbre of the pipes. There are other important characteristics, however, which determine the overall effect of any given rank of pipes. In addition to the loudness and the efficiency of a rank, the nature of the attack is very important. We will consider the attack time (which determines how promptly a pipe speaks) and the number of attack transients (which determines the smoothness of the attack). Since a vibrating reed emits air in a series of puffs there is little wasted energy. Flue pipes, on the other hand, are much less efficient: there is a continuous stream of air through the pipe. This hissing "wind noise" is often audible in the tone of the pipe.

The attack time of a reed pipe is usually one-half to one-fifth the attack time for a flue pipe of the same pitch (Nolle and Boner, 1941). In a flue pipe the tone "builds up" as the air within the pipe begins to vibrate; since a reed pipe is dominated by the vibrating brass tongue, it attains its full vibration almost immediately. In almost all nonpercussive instruments, the low tones have longer attack times than higher tones, as we discovered in section 6.A. Organ pipes are no exception to this rule. As you descend in pitch through any rank, the attack time increases. Experiments (Nolle and Boner, 1941) indicate that for any rank of flue pipes the attack time is approximately proportional to the wavelength. Also, the greater the mass of air contained within a pipe, the larger the attack time will be; particularly long attack times can be observed for 16' or 32' open-open pipes of large scale. The brass tongues of 16' or 32' chorus reeds sometimes have a small weight attached in order to increase the mass, and thus lower the frequency, without changing the stiffness of the tongue. The extra mass lengthens the attack time substantially, sometimes to such an extent that the usefulness of the stop is greatly reduced. George Plitnik, one of the authors of this book, owns a 32' chorus reed with weighted tongues: the attack time of the lowest notes exceeds two seconds! Obviously such a stop can be used effectively only for slowly moving pedal passages.

The nature of the attack of a flue pipe is dependent upon various details of the design and the voicing of the pipe. A small-scale imitative string pipe, for example, is voiced to yield an abnormally full compliment of upper partials. The techniques used tend to encourage the pipe to speak its second harmonic rather than the fundamental. A cylindrical "bridge" placed in front of the mouth, parallel to the flue opening, has been found to eliminate this problem without adversely affecting the desired spectrum. The voicing device which has the greatest influence on the nature of the attack is termed *nicking*. Nicking is the process of cutting narrow, vertical grooves along the front edge of the languid and the inner side of the lower lip, at right angles to the flue. Nicking increases the width of the air stream, securing a softer, smoother attack with fewer transients. When flue pipes have been voiced with heavy nicks we say that the voicing is "romantic." When only very light nicking (called "feathering") is used, the voicing is baroque, because nicking was an unknown technique during the baroque era. Since pipe organs are generally located in buildings with fairly long reverberation times, and since the pipe organ is played primarily as a legato instrument, baroque voicing of flutes and principals is considered to be highly desirable. The "percussive consonant" preceding the "vowel" of steady-state sound helps to separate succeeding tones and adds a clarity to the voice lines of polyphonic music which cannot be achieved with heavily nicked pipes. In the performance of romantic and popular music it is not essential to separate voices; rather, a mass of tone color is desired. The smooth attacks of heavily nicked pipes are considered to be conducive to producing such an effect.

Analysis of Tonal Qualities

We now consider the steady-state spectra of typical representatives of the pipe families given in Table 6.F-1. While such spectra are useful in order to understand the physical basis of the tonal families and to compare members of the same family, we must remember not to attach undue importance to these spectra. As we have seen, the steady-state spectrum is only one factor in the overall tonal effect of any pipe. Furthermore, the spectrum measured for an open-open pipe will vary considerably depending on whether the recording microphone is placed near the mouth, near the open end, or some distance from the pipe. Since organ pipes are almost never heard in close proximity, we have recorded the sound of various pipes at a distance from 2 to 4 meters (in their natural environment) so that an overall spectrum, characteristic of the pipe, would be received. Fig. 6.F-7 is the spectra of typical examples of the flue pipes shown in Fig. 6.F-5. Fig. 6.F-8 is the spectra of typical examples of the reed pipes pictured in Fig. 6.F-6.

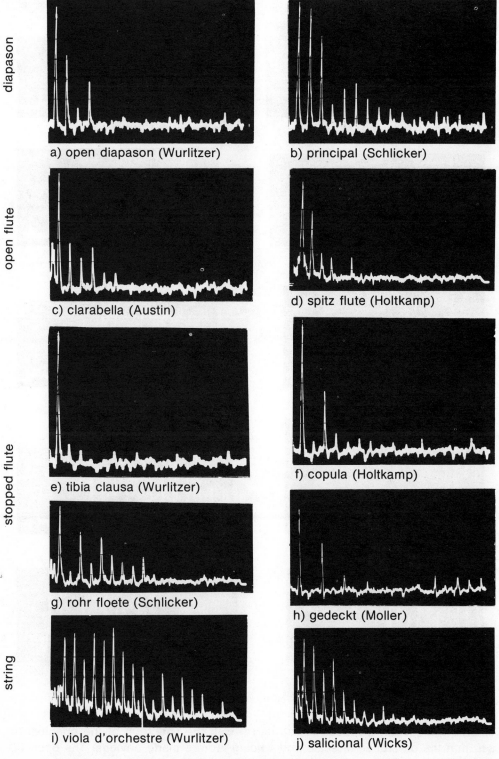

a) open diapason (Wurlitzer)

b) principal (Schlicker)

c) clarabella (Austin)

d) spitz flute (Holtkamp)

e) tibia clausa (Wurlitzer)

f) copula (Holtkamp)

g) rohr floete (Schlicker)

h) gedeckt (Moller)

i) viola d'orchestre (Wurlitzer)

j) salicional (Wicks)

diapason

open flute

stopped flute

string

Fig. 6.F-7. Spectra of flue pipes

These spectra were recorded in many different environments, but unless otherwise stated, the note being analyzed is middle C (C4). (A tape of all the sounds analyzed in Figs. 6.F-7 and 6.F-8 is present in section 6.F of the tape *Descriptive Acoustics Demonstrations*.) A brief discussion of these spectra follows. Although each spectrum shows the relative proportions of each of its ingredients, direct comparisons cannot be made between spectra since they were not recorded under uniform conditions or at the same loudness level.

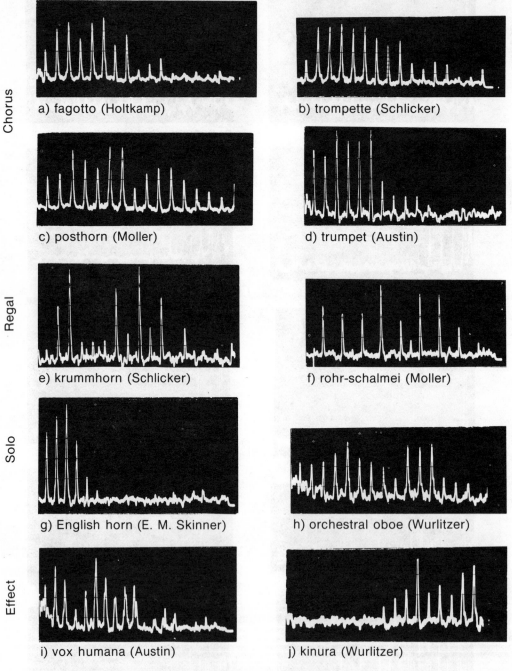

Fig. 6.F-8. Spectra of reed pipes

Comparing the open diapason spectrum (a) with the spectrum of the principal (b), we see that the difference in scale and voicing is here made obvious. The open diapason has only four harmonics, while the principal shows nine or ten. The spectra of

(c) and (d) are for two open flutes; the clarabella and the spitz flute. The clarabella (c) is a fairly large-scaled rectangular flute. You may notice that the spectrum is not too different from that of the open diapason, except that the clarabella gives more emphasis to the fundamental. The spitz flute (d), which is tapered (see Fig. 6.F-5d), has an increased number of partials present. The taper helps to increase the strength of the second partial without decreasing the first partial: hence the brighter sound. The next four spectra are various types of stopped flutes which, of course, emphasize the odd harmonics. The tibia clausa (e) is a very large-scale flute; so we should not be surprised to discover that the spectrum shows only the fundamental (the third harmonic is of smaller amplitude than even the second harmonic). The copula (f) is a capped metal flute of somewhat smaller scale; hence we note the prominent third harmonic. The gedeckt (h) is an even smaller scale, and we see that the fifth harmonic is now apparent, although small. The rohr flute (g) is basically a stopped flute, which is apparent by the emphasis of odd partials. The small chimney is usually of such a length to emphasize the fifth or sixth harmonic. In this case, the fifth harmonic has been made somewhat more prominent. Finally, the spectra of two string pipes are presented. The viole d'orchestre (i), which is of very small scale (the diameter of the 8' pipe is less than 4 cm), shows many harmonics, the first six being especially prominent. It is also interesting to note that the second and third partials are both somewhat stronger than the fundamental. This is because the pipe is first voiced to overblow and sound the second partial; a small wooden dowel (the beard) is then fitted near the mouth to cause the pipe to again sound the fundamental. Needless to say, the above process gives an undue emphasis on the second and third partials, which is deemed to be desirable for a "realistic" orchestral string effect. The salicional spectrum (j), on the other hand, is somewhat bland in comparison to the orchestral string. Although there are more harmonics present than for a small-scale flute, the lessened harmonic development results in a more refined tone.

The spectra of reed pipes given in Figs. 6.F-8a-d are for chorus reeds. The first three, all built within the past 10 years, show a high development of partials with the fundamental being somewhat attenuated. The trompette (b) is voiced to be somewhat brighter than the fagotto (a), although they have very similar tonal qualities. The posthorn (c) is voiced to be extremely bright and loud; the spectrum shows sixteen prominent partials and there were undoubtedly quite a few more beyond the range of the spectrum analyzer. The trumpet (d) was made about fifty years ago when chorus reeds were of larger scale and were not voiced to be as bright as is the general practice today. This is obvious from the fact that the harmonics higher than the sixth are not very prominent. The krummhorn (e) not only shows a reduced fundamental; the fundamental is missing entirely! Since the krummhorn consists of a cylindrical resonator (albeit of very small scale) we would expect to find mostly odd harmonics. Except for the very prominent second harmonic and the missing fifth harmonic, this is essentially true. The rohrschalmei (f) (which is C5 instead of C4) consists of two cylindrical tubes of approximately the same length. We note that although all partials are present in the spectrum the fourth harmonic is the most prevalent. The English horn (g), consists of a small-scale conical resonator with a bulge on the end which resembles the bulge of its orchestra prototype. Since the end is essentially closed with a relatively small opening for the sound to escape, the upper harmonics are greatly attenuated. One will notice, however, that the second partial is very prominent. The orchestral oboe (h) has a very small-scale capped resonator. This stop, although fairly soft, is voiced to have many harmonics of about equal intensity present. The vox humana (i) utilizes a capped cylindrical resonator having a length of only one-eighth that of a full-length conical resonator. The resonator, being very short, will not control the reed, but rather it will serve only to emphasize certain frequency regions and attenuate other regions (in a manner similar to the way the vocal tract modified the vocal cord energy). The kinura (j) consists of an extremely short conical resonator (8 cm long at 8' C), which exerts no influence at all on the reed but serves somewhat as an impedance-matching device from the reed to the outside air. That it fails miserably in this regard for the low partials is evidenced by the fact that the first eight partials are completely absent! The higher partials are present, including many which are beyond the range of the spectrum analyzer. You may have noticed that for reed pipes there is almost no limit to the type of spectrum (and its associated tone quality) which can be produced by combining unusually shaped resonators with certain types of shallot.

The characteristic ensemble effect of pipe organs, which has not yet been satisfactorily reproduced by their electronic imitators, occurs when more than one set of pipes play at the same time. Even though the pipes may have been tuned to the same frequency, they will be very slightly out of tune. This gives rise to beating among the various partials and adds warmth to the tone. The ensemble or choral effect is produced by beating when more than one voice is sounded. The ensemble effect is missing on most electronic organs because the voices are exactly in tune; there is no beating, and the tone lacks the warmth of a pipe organ tone. (See section 6.A.)

It is not uncommon, on many organs, to have a rank of pipes which is purposely tuned slightly sharp. This stop, usually called the voix celeste, is a member of the string family. When it is played with a similar string (such as the salicional) very noticeable beats are produced, giving a warm sound reminiscent of the orchestral strings. Although the string family, because of their many partials, are best adapted to this purpose, very beautiful flute celestes have also been produced.

A vibrato is produced in organ stops by a mechanical device called the tremolo. The *tremolo* (for pipe organs) sets up a periodic fluctuation of the air pressure in the windchest. Since the pressure is alternately increasing and decreasing around its average value, a fluctuating pitch change is introduced into the tone.

Exercises

1. Compare the advantages and disadvantages of a tracker action and an electric action pipe organ. (Hint: Consider such items as cost, flexibility, and keyboard touch.)

2. What would be the effect on the tone of a reed pipe if a thin sheet of soft leather were put on the shallot face where the tongue rests? Explain.

3. Several different types of shallot are used in reed pipes. What difference in volume and tone would you expect for a shallot face with a large opening compared to one with a smaller opening? Explain.

4. Chorus reeds are sometimes found with half-length conical resonators (i.e., an 8' pitched pipe will have a resonator approximately 4' in length). What effect would using a half-length resonator have on the tone quality? On the amplitude of the fundamental? On the perceived pitch? Why?

5. Would you expect a typical open diapason pipe to be nicked or unnicked? Why? What about a typical principal pipe? Why?

6. Compare an open diapason and a nonimitative string pipe. Consider scale, harmonic development, strength of the fundamental, and loudness.

7. How would you expect the timbre of a very large-scale open flute to compare to the timbre of a very large-scale stopped flute? Explain.

8. Pipe organs almost always have a diapason chorus and/or a flute chorus, but seldom does one find a complete string chorus. What might be some reasons for this? Explain.

9. The purpose of a diapason chorus is to reinforce the harmonics of the 8' diapason stop. Would large-scale open diapasons or the smaller-scale principals fulfill this purpose better? Explain.

10. In small pipe organs the 16' pedal pipes are almost always stopped wooden pipes. Why?

11. Which families or organ stops have the greatest upper harmonic development? The least? Why?

12. An organ stop called the French horn is a fairly loud reed with a very strong fundamental and little harmonic development which does not blend well in ensemble. With which family would you classify this stop? Why?

13. The translation of the Latin words *vox humana* is "human voice." The typical vox humana utilizes a one-eighth length cylindrical resonator which is partially closed on top. How is this similar to the human vocal apparatus? How is it different?

14. An organ stop labeled "32' Resultant" is often found in the pedal division of small or medium-sized organs. The bottom octave of this stop is a 16' tone and a 10 2/3' tone (the fifth above) played simultaneously. By referring to the discussion of difference tones in section 3.C, explain how a 32' pitch is produced by this effect.

15. For an imitative orchestral string pipe would it be more effective to use nicking or not? Explain.

16. Will a diapason and a string pipe of the same pitch have exactly the same length? Explain.

17. Why is a voix celeste an effective stop to help produce an orchestralike quality of string tone?

18. In order to increase the loudness of an organ ensemble it is more effective to add stops of higher pitch than to add more 8' stops. Explain in detail why this is true.

*19. Arrange to visit a pipe organ and have someone explain the workings of the instrument and the tonal families.

Further Reading

Anderson, P. G. 1969. *Organ Building and Design* (Oxford University Press, New York).

Barnes, W. H. 1964. *The Contemporary American Organ,* 8th ed. (J. Fisher, New Jersey).

Bonavia-Hunt, N., and H. W. Homer. 1950. *The Organ Reed* (J. Fisher, New Jersey).

Fletcher, H., E. D. Blackham, and D. A. Christensen. 1963. "Quality of Organ Tones," J. Acoust. Soc. Am. *35,* 314-25.

Fletcher, N. 1974. "Nonlinear Interactions in Organ Flue Pipes," J. Acoust. Soc. Am. *56,* 645-52.

Fletcher, N. 1976. "Jet-drive Mechanism in Organ Pipes," J. Acoust. Soc. Am. *60,* 481-83.

Ingersler, F., and W. Frobenius. 1947. "Some Measurements of the End Corrections and Acoustic Spectra of Cylindrical Open Flue Organ Pipes," Trans Danish Acad. Tech. Sciences *1.*

Jones, A. T. 1941. "End Corrections of Organ Pipes," J. Acoust. Soc. Am. *12,* 387-94.

Klotz, H. 1961. *The Organ Handbook* (Concordia Publishing House).

Nolle, A. W., and C. P. Boner. 1941. "Harmonic Relations in the Partials of Organ Pipes and of Vibrating Strings," J. Acoust. Soc. Am. *13*, 145-48.

Nolle, A. W., and C. P. Boner. 1941. "The Initial Transients of Organ Pipes," J. Acoust. Soc. Am. *13*, 149-54.

Audiovisual

1. *Instruments Using a Bellows* (3 3/4 ips, 2 track, 30 min, UMAVEC)
2. *Organ Tones, Descriptive Acoustics Demonstrations,* Section 6.F (GRP)
3. *Organ,* Part 1 (19 min, color, OPRINT)
4. *Organ,* Part 2 (22 min, color, OPRINT)
5. *Vibrating Strings and Air Columns* (30 min, 1957, EBEC)
6. *Wind and Percussion Instruments* (30 min, 1957, EBEC)

Demonstrations

1. Various types of pipes connected to laboratory air supply

6.G. BOWED STRING INSTRUMENTS

The vibrating string, which we considered in section 2.D, forms the basis of this section. Strings may be made to vibrate by bowing, striking, or plucking. We will consider the resultant acoustical effect of each type of excitation—bowing in this section and striking and plucking in the next. In addition we will consider the different types of resonators usually associated with these strings and the spectra produced by strings under various conditions. Finally we will discuss briefly the history and development of the violin, and current research on the structure and tone of violins.

The violin is representative of bowed string instruments. Other common members of this class are viola, cello, and double bass. They have the following common features:

1. A string approximately fixed at both ends is the primary vibrator and as such determines the frequencies produced.
2. All partials are present, with the exceptions noted below.
3. Most of the energy is radiated by the body of the instrument and a smaller amount by the string.
4. The body of the instrument imposes its own resonance features on the radiated sound.
5. The bowed strings have fairly long attack times.

Construction of the Violin

A sketch of the main parts of the violin body is given in Fig. 6.G-1. The strings which provide the original vibration are attached to adjustable pegs at one end and to the tail piece at the other end. The vibrations of the strings are transmitted to the body of the instrument through the bridge. Directly under the feet of the bridge, on the underside of the top plate, are the bass bar and the sound post. The bass bar stiffens the top plate and the sound post transfers some of the vibration energy to the back plate so that it also vibrates. The fingerboard provides a support against which the strings can be stopped to shorten them. The holes in the top plate, called the f-holes, enable the air inside the body to participate in the vibration (as a Helmholtz resonator) so that an important air resonance is present. The graceful curves in the sides of the instrument, called the C-bouts, enable the violinist to bow the outside strings without rubbing the side of the instrument.

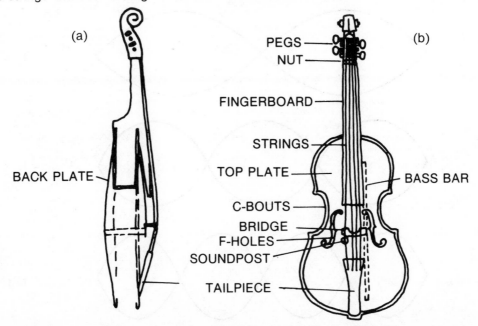

Fig. 6.G-1. Construction aspects of violin: (a) side view; (b) top view.

The four open strings of the violin are G3 (G below middle C), D4, A4, E5. The unstopped strings have their effective length for vibration defined between the bridge of the instrument at one end and the nut at the other. By stopping the strings with the fingers of the left hand, one can easily obtain a range of over three octaves.

Role of the Strings

The laws governing the frequencies of the various modes of a vibrating string which is fixed at both ends have been discussed previously in section 2.D. In summary, the frequencies of the modes of a vibrating string are (1) inversely proportional to the wavelengths of the modes, which in turn are proportional to the length of the string, and (2) directlly proportional to the wave speed in the string, which in turn is proportional to the square root of the tension on the string and inversely proportional to the square root of the mass per unit length of the string. It is then obvious that for any given material the mass per unit length will be dependent on the square of the diamater of the string, and that therefore thicker strings create lower frequencies. Also, doubling the length of a string lowers its modal frequencies by an octave, while halving the length of a string raises them by an octave. To raise the frequencies of a stringed instrument, one need only increase the tension, or decrease it to lower the frequencies (as is typically done in tuning all stringed instruments). The bass strings on violins, harps, and pianos are of a larger diameter than the treble strings, which enables the achievement of lower pitches without requiring strings of inordinate length.

Assuming a string of constant cross-sectional area (and thus constant mass per unit length) and constant tension, recall that the natural modes of the vibrating string of length l are all integer multiples of the fundamental frequency f_1, as shown below in Fig. 6.G-2. Note that all of the odd-numbered modes have antinodes (maximum

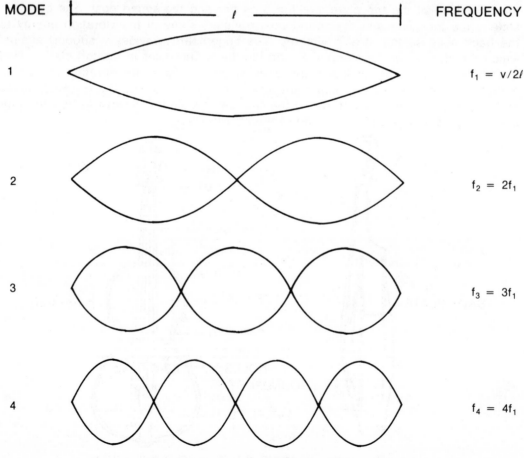

MODE		FREQUENCY
1		$f_1 = v/2l$
2		$f_2 = 2f_1$
3		$f_3 = 3f_1$
4		$f_4 = 4f_1$

Fig. 6.G-2. First four natural modes of a string.

276

displacement) at the center of the string, while all the even modes have nodes (zero displacement) at the center of the string. Now suppose that we bow a string at its center. Any natural modes with nodes at the center of the string will not be excited because the point where we bow the string is a point of maximum vibration, or an antinode. Thus, odd modes are excited and even modes are not excited for a string bowed at its center. Suppose a string is excited (by bowing, plucking, or striking) at a point 1/3 of its length from one end. It can be seen that any partial having a node 1/3 of the distance from the end of the string is not possible; hence the third, sixth, ninth, etc. modes are not excited. In general, the partials that are missing are those having a node at the point of excitation.

In practice, the bow has a finite width and is not maintained at a fixed position relative to the end of the string, which means that particular partials will not be completely missing from the spectrum. However, the position of the bow in relation to an end of the string will have a strong influence on the spectrum, and the tonal quality can be made quite different by changing the bow position on the string. Any given partial will be made to sound more strongly as the bow is placed more closely to an antinode of the mode associated with the partial; it will sound weaker as the bow is placed closer to a node of the associated mode.

Action of the Bow

When the horsehair bow is drawn across a violin string, the frictional force exerted on the string by the bow displaces the string until the restoring force in the string is great enough to overcome the frictional force of the bow and the string snaps back toward its rest position. Frictional forces of the bow then displace the string again, and the process is repeated periodically, thus resulting in vibration of the string, as shown in Fig. 6.G-3.

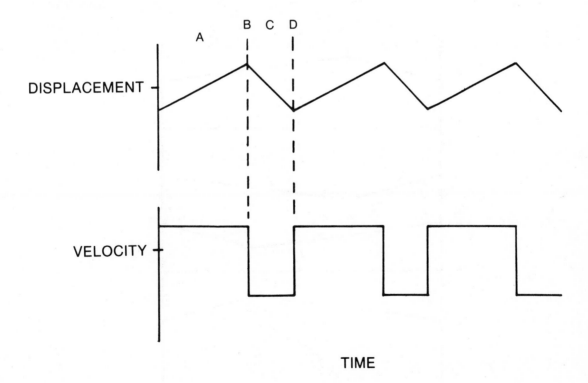

Fig. 6.G-3. Idealized displacement and velocity waves in a string. The string is moving with the bow in region A; the string begins slipping at point B; the string is slipping past the bow in region C; the bow "captures" the string at point D. (After Schelleng, 1974.)

The time the string is free from the bow is substantially shorter than the time the bow is displacing the string. (These times could be equalized by bowing the string at its center.) It is also apparent from the figure that the string does not merely return to its equilibrium position, but overshoots this position—a characteristic of oscillatory motion. The time required for the string to slip back is governed by the sliding friction between the string and the bow; since sliding friction is always less than static friction, the return time is shorter than the time required for displacement. The motion of the string is then a complex type of vibration which will contain all the harmonics of the fundamental frequency. The above picture is somewhat oversimplified, however, because we might expect the period of the waves on the string to depend on the bowing force exerted by the hand. In actuality the period of the string is remarkably constant over a wide range of bowing forces (Schelleng, 1974). The explanation is found in the time it takes the disturbance on the string to travel to the near end of the string and then to the far end and back to the point of excitation. The appearance of the string at different instants in the cycle is shown by the straight line segments in Fig. 6.G-4. If the A-string, for example, is bowed the disturbance makes 440 complete trips to the near end, then to the far end, and finally back to the excitation point, every second (since the frequency is 440 Hz). When the bow changes from ''up bow'' to ''down bow'' the direction of circulation of the disturbance also changes. When the disturbance arrives back at the point of excitation a maximum displacement is produced, which means that the restoring force on the string (the force tending to move the string in the direction of its rest position) has its largest value directed oppositely to the frictional force of the bow, and this causes the string to break away from the bow. When the wave reflected from the near end arrives back at the bow a maximum negative displacement occurs at the bow position,

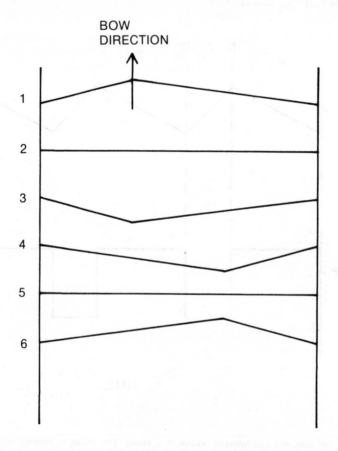

Fig. 6.G-4. Shape of violin string at different times during one cycle of vibration. String slips from bow at (1) and catches bow at (3). Compare Fig. 6.G-3. (After Schelling, 1974.)

which results in the maximum restoring force in the same direction as the bowing force and causes the string to stick to the bow again. In this way the standing waves on the string cause the string to stick to the bow and to break away from the bow at periodic intervals.

Although the frequency of the tone emitted by bowing does not depend strongly upon the bowing force, the quality of the emitted sound is dependent upon the force, as well as the position of the bow on the string and the speed of the bow. The bow force, however, is more effective in modifying tone than is the bow speed; increasing the bow force tends to increase the intensity of the higher harmonics. The point at which the bow is applied to the string also greatly modifies the tone quality. Bowing close to the bridge increases the number of higher harmonics, while bowing farther away from the bridge reduces the higher harmonics and gives the sound a softer quality. The amplitude of the string vibration at a particular bowing position depends on how rapidly the bow is drawn over the string—the greater the bow speed, the louder the resulting sound. The amplitude of vibration, however, does not depend on bowing force (Saunders, 1962).

For a constant bowing speed, there is a maximum bow force and a minimum bow force which will give the required frequency. The range of acceptable forces depends upon the distance from the bridge, as shown in Fig. 6.G-5. Note that the distance between maximum and minimum bow force increases as the distance from the bridge increases, and that there is a point (very close to the bridge) where the maximum and minimum forces are equal. When the maximum force is exceeded, the discontinuity in the string due to the traveling waves can no longer provide the necessary force to trigger the slipping, and a raucous, erratic vibration results. If less than the minimum force is provided, the static friction is insufficient to hold the string for its full displacement and the string slips sooner, thus replacing the fundamental tone by the octave.

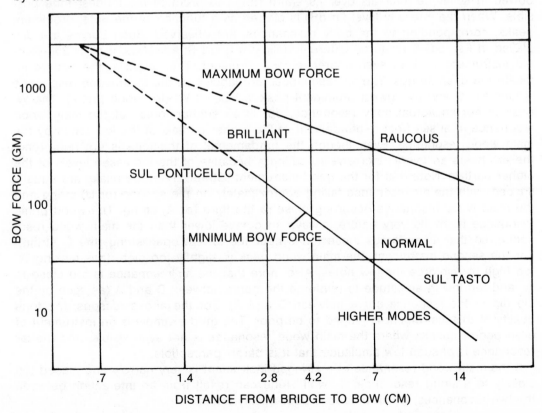

Fig. 6.G-5. Range of acceptable bow forces as related to bow position for a cello bowed at a speed of 20 cm/sec. (From Schelling, 1974. Copyright by Scientific American, Inc. All rights reserved.)

As mentioned previously, as the bow is moved away from the bridge the proportion of higher harmonics decreases, so that the timbre has the gentle characteristic designated by *sul tasto* (bow over the fingerboard) and a wide degree of latitude exists between maximum and minimum bow forces. As one bows closer to the bridge, larger bow forces are required and there is a less generous latitude between maximum and minimum force, but the tone quality is greatly enriched by the increase of the higher partials. An experienced hand is required to bow successfully in this region; amateurs find it prudent to confine their bowing to the region closer to the fingerboard. As one attempts to bow even closer to the bridge, the bow force mounts even more and the fundamental tone disappears, leaving only a swarm of high harmonics; composers call this the *sul ponticello* (bow over the little bridge).

Influence of the Body

A vibrating violin string transfers most of its energy to the body of the instrument through the bridge; a small amount of energy is radiated directly by the string, but without the violin body the sound would be very weak. The body of the violin, however, modifies the sound of the strings because the body itself has a complex set of resonances. Thus the radiated sound is the sound produced by the strings as "filtered" by the body.

Research has shown that one of the main determining factors of good-quality violins is having the resonances placed so that the air and main wood resonances reinforce partials in the vicinity of the D and A strings respectively. The air resonance (which results from the air contained in the body) can be adjusted by changing the air volume of the body and the size of the openings (f-holes) in the body. The main wood resonance (which is the lowest fundamental mode of the coupled top and back plates) can be adjusted by proper construction of the plates.

In one method for determining where the resonances are located, the violin is bowed in semitone intervals over its entire range to produce the loudest tone possible. When the intensity level (in dB) is plotted as a function of the note, prominent peaks, corresponding to the body resonances, are observed. Such curves are displayed in Fig. 6.G-6 for three different violins: a good Stradivarius, a poor 250-year-old instrument, and an even poorer, older instrument. The letters at the bottom indicate the open strings. The air resonances are labeled A, the main wood resonance is labeled W, and the lowest prominent peak is labeled W' (for wood prime). The W' peak is not an actual body resonance, but is a "subharmonic" of the main wood resonance; it arises from reinforcement of the higher partials of the low tones by the main wood resonance. Even though the fundamental of the tone is not reinforced, the ear hears an overall increase in loudness because of the increased levels of the higher partials. Note that for the good instrument (Stradivarius) the peaks are equally spaced, with the air resonance falling approximately on the second (or D) string and the main wood resonance occurring close to the third (or A) string. The wood prime resonance is, by its very nature, always an octave lower than the main wood resonance, so that it reinforces those tones near the lowest open string (the G string). For the second instrument, the main wood peak is slightly too high, thus making W' too high to reinforce the low notes. Also, note that the air resonance is too close to W' and too low in amplitude to reinforce the notes between D and A (as seen by the big dip in the response curve between D and A). For the above reasons, the tone quality of this instrument is judged to be poor. The third example is an instrument of even poorer quality, where the main wood resonance is not even visible, and the air resonance is of such low amplitude that it is barely perceptible.

If the main wood resonance of a stringed instrument is very narrow and placed too closely to a string resonance, a "wolf" tone can result from an interaction between the two resonances, which produces poor-quality tones.

Additional Factors in Violin Tone

Recently many other characteristics of violins have been investigated to determine

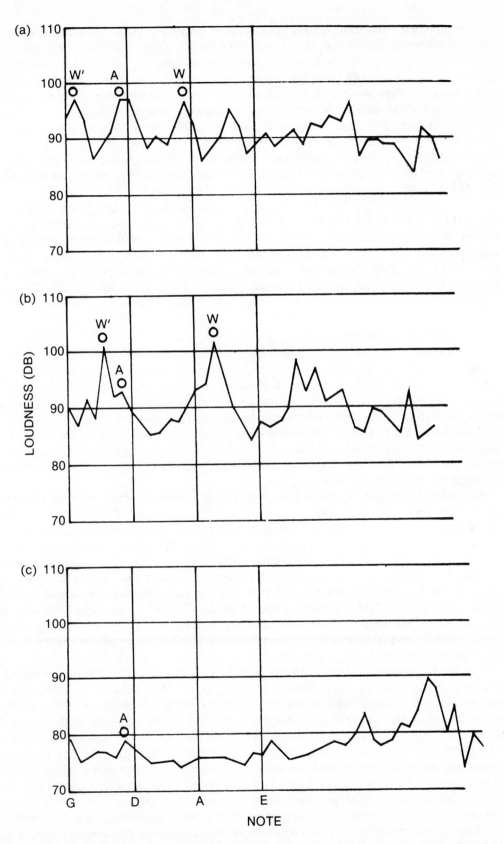

Fig. 6.G-6. Violin "loudness curves" compare maximum sound levels of (a) a good 1713 Stradivarius, (b) a poor 250-year-old instrument, and (c) a poorer Guarneri. Letters at bottom indicate open string tuning. Air resonances are marked A; wood resonances W; and "wood prime resonances" W'. (From Hutchins, 1962. Copyright by Scientific American, Inc. All rights reserved.)

their influence on the tone quality. The type of wood and how the wood is cut in relation to the grain play a role in the final tone quality (Abbott and Purcell, 1941). The effect of aging on violins is being studied; the preliminary results seem to indicate that aging in and of itself does not improve tone quality (Saunders, 1962). It does appear, however, that playing an instrument can eventually improve the tone quality to some extent. One reason is that continued playing loosens the glue joint between the top plate and the sides, thus making the top plate more flexible (and hence more responsive). It has been discovered (Hutchins, 1962) that the purfling—a narrow groove cut around the outside edge of the top and back plate, with thin strips of wood glued inside—increases the flexibility of the plates as the glue ages and begins to crack slightly.

The excellent tonal quality of the Stradivarius violins has often been attributed to the varnish used, since the formula was a secret, shared only by the pupils of the master, and it has been lost for about 200 years (Farga, 1969). Recent acoustical investigations have shown, however, that while the response peaks of a violin are lowered slightly after the varnish has been applied, the change is so small that it could not ordinarily be detected by ear (Meinel, 1957; Saunders, 1962). The main function of the varnish seems to be to seal the pores and thus to protect the wood.

History and Recent Developments of the Violin

In the violin, the art of the eighteenth-century craftsmen achieved its greatest triumph in terms of tonal beauty, construction, and appearance. Around the year 1550 the violin, in the form we know it today, emerged from several immediate ancestors which flourished during the renaissance. In particular, the lyra da braccio, although slightly larger, was remarkably similar to the violin in bodily outline, but utilized more strings. It is often erroneously assumed that the viol family, because it preceded the violin by about 100 years and has now become obsolete, was the ancestor of the violin. Actually, the viols represent a separate class of instruments with different construction and stringing.

The violin developed rapidly in Italy, starting with Andrea Amati, who, in the middle of the sixteenth century, founded the great Cremona school of violin makers. Within 150 years of his death, his descendents and his pupils' descendents (notably Antonio Stradivari and Guiseppe Guarneri) had brought the art of violin making to such a state of perfection that their instruments are only now being equaled. By the end of the seventeenth century Antonio Stradivari had developed a classical model of violin which has become the standard for succeeding generations. About 1800 the violin underwent its last important transformation. Greater power and brilliance were needed to fill the larger music halls and were achieved by greater string tension, higher bridge, and greater playing length of the strings, all of which resulted in further internal changes in the instrument.

In order to determine the placement of the main wood resonance in a finished instrument, violin makers have traditionally employed "tap tones"; i.e. they tap the unassembled front and back plates and listen to the pitch produced. The ability to correctly judge the resonances of the top and back plates was part of the art of violin making. Recent research by Hutchins (1962) has made the determination of the free plate resonances accessible to scientific analysis. By supporting the plate at its nodes and driving it with a variable oscillator, one can plot the response of the plate as a function of frequency. Good instruments result when adjacent peaks of a plate are about a semitone apart and when the front and back plates have peaks which alternate and do not coincide in frequency. Then, when the instrument is assembled, the main wood resonance is about seven semitones above the average frequencies of the tap tones from the front and back plates. Thus, by accurately predicting the tap tones and placing them properly, it is possible to place the main wood resonance where it will do the most good for the tone of the instrument.

Although the violin design became more or less optimized under the skilled hands of the Cremona craftsmen, the same development has never been achieved for the other members of the string family. The air and wood resonances are generally too high for really excellent tone quality to be achieved on violas and cellos. With this in mind, Hutchins (1967) has applied the principles of good violin design to designing and constructing a family of eight stringed musical instruments covering a frequency range from low E (E1) to E6. If the instruments were scaled to the violin in a linear manner, the large bass would have a body length 2.1 m long (six times that of the violin) and the small treble violin would be only about 16 cm in length. Table 6.G-1 gives the actual body lengths chosen for the instruments, as well as their theoretical lengths. The scaling actually used makes the body of the smallest instrument 27 cm long and the large bass body is a more easily managed 130 cm. Given a practical size, the philosophy of design was to then determine the other dimensions so that the air resonance and main wood resonance occurred near the two central strings respectively. In order to get the air resonance of the trebel violin high enough without having unduly small ribs, a set of six holes was made in the ribs. The viola was redesigned to have a slightly longer body length (52 cm as opposed to the standard 41 cm), which necessitates having small players play it vertically on the floor like a cello. The new baritone instrument is similar to the cello in length but has slightly longer strings. The tone quality, however, has been judged to be superior to that of a conventional cello. The small bass is a newly constructed instrument, being slightly smaller than the conventional bass but having a desirable singing quality of tone and ease of handling. The new contrabass is 1.3 m high, and, although somewhat hard to handle, it has an exceptional depth of tone. The overall evaluation of these instruments is that they are musically successful, a primary weakness being a lack of musical literature for some of them.

Table 6.G-1. Data on new bowed string family of instruments. Length scaling is relative to conventional violin (mezzo). Multiply length ratios by 36 to get lengths in cm. (Data from Hutchins, 1967.)

Instrument	Strings	Body length ratios	
		Theoretical	Actual
Contrabass	EADG	6.0	3.60
Small bass	ADGC	4.0	2.92
Baritone	CGDA	3.0	2.42
Tenor	GDAE	2.0	1.82
Alto	CGDA	1.5	1.44
Mezzo	GDAE	1.0	1.07
Soprano	CGDA	0.75	0.87
Treble	GDAE	0.50	0.75

Another area of research has been in the application of electronics to the string instruments in order to simulate the acoustical properties of the body with electronic filters. The sound is produced by bowing the strings in the usual fashion but is picked up by electromechanical transducers, fed to the body-simulating filters, and finally radiated by a loudspeaker mounted either externally or internally in the instrument body. Demonstrations of violins (Mathews and Kohurt, 1973) and violas (Gorrill, 1973) so constructed have been conducted.

Exercises

1. Some partials are (approximately) not present in a string tone. As an example, consider wave 6 from exercise 8 of section 2.E as a string bowed at its center. What partials are present? What partials are absent?

2. What partials are missing for a string excited 1/4 the distance from one end? For 1/5? For 1/6? For 1/7?

3. Why is the spectrum of radiated sound for violins different from the spectrum of the string waveshape?

4. Where is the violin bowed? What partials (approximately) are missing?

5. Name other instruments with characteristics similar to those of the violin.

6. What are the approximate frequencies of the air resonance and the main wood resonance of the violin?

7. What would the air and main wood resonance frequencies be for the new treble violin, which is scaled an octave higher than the violin? For the new tenor violin, which is scaled an octave lower?

8. How do the standing waves on a violin string affect the interaction between the string and the bow?

9. Why does shortening a string by fingering increase its frequency?

10. If you owned a violin of poor quality and research showed that the wood resonance was located too high, what would you do to lower the frequency of the resonance? How would you modify the instrument in order to change the frequency of the air resonance? (Better get expert help if you actually do it.)

11. In the viola and cello, the air and wood resonances are generally three or four semitones higher with respect to the frequencies of the open middle strings than for the violin. Can you think of any reasons why this might be expected? What can you predict about the tone quality of these instruments?

*12. The spectra below resulted from a violin string excited by plucking at its midpoint and at a distance 1/8 the length of the string from one end.
   a. What are the relative strengths (in dB) of the first ten partials in each case? (There are some "extraneous peaks" in the spectra. The fundamental was in each case about 200 Hz.)
   b. What are the strengths of the partials relative to the strongest partial in each case?
   c. What observations can you make from these spectra about the body resonances?

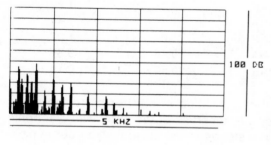

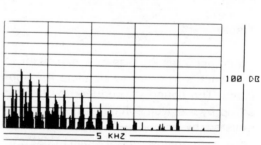

*13. a. Decide on the gross features of a string instrument that you wish to construct.
   b. By consulting appropriate sources, determine the physical properties of your instrument (such as string length, string density, fret positions if any, etc.) so that it will produce the desired frequencies.
   c. Construct the instrument.
   d. Take the instrument into the laboratory and make measurements to determine how well you have achieved your design goals.

Further Reading

Backus, chapter 10

Benade, chapters 23, 24

Farga, F. 1961. *Violins and Violinists* (Frederick A. Praeger, N.Y.).

Gorrill, W. D. 1973. "Viola Tone Quality Study Using an Instrument with Synthesized Normal Modes," J. Acoust. Soc. Am. *54,* 311(A).

Hutchins, C. M. 1975. *Musical Acoustics, Part I: Violin Family Components* (Dowden, Hutchinson, and Ross, Stroudsburg, Penn.).

Hutchins, C. M. 1975. *Musical Acoustics, Part II: Violin Family Functions* (Dowden, Hutchinson, and Ross, Stroudsburg, Penn.).

Hutchins, C. M. 1967. "Founding a Family of Fiddles," Physics Today (2), 23-27.

Hutchins, C. M. 1962. "The Physics of Violins," Scientific Am. *207* (5), 78-93.

Mathews, M. V., and J. Kohut. 1973. "Electronic Simulation of Violin Resonances," J. Acoust. Soc. Am. *53,* 1620-26.

Meinel, H. 1957. "Regarding the Sound Quality of Violins and a Scientific Basis for Violin Construction," J. Acoust. Soc. Am. *29,* 817-22.

Saunders, F. A. 1962. "Violins Old and New, an Experimental Study," Sound *1* (4), 7-15.

Schelling, J. C. 1974. "The Physics of the Bowed String," Scientific Am. *230* (1), 87-95.

Audiovisual

1. *Bowed String Instruments and Zithers* (3 3/4 ips, 2 track, 30 min, UMAVEC)
2. *Real and Synthetic Bowed String Tones,* Section 6.G., Descriptive Acoustics Demonstrations, (GRP)
3. *The Cello,* Part 1 (24 min, color, OPRINT)
4. *The Cello,* Part 2 (23 min, color, OPRINT)
5. *The Violin,* Part 1 (22 min, color, OPRINT)
6. *The Violin,* Part 2 (26 min, color, OPRINT)

Demonstrations

1. Effect of bowing string in different places
2. Effect of loading bridge, covering f-holes, and loading top plate of violin

6.H. PERCUSSIVE STRING INSTRUMENTS

We have previously discussed two classes of vibrators: free and forced. The percussive strings are in the free category, whereas the bowed strings lie in the forced category. Percussive strings share many common features with bowed strings: a string fixed at both ends is the primary vibrator, all partials are present with some exceptions, most of the energy is radiated by the body, and the body imposes its own resonance features on the radiated sound. However, the features of inharmonic partials is quite apparent for some free strings, but is not observed for bowed strings.

The piano (see Fig. 6.H-1) will be considered in some detail because of its great importance. Other percussive strings will be considered more briefly.

Fig. 6.H-1. View of grand piano. Note "overstringing" of the bass strings.

Inharmonicity in Percussive Strings

In our discussion so far, we have assumed that strings are all perfectly flexible and that the only restoring force is due to the tension applied to them. Actual strings have some stiffness which provides restoring force (in addition to the tension), slightly raising the frequency of the string partials. This additional restoring force is greater for the higher partials of the string, which means that the higher partials are increased in frequency by a larger fraction than the lower partials (Fletcher, 1964). Calculations indicate that the frequency increase is related to the square of the partial frequency, so that the second harmonic will be shifted four times as much as the fundamental. The fact that the partials of a piano tone become successively sharper has an important bearing on what we consider to be a normal piano tone. Fletcher et al. (1962) synthesized piano tones and played them for a jury consisting of both musicians and nonmusicians. When the synthetic tones consisted of all harmonic partials, they were unconvincing and were described as "lacking warmth," even though the attack and decay of the waveforms duplicated those of real piano tones. When tones were synthesized with inharmonic partials, however, the jury could usually not distinguish the real tones from the synthetic tones. Fig 6.H-2 shows the stretching of the partials for a piano tone.

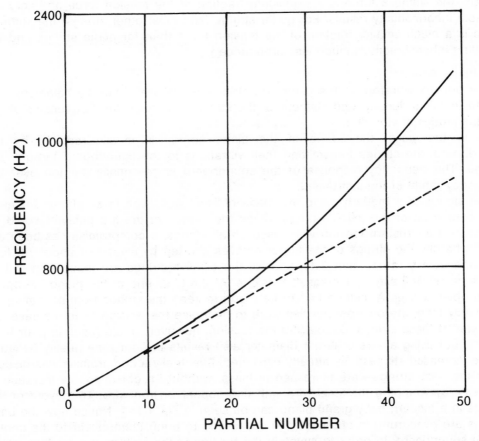

Fig. 6.H-2. Inharmonicity of real piano tones (solid line) compared to harmonic tone (dashed line). (From Blackham, 1965. Copyright by Scientific American, Inc. All rights reserved.)

While a certain amount of inharmonicity is desirable in piano tone, too much inharmonicity ruins the tone. It has been found, for example, that the bass tones on a piano are better if the strings are made longer. When short, thick strings are used for the bass notes of a piano (as in the typical spinet piano) the inharmonicity is much greater because of the greater stiffness of the strings. Since the fundamental of low-frequency piano tones has very little energy, our ears rely on the difference tone produced by the upper partials to determine the pitch. The increased inharmonicity of short bass strings, however, causes the partials to be stretched to such an extent that a single clear difference tone is no longer produced. The sound is then a rough tone with poorly defined pitch. This is why spinet pianos sound poor

287

in the bass range, while the notes of a grand piano in the same range are more clear and sonorous.

The stretching of the partials of a piano tone plays an important role in piano tuning. Since pianos are tuned by eliminating beats between octaves and fifths, the octaves will all be stretched slightly. This occurs because a piano tuner, when tuning an octave, will eliminate beats between the second partial of the lower tone and the first partial of the upper tone. But since the second partial of the lower tone is slightly stretched, when the beats are eliminated the octave will be in tune on the piano, but stretched slightly from twice the fundamental frequency of the lower note. The piano then, strictly speaking, is not tuned to the equal tempered scale, but is stretched somewhat from this scale. In fact, a piano tuned exactly to the equal tempered scale sounds quite flat in the upper register.

The actual amounts of inharmonicity depend on how big the "stiffness force" is relative to the "tension force" for restoring the string to its rest position. For piano strings, the stiffness force is a significant fraction of the tension force and very noticeable inharmonicity results. For guitar strings (and most other strings) the stiffness force is a much smaller fraction of the tension force than for piano strings and the resulting inharmonicity is much less pronounced.

The Piano Action

The primary vibrators of the piano are strings which are struck by hammers. The length, tension, density, and stiffness of the strings determine the frequencies of the various modes of vibration.

As in the violin, the sound of the piano strings is mostly radiated by the soundboard, the strings transmitting their vibrations to the soundboard through the bridge. The frequency response of the soundboard is essentially uniform over the frequency range of the instrument.

The strings of a modern piano are made of steel wire, with three strings for each note over most of the piano's range. Since the strings of the top octaves would be too short if a constant tension were used on all strings, a compromise has been devised whereby the strings of the upper octaves shorten by a ratio of about 1.9 to 1 for each octave rather than 2 to 1. Also, the smaller strings are made with reduced diameters and slightly increased tensions. At the bass end of the piano the opposite problem exists. A method had to be found to keep the strings from becoming inordinately long; the only possibilities were to decrease the tension or to increase the diameter of these strings. Decreasing the tension, however, would result in poor tone quality, but using a wire of larger diameter also results in poor tone quality because of the increased stiffness, as already explained. Eventually a compromise was discovered: the bass strings were increased in mass, without an objectionable increase in stiffness, by wrapping the steel wire with fine copper or iron wire. The layout of the strings in a modern baby grand piano can be seen in Fig. 6.H-1. Notice how the bass strings are overstrung in order to save space and to bring them closer to the center of the soundboard. In order to maintain the tension in the strings of a modern grand piano, the tuning pin which keeps each string in tune is driven into a pinblock consisting of about a dozen cross-grained layers of hardwood.

The moving parts of the piano which are involved in striking the string are called the action. Although nearly all modern grand piano actions are descendants of the original action invented in 1709 by Cristofori, modern actions are of greatly increased complexity, consisting of about 7000 separate parts. A comparison of Cristofori's action and that of a modern grand piano is presented in Fig. 6.H-3. In the Cristofori action, shown in Fig. 6.H-3a, the key 1) is seen to be pivoted at point (2). The escapement, or jack (3), then actuates the intermediate level (4), which delivers an impulse to the hammer (5). The back check (6) prevents the hammer from bouncing off the mechanism and hitting the string a second time. In the modern grand action, illustrated as Fig. 6.H-3b, the key (1) pivots about point (2), raising the capstan (3), which causes the wippen (4) to rotate about point (5). The rotating wippen raises the jack (6), which is pivoted at the other end of the wippen. The upper end of the jack pushes on the roller (7), which is attached to the hammer shank (8), thus causing the hammer (9) to rise toward the string. Just before the hammer encounters the string, the lower end of the jack strikes the jack regulator (10), which causes the jack to rotate so that its upper end moves out from under the roller and hence no longer pushes on it. Since it is inertia which carries the hammer toward the string, the hammer rebounds and falls back until the roller rests on the repetition lever (11). The back-check (12) prevents the hammer from striking the string a second time. When the key is lifted slightly, a spring pulls the jack back under the roller, so that pressing the key again allows the note to be repeated. The ability of the grand piano action to allow for such rapid repetition of notes, without having the key return to its starting position, is the real reason for the superiority of the grand action to the upright action. A careful consideration of the action of an upright piano will show that a note cannot be repeated unless the key returns to its starting position.

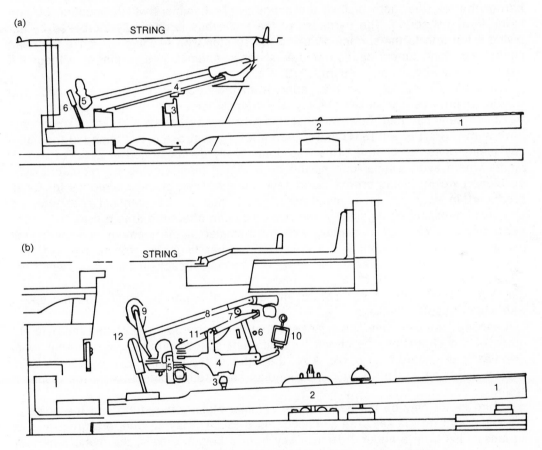

Fig. 6.H-3. Piano actions: (a) Cristofori and (b) modern grand.

The sustaining pedal allows one to raise all of the dampers off all of the keys so that the tone may be sustained. The soft pedal, operated by the left foot, shifts the entire action (in a grand piano) to the right or to the left so that the hammer strikes only two of the three strings. In the earlier pianos, when there were only two strings per note, the soft pedal caused the action to strike only one of the strings; hence the term *una corda,* which is still used to indicate the use of the soft pedal in musical scores. In upright pianos the soft pedal moves the hammers closer to the strings and thus reduces the volume of sound produced by decreasing the distance the hammer travels before encountering the strings.

Piano Tone Quality

We have mentioned that a partial having a node at the point of excitation is expected to be missing or very weak. The hammers of a piano are arranged so that they all strike their respective strings at a distance of about one-seventh the length of the string from one end. It has frequently been asserted that this is done in order to eliminate the seventh harmonic from the tone. However, recent research (Fletcher et al., 1962) has shown that the seventh partial is present with an intensity comparable to the other partials (see Fig. 6.H-9). Since the hammers do not strike the string at a point, but rather make contact with a certain length of string, the above conditions for missing partials do not apply exactly. The point where the hammer strikes the string has been chosen to be one-seventh of the length of the string because trial and error have determined that the resultant tone quality is then good.

To achieve a clear sound on a piano, it is important that the hammer heads be properly voiced; that is, they must be smooth and uniformly shaped. The degree of hardness or softness of the hammers is usually dictated by personal tastes—the harder the felt, the more brilliant the sound—while hammers which become too soft result in a dull sound. The hardness of the felt heads is generally increased by applying a hot iron to the surface. Pricking the surface with needles will make the head softer, but piano tuners caution amateurs not to attempt these changes without the benefit of expert guidance (Kuerti, 1973).

It has often been asserted that, apart from the force of the hammer impact, the way in which a pianist strikes the key is extremely important to the resultant tone. It has been demonstrated, however, that this opinion has no basis in fact. Experiments have shown (Hart et al., 1934) that if the force of the stroke is not changed the musical quality does not change. This was dramatically demonstrated in a set of experiments where every single tone played by a skilled pianist could be recreated by a cushioned weight falling on the piano key. A comparison of the waveforms for tones produced in each of these manners showed them to be identical. Although no amount of pressing or wiggling of the piano key can affect the tone quality (since the performer is totally out of contact with the hammer at the moment of impact), the pianist can and does control the speed of his/her finger in striking the key. The greater this speed, the greater the force of hammer impact on the string, which increases the development of upper partials, thus resulting in a richer, as well as louder, tone. Thus a pianist can control the quality of sound by controlling the speed with which to strike the keys.

A second factor of some importance is that when chords are sounded the different notes of the chord may be struck with different speeds, thus giving different tonal qualities to the chord as a whole. Also, the notes of the chord may not be initiated at exactly the same instant of time, thus lending still another difference to the resultant sound.

A third factor may be the noise of the hammer striking the string and the noise of the finger striking the key. The so-called "hard touch" may be due to a great deal of surface noise from a finger striking a key from a height. Finally, the piano tone may be influenced at cessation by the manner in which the pianist releases the note. A sudden dropping of the felts may dampen the vibrations differently than releasing the key gradually, although more research would be needed to ascertain if this effect is really perceptible.

A discussion of piano tone quality would be incomplete without mentioning the possible differences in timbre of pianos produced by different manufacturers. A preliminary experimental investigation (Culver, 1956) indicates that comparable pianos by different manufacturers do indeed exhibit quite different tonal qualities, as shown in Fig. 6.H-4a. The spectra shown are for one note of each of the different pianos (7-foot grand) by well-known manufacturers. Even more surprising is the fact that a grand piano and a spinet piano by the same manufacturer had remarkably similar spectra (in the middle and upper register, as indicated by Fig. 6.H-4b), although the dynamic range of the small upright was much more limited.

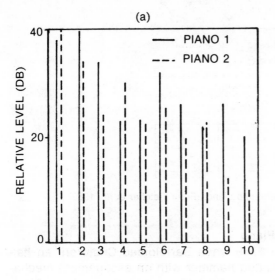
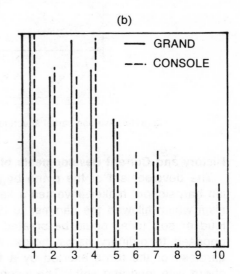

HARMONICS

Fig. 6.H-4. Comparison of piano spectra: (a) two different makes; (b) grand and console of same make. (After Culver, 1956.)

The tone of a piano is without a steady state, but consists entirely of a brief attack and a decay which may be fairly long, as illustrated in Fig. 6.H-5. Furthermore, the spectrum of piano tones is very dependent upon pitch, the lower tones having many more partials than the higher notes. This is clearly shown in Fig. 6.H-6, where the spectra of seven C's on a studio piano are displayed. Note the decrease in the number of the harmonics for the higher notes and also the rapidity with which the upper harmonics decrease in intensity for the higher notes compared to the lower. Fletcher et al. (1962) give an acceptable range of about a 2-dB decrease per partial for low tones to as much as a 40-dB decrease per partial for high tones. It is also apparent that the seventh partial is present in most of these tones.

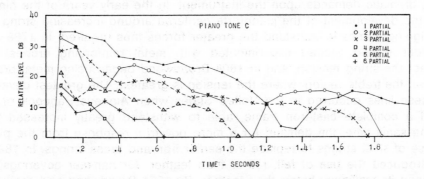

Fig. 6.H-5. Relative level vs time for the first six partials of the tone C. (From Fletcher et al., 1962.)

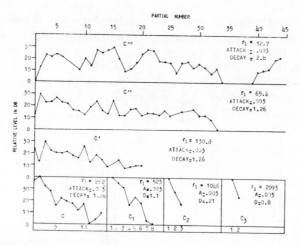

Fig. 6.H-6. Average partial structure of the studio piano. (From Fletcher et al., 1962).

History and Current Developments of the Piano

The development of the piano began in 1709 when Bartolomeo Cristofori, an Italian harpsichord maker, invented a key-actuated hammer with an escapement mechanism which allowed the hammer to bounce away from the string after striking. Since loud or soft tones could be created at will by striking the key lightly or heavily—as opposed to ordinary harpsichords, which produce tones at constant loudness, regardless of the force exerted by a finger—Cristofori called his invention "a harpsichord with loud and soft." The invention came to be known as the piano forte, from *piano* ("soft") and *forte* ("loud").

By 1726 Cristofori had arrived at all the essentials of the modern pianoforte mechanism, but it suffered from one severe disability: the frame did not have sufficient strength to withstand the string tension required to give musical effect to the force the mechanism was capable of exerting on the string. Consequently, the instrument languished for almost 70 years—little interest was shown in it by composers or musicians. About 1770, however, Johann Stein developed what came to be known as the German action. Also, about this time some composers began playing upon, and thus helping to popularize, the pianoforte, Johann Christian Bach being one of the foremost. It was left for Mozart, however, to start the pianoforte on its next stage of development. During the latter part of the eighteenth century he evolved a style of keyboard composition for the piano which could not be adequately performed on a harpsichord. Meanwhile, the double action, developed by Cristofori, was being produced in England. It is the standard piano action used today, having replaced the German action.

The year 1800 may be taken as a major turning point in the development of the piano. About this time Beethoven began composing his piano pieces, which made increasing dramatic demands upon the instrument. In the early years of the nineteenth century the development of the pianoforte centered around increasing string tension and designing frames to withstand the greater forces thus required. In 1788 an English builder named Stodart experimented with metal reinforcing arches to help counteract the string tension, and in 1808 Broadwood began using metal bracing to strengthen the treble region, where the tension is greatest. The greatest development along these lines occurred in the United States when Alpheus Babcock of Boston patented a complete cast-iron frame, able to withstand greatly increased tension. About the same time, the demand for a more powerful response from the piano led to the use of steel strings to replace the earlier iron and brass strings. In 1826 Henri Pope introduced the use of felt, rather than leather, for hammer coverings, as the felt retained its resiliency better than leather. By 1851 Broadwood had produced his first completely iron-framed grand piano, and in 1855 the Steinway metal frame (which has been the model for all successful frames ever since) was introduced. By 1859 Steinway had also perfected the system of overstringing, in which the bass strings run at an angle to the treble strings, thus shortening the piano case and also distributing the tension more evenly over the frame.

Meanwhile, the keyboard had been extended to over six octaves (C1 to E7) by 1840 and become standardized at seven octaves (A0 to A7) by 1855. By this time the piano had essentially achieved its modern form. The double action had evolved into the double escapement, or repetition, action, which is the direct ancestor of all modern grand piano actions. The middle pedal, which allows a given chord to be sustained, was invented in 1862 by C. Montal, the other two pedals, the soft pedal and the sustaining, having come into almost universal vogue by 1790. Soon after 1880 the three top notes, which extended the piano compass to C8, were adopted, so that by 1885 the grand piano had achieved its modern form in all particulars.

The upright and spinet pianos came into vogue after 1890, in part to conserve space and in part to reduce production costs and increase sales. The piano continues to evolve, primarily in terms of materials for construction, with plastic, teflon, nylon, and aluminum being used in place of earlier materials.

A recent interesting electronic innovation involves replacing the soundboard by electromechanical transducers, filters, amplifiers, and loudspeakers (Martin, 1970). The strings are excited in the usual way by striking, but the string vibration is transduced into an electrical signal, which is modified and amplified in a manner similar to that of the soundboard before being radiated by loudspeakers. Such an ''electronic soundboard'' has many advantages and gives the performer a much greater flexibility and control of the instrument.

Plucked String Instruments

Plucked string instruments share the possibility of inharmonic partials because they are free vibrators and have short attack times, no steady state, and variable length decays, in common with the struck string instruments. The plucked string instruments we will consider are the guitar, the harpsichord, and the dulcimer.

First consider the guitar. Most of the energy of the vibrating strings is coupled via the bridge to the body of the guitar, from which the sound is radiated. The guitar body ''colors'' the string spectrum in much the same way as a violin body influences the spectra of its strings.

The spectrum produced on a guitar string depends on where the string is plucked. An example of the spectrum for a guitar string plucked one-quarter the length of the string from one end is shown in Fig. 6.H-7a. Note that the fourth partial is conspicuously absent. The spectrum of the radiated sound is shown in Fig. 6.H-7b. Note that the first and fifth partials are attenuated substantially in relation to the string spectrum shown in Fig. 6.H-7a.

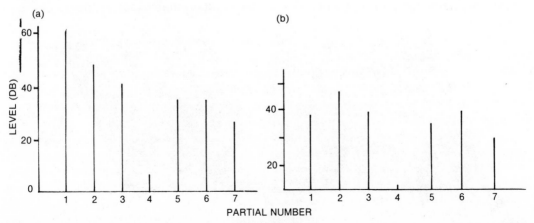

Fig. 6.H-7. Spectra of (a) guitar string plucked one-quarter of its length from one end and (b) radiated sound.

293

The difference in the above two spectra is due to the influence of the guitar body. This influence is quite substantial for all of the string instruments, because without the body of the instrument to radiate the sound, the strings would be almost inaudible. The guitar body consists of a wooden cavity with a circular opening. The wooden plates of the instrument have a complicated spectrum of resonances, the lowest resonance being called the W (or wood resonance). The placement of this resonance is crucial for well-designed instruments. The enclosed cavity and opening constitute a Helmholtz resonator, which gives rise to the air resonance. The air resonance is always found at a lower frequency than the main wood resonance, and its placement (in frequency) is just as important as the placement of the wood resonance. Approximate resonances of a guitar body, obtained by taking the differences between the spectra of Figs. 6.H-7a and b, are given in Fig. 6.H-8.

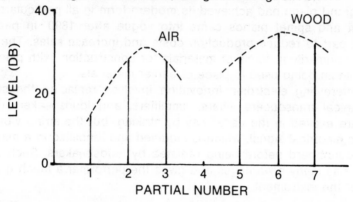

Fig. 6.H-8. Approximate resonances (dotted lines) of a guitar body. The first partial frequency was about 145 Hz.

Forced strings, as in the bowed violin, exhibit harmonic spectra, whether or not the strings have appreciable stiffness. Free strings, as in the percussive strings, exhibit harmonic partials if the strings have no stiffness, but they exhibit significant inharmonicity among the partials when the strings have appreciable stiffness. The piano exhibits significant amounts of inharmonicity. The guitar exhibits very small, but measurable, amounts of inharmonicity, the higher partials being slightly higher in frequency than harmonic partials would be.

Percussive sounds are characterized by an abrupt attack transient followed by a more or less extended decay transient. The information in the decay transient is important in our perception of the tone. The time-evolving spectra shown in Figs. 6.H-9 and 10 illustrate some of these features of the decay transients for guitar and banjo. Note that time increases from back to front in the figure, frequency increases from left to right, and sound level appears in the vertical. The guitar tone spectrum exhibits almost harmonic partials that have a longer duration than those of the banjo.

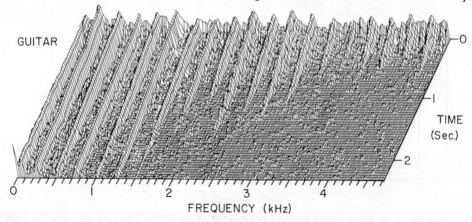

Fig. 6.H-9. Time evolving guitar spectrum. (Courtesy of E. P. Palmer and I. G. Bassett.)

294

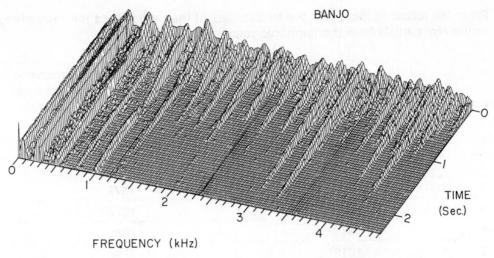

BANJO

FREQUENCY (kHz)

TIME (Sec.)

Fig. 6.H-10. Time evolving banjo spectrum. (Courtesy of E. P. Palmer and I. G. Bassett.)

Exercises

1. Why are violin partials harmonic and guitar and piano partials inharmonic?

2. What would happen to the violin partials if the instrument were plucked instead of bowed? Explain.

3. Where is the piano string struck? What partials should be missing? Are they? Why?

4. What are the resonance characteristics of a piano soundboard? How does this compare to the violin and guitar?

5. Suppose a piano were built with all strings made of the same material, having the same diameter and subjected to the same tension. If the top note (C8) had a string length of 8 cm, compute the length of the lowest C string (C1). Is this practical? Why?

6. How long would the string in exercise 5 have to be if mass per unit length were increased by a factor of four? If the tension were reduced by a factor of four? If both of the above changes were used?

7. When a chord is played on the piano while the sustaining pedal is depressed, the tone is richer than when the same chord is played without using the sustaining pedal. Explain.

8. What is the approximate frequency of the air resonance for the guitar?

9. Superimpose the guitar body resonance curve given in Fig. 6.H-8 on the spectra of guitar strings given below to show the spectrum of the radiated sound in each case.

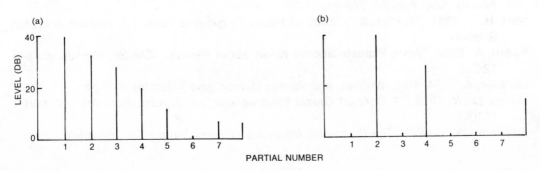

PARTIAL NUMBER

295

10. Fill in the following table with the frequencies of the partials and the "stretching" of the real partials from the harmonic value.

Partial	"harmonic" guitar	real guitar	stretching of guitar partials	"harmonic" piano	real piano	stretching of piano partials
1	145	145		393	393	
2		292			785	
3		436			1180	
4		—			1577	
5		728			1976	
6		871			2380	
7		1018			2798	
8		—			3174	
9		—			—	
10		—			—	

11. Name other struck string instruments.

12. Name other plucked string instruments.

*13. Place your finger on a piano key and, keeping contact between finger and key, depress the key with maximum speed. Now, continually reduce your speed until depressing the key no longer produces a tone. Describe what is taking place in the action when this occurs.

*14. Try placing weights on a piano key to determine the force necessary to depress it. Do different keys on the piano require different weights? If possible, repeat the experiment on different pianos of different makes and draw some general conclusions.

Further Reading

Backus, chapter 13
Benade, chapters 7, 8, 9, 17, 18
Culver, chapter 11
Kent, part I
Olson, chapters 5, 6
Blackham, E. D. 1965. "The Physics of the Piano," Scientific Am. *99* (6), 88-89.
DiLavore, P., ed. 1975. *The Guitar* (McGraw-Hill, N.Y.)
Fletcher, H. 1964. "Normal Vibration Frequencies of Stiff Piano Strings," J. Acoust. Soc. Am. *36*, 203-9.
Fletcher, H., E. D. Blackham, and R. Stratton. 1962. "Quality of Piano Tones," J. Acoust. Soc. Am. *34*, 749-61.
Hart, H. C. 1934. "A Precision Study of Piano Touch and Tone," J. Acoust. Soc. Am. *6*, 80-94.
Kuerti, A. 1973. "What Pianists Should Know about Pianos," Clavier, May-June 1973, 12-22.
Loesser, A. 1954. *Men, Woman, and Pianos* (Simon and Schuster, N.Y.).
Martin, D. W. 1970. "A Concert Grand Electropiano," J. Acoust. Soc. Am. *47*, part 1, 131(A).
McFerrin, W. V. 1971. *The Piano—Its Acoustics* (Tuners Supply Co., Boston).

Audiovisual

1. *Plucked String Instruments* (3 3/4 ips, 2 track, 30 min, UMAVEC)
2. *Real and Synthetic Piano Tones,* section 6.H, Descriptive Acoustics Demonstrations (GRP)
3. *The Guitar* (23 min, color, OPRINT)
4. *The Piano*, Part 2 (21 min, color, OPRINT)

6.I. PERCUSSION INSTRUMENTS

The percussion instruments may be divided into four classes on the basis of their vibrator type: membranes, bars, plates, and bells. They can be further subdivided into definite pitch percussives (those producing tones with a definite pitch) and indefinite pitch percussives. The following general observations can be made concerning the percussion instruments:

1. The natural mode frequencies are usually inharmonic.
2. Compared to percussive strings, the vibrating area is generally large; so the energy is radiated directly.
3. The excitation of percussive instruments is usually impulsive.

Membrane Percussives

Membrane percussives employ as their vibrator a membrane stretched in two dimensions. Two-dimensional waves are set up in the membrane, which can give rise to rather complex standing wave patterns. Indefinite pitch membrane percussives include bass drum and snare drum, each of which employs two membranes on opposite ends of a short cylindrical tube. The kettledrum is a definite pitch percussive and employs a single membrane.

Kettledrums (timpani) consist of circular membranes stretched tightly over the open sides of large hemispherical bowls. The player sets the membrane into vibration by striking it with a felt-covered stick. The drum also has a mechanism for changing the tension on the membrane, which changes the frequency of vibration. The kettledrum is the only percussion instrument utilizing a stretched membrane which has a tone of definite pitch. This pitch is determined by four factors: (1) the diameter of the membrane, (2) the volume of the bowl, (3) the tension on the membrane, and (4) the density (mass/area) of the membrane. The kettledrum is usually struck at a distance several centimeters from one edge, so that many natural modes of vibration are excited.

Many different modes of vibration are possible on a two-dimensional membrane. We considered some of these for a square membrane in section 2.D. Circular membranes are more commonly used for musical percussion instruments, however, and we now consider their natural modes. The six lowest-frequency natural modes for a circular membrane fixed around its circumference are shown in Fig. 6.I-1. The circular and straight lines represent nodal lines of zero displacement for the membrane. We note that the entire membrane moves as a unit in the first mode. It moves as two opposing parts in the second mode and as four parts in the third mode. The fourth mode involves the motion of two opposing parts, but they are concentric with each other. The fifth mode involves six parts separated by diametrical nodal lines. The sixth mode involves four parts separated by diametrical and circular nodal lines.

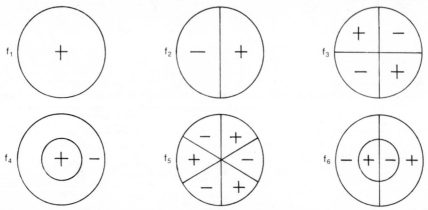

Fig. 6.I-1. Six lowest frequency modes of a circular membrane. The + and − signs show relative direction of motion.

298

The natural frequency ratios for the six modes of a circular membrane are listed in Table 6.I-1. The frequency ratios in the second column show how the frequency of each mode is related to that of the lowest mode. We note that in general the modes bear a very inharmonic relationship to the lowest mode. The frequency ratios in the third column show how the frequency of each mode is related to that of the second mode, and again we see very inharmonic relations.

Table 6.I-1. Frequency ratios of the first six modes of circular membrane "instruments."

| Mode | Freq. Ratio to First Mode | Freq. Ratio to Second Mode | | |
	Membrane Alone	Membrane Alone	Membrane and Air	Actual Kettledrum
1	1.00	—	—	—
2	1.59	1.00	1.00	1.00
3	2.14	1.35	1.50	1.56
4	2.30	1.45	1.83	1.76
5	2.65	1.67	1.92	2.0
6	2.92	1.84	2.33	2.2

We are confronted with the question of how a kettledrum can have a definite pitch if the modes are so inharmonically related. We find part of the answer by considering the effect the air in the "kettle" has on the different modes. Imagine the excitation of the lowest mode of the membrane in which the membrane moves as a unit. The motion of the membrane will tend to compress and expand the air in the "kettle," which will result in more stiffness for the mode and thus a higher frequency. However, two features come into play that make the lowest mode of limited importance for the kettledrum. The small hole in the bottom of the "kettle" lets air move in and out as the membrane tries to expand and compress the air; this produces a lot of resistance that damps the mode. In addition, the lowest mode radiates sound fairly efficiently, which also helps to damp the mode. Thus, any excitation of the lowest mode will die out rather quickly and will contribute mostly an initial "thump" to the sound. The time-evolving spectrum in Fig. 6.I-2 shows this quick dying out of the lowest mode.

KETTLEDRUM

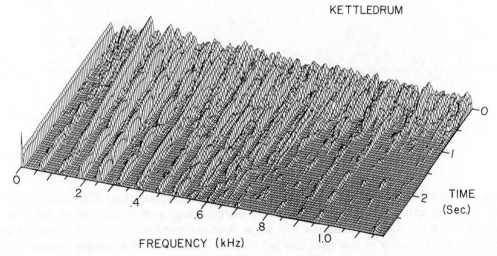

Fig. 6.I-2. Time evolving kettledrum spectrum. (Courtesy of I. G. Bassett.)

The second mode turns out to be the first musically useful mode. Because the membrane moves as two opposing parts, this mode does not tend to compress and expand the air inside, but rather to push it back and forth from one side of the

299

"kettle" to the other. The air in this case can be thought of as adding mass to the vibrator, thus decreasing the frequency. This mass loading due to the air affects most of the higher modes, but to a much smaller extent than the second mode. Hence, the second mode is lowered in frequency by a greater factor than the higher modes, which results in the frequency ratios shown in column 4 of Table 6.I-1 for an air-loaded membrane. (The assumptions made for the calculations were that the air loading increased the effective membrane surface density by 60%, 29%, and 20% for modes 2, 3, and 5 respectively. Modes 4 and 6 were assumed to be unaffected.) Frequency ratios actually measured for a kettledrum are shown in the last column. A comparison of the two shows that the frequency-ratio differences between a plain membrane and an air-loaded membrane can be explained reasonably in terms of an increased mass due to the air loading.

The frequency ratios for the kettledrum are seen to be more "musically interesting" than the corresponding frequency ratios for a plain membrane. Note in the last column of Table 6.I-1 that if we ignore the first mode the frequency ratios of the second and fifth modes are roughly 1.0 and 2.0, which corresponds to the fundamental and second harmonic partial. Furthermore, the frequency ratios of modes 2, 3, and 5 are about 1.0, 1.5, and 2.0. But this is just an appropriate collection of frequencies to be the harmonics of a missing fundamental of 0.5. Either of these combinations gives rise to the perception of a definite pitch which we have come to expect from a kettledrum. Apparently the fundamental, having a relative value of 1.0, is the one commonly associated with the pitch of the instrument.

The bass drum does not produce sounds of definite pitch, and its vibrations are more complex than those produced by the kettledrum. This is because it has two heads tuned to slightly different frequencies, which gives rise to beating in the tone. When a drummer strikes one head, the vibrating air inside the drum causes the other head to vibrate, but the vibrations of the two heads alternately add constructively and destructively. The time-evolving spectrum shown in Fig. 6.I-3 illustrates the inharmonic modes of the bass drum. The effects of beating can be seen with some of the modes. (Note in Fig. 6.I-3 that time increases from front to back.)

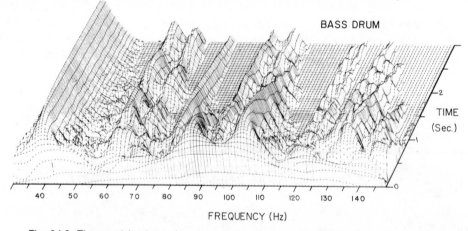

Fig. 6.I-3. Time evolving bass drum spectrum. (Courtesy of H. Fletcher and I. G. Bassett.)

An additional feature of interest with the bass drum is that when it is struck a hard blow all of its natural frequencies are a few percent higher than normal at the beginning of the tone. This is because the large displacement caused by the hard blow produces an increase in the effective tension on the membrane, which results in an increase of all the natural frequencies.

Drum heads are usually made of calfskin, but in recent years manufacturers have been providing plastic heads which are not as readily affected by temperature and humidity changes. The tone of these heads, however, is different, and many players do not care for the plastic heads.

Percussive Bars

Instruments belonging to the percussive bar family include the glockenspiel, marimba, xylophone, vibraphone, chimes, and triangle. They employ "bars" as their vibrating element, in which the restoring force is due to internal stiffness of the bar and not to any externally applied tension. The vibrating bars can be considered as one-dimensional vibrators, somewhat analogous to strings. (See Rossing, 1976, for a discussion of torsional, longitudinal, and edgewise modes not considered here.) When we discussed piano strings in section 6.G we observed that the partials were stretched and made inharmonic because of the stiffness of the string. We might expect a much more exaggerated inharmonicity for vibrating bars.

The shapes of the first four modes of a uniform bar free at both ends are shown in Fig. 6.I-4. Corresponding natural mode frequency ratios are listed in column two of Table 6.I-2. We note that the frequency ratios are very inharmonic, which gives rise to an indefinite pitch sound. However, several things can be done to adjust these ratios and to produce a definite pitch sound.

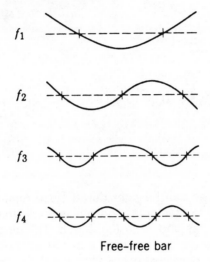

Free-free bar

Fig. 6.I-4. Bar shapes (exaggerated) for the first four modes of a uniform free-free bar.

Table 6.I-2. Frequency ratios for first six modes of bar vibrators. (After Rossing, 1976.)

Mode	Uniform Bar	Glockenspiel	Xylophone	Chime	
1	1.00	1.00	1.00	1.00	
2	2.76	2.71	2.97	2.79	
3	5.40	5.15	5.9	5.45	
4	8.93	8.43	—	8.95	(2.0)
5	13.34	12.21	-	13.20	(2.95
6	18.64	—	—	18.16	(4.06)

The glockenspiel (orchestra bells) employs nearly uniform rectangular metal bars, as shown in Fig. 6.I-5a. No attempt is made to adjust the frequency ratios of the higher modes because they die out rather quickly, leaving the fundamental mode to linger on. The pitch is associated with the frequency of the lowest mode. The higher modes contribute primarily to the initial decay transient of the instrument. The decaying partials can be seen in Fig. 6.I-6, which shows a time-evolving spectrum for a note of a glockenspiel.

(a)   (b)

Fig. 6.I-5. Vibrating bars: (a) uniform; (b) undercut.

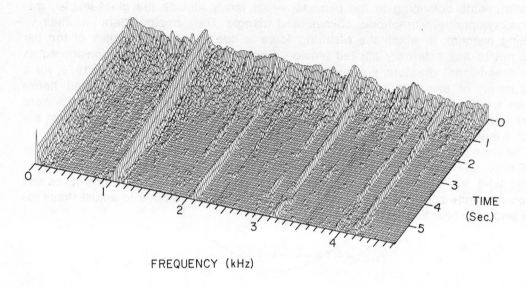

FREQUENCY (kHz)

Fig. 6.I-6. Time evolving glockenspiel (orchestral bells) spectrum. (Courtesy of I. G. Bassett.)

The marimba and xylophone can be considered close relatives in that both employ wooden bars as their vibrating elements. The bar is made thinner in the middle than at the ends by undercutting, as shown in Fig. 6.I-5b. Thinning a bar at a nodal point results in little change in frequency. Thinning a bar at antinodal points can decrease the frequency of the bar by decreasing the stiffness if the thinning is done in a region of maximum bending (e.g., mid-point of bar for mode 1). If the thinning is not at a point of bending, the thinning reduces the mass and raises the frequency. Thinning results in a modification of the nodal points of the various modes and in a significant modification of the frequency ratios of the modes. Typical frequency ratios appear in column 4 of Table 6.I-2 for a xylophone. Note that the second mode frequency is about three times that of the first mode and when sounding with the first will produce a pitch associated with the first mode. The xylophone and marimba also employ cylindrical resonators (located beneath the bar) tuned to the fundamental frequency so as to amplify the sound. The time-evolving spectrum of a typical xylophone tone appears in Fig. 6.I-7. Note that the fundamental is the strongest component and that the next most prominent component has a frequency about three times that of the fundamental. Similar spectra for marimba tones show the second prominent partial to be tuned to a frequency nominally four times that of the fundamental.

The vibraphone employs undercut metal bars in which the second mode is tuned approximately to the octave of the first mode. A unique feature of the vibraphone is a set of motor-driven discs which rotate over the resonator. By alternately opening and closing the top of the resonator a tremolo effect is produced.

Chimes employ cylindrical metal tubes as their vibrating elements. A plug is placed in one end of the tube, which modifies some of the natural frequencies. Frequency ratios for a chime are listed in column 5 of Table 6.I-2. If we assign the fourth mode a relative value of 2 (in parentheses) we see that the fifth and sixth modes have relative values nearly equal to 3 and 4 respectively. These three modes produce frequen-

302

XYLOPHONE

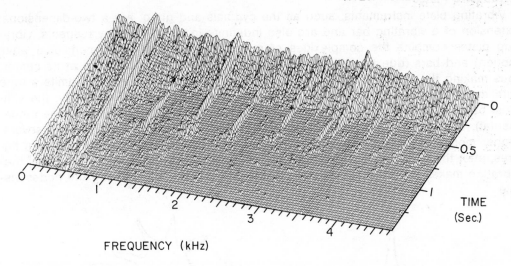

FREQUENCY (kHz)

Fig. 6.I-7. Time evolving xylophone spectrum. (Courtesy of I. G. Bassett.)

cies having relative values of 2, 3, and 4, which give rise to a perceived pitch for the strike tone having a relative value of 1, i.e., an octave lower than the fourth mode. The first three modes contribute primarily to tonal quality, but not to the perceived pitch. We can observe these features in the time-evolving A4 chime spectrum shown in Fig. 6.I-8. Note that the first partial dies out rapidly, the second partial lasts longer, and the third component is very long lasting. Note that the fourth, fifth, and sixth partials have frequencies of about 880 Hz, 1300 Hz, and 1750 Hz, which give rise to a "pitch frequency" of 440 Hz.

CHIME A$_4$ microphone 4

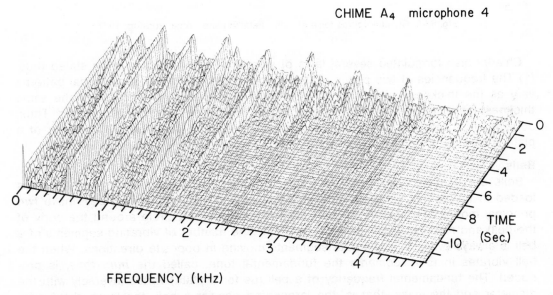

FREQUENCY (kHz)

Fig. 6.I-8. Time evolving chime spectrum. (Courtesy of H. Fletcher and I. G. Bassett.)

Triangles are of indefinite pitch because several of the inharmonic modes of a uniform bar are excited at significant levels.

Percussive Plates

Vibrating plate instruments, such as the cymbals and gong, are a two-dimensional extension of a vibrating bar and are also indefinite in pitch. In many respects, vibrating plates combine the complexities of membranes (due to two-dimensional wave motion) and bars (due to stiffness). If a solid circular plate is clamped at its center, thus making the center a nodal point, and then bowed on the edge, it emits a tone with many inharmonic partials. The plate vibrates in a series of segments, the simplest being when there are four segments and four nodal lines (producing the fundamental). The series of sketches in Fig. 6.I-9 shows this type of vibration and several higher modes of vibration for such a plate. These figures are known as Chladni figures, after the man who first obtained such figures by placing sand on the plate; the vibration makes the sand accumulate on the nodal lines, thus making the nodes visible.

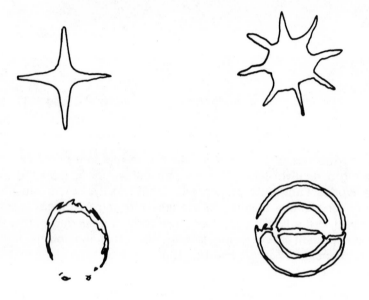

Fig. 6.I-9. Chladni figures for a circular metallic plate. (After Rossing, 1977.)

Chladni also formulated several laws of vibrating plates which can be stated thus: (1) The frequencies of two plates of like shape and showing the same nodal patterns vary as the thickness of the plates. (2) The frequencies of two plates of the same thickness and same nodal pattern vary as the inverse square of the diameter. Thus, doubling the plate thickness doubles the frequency and doubling the diameter of a plate lowers the frequency by a factor of 0.71.

Bells

Bells are acoustically related to vibrating plates, but they are curved and mass loaded in the center. Like vibrating membranes and vibrating plates, bells have two principal modes of vibration: circular (having nodes which circle around the body of the bell) and meridian (having vertical nodes). The number of vibrating segments of a bell is always even, with alternate segments moving in opposite directions. When the bell vibrates in four segments the fundamental tone, called the *hum tone,* is produced. The fundamental frequency of a bell (as for a plate) varies inversely with the diameter and the mass—that is, the larger and heavier a bell, the lower the fundamental frequency.

There are four meridian nodes for the hum tone, meaning that alternate quarters of the bell are moving outward and inward. The next mode is called the prime mode because it has a frequency close to what you hear when the bell is struck (the so-

called "strike tone"). The *prime tone,* which also has four meridian nodes, is tuned to one octave above the hum tone (which is perceived as a lower-pitched hum following the initial strike tone). The next mode above the prime tone is called the third because it is tuned to sound a minor third above the prime tone. The next mode is termed the fifth because it is tuned a perfect fifth above the prime. Both the third and fifth have six meridian nodes. The next higher mode is the nominal (or octave), which has twice the frequency of the prime tone. The octave has eight meridian nodes. The strike tone, alluded to earlier, has traditionally been associated with the perceived pitch of the bell. The perceived pitch of the strike tone is determined by the nominal and two additional upper modes: the twelfth and the superoctave. These three modes are tuned to a 2:3:4 frequency ratio, and the resulting strike tone is adjusted so as to have a frequency close to that of the prime tone.

When a bell is thinned along a nodal line the frequency of the associated mode is affected very little. When a bell is thinned at antinodal points the frequency of the associated mode may be increased or decreased (in a manner similar to bars), depending on whether the mass or stiffness for a particular mode is affected.

After the bell has been cast, the various partials are tuned by machining metal from the inside surface with a lathe. By knowing the vibration pattern of the bell and by carefully thinning the bell at the appropriate place one can bring the bell into tune with itself. Fig. 6.I-10 shows the vibration pattern of the first five modes, as well as the areas on the interior surface which are important for tuning each mode.

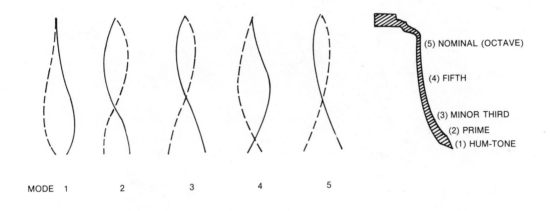

Fig. 6.I-10. The first five modes of a carillon bell. (After Grutzmacher, et al., 1966.)

It has long been considered (see Schad and Warlimont, 1973) that the pitch associated with a bell is that of the strike tone. This initial subjective perception of pitch is then reinforced by the well-tuned prime tone. Table 6.I-3 lists the measured frequencies (and frequency ratios) for the first five modes of three carillon bells. The carillon, located on the Brigham Young University campus, was built by the I.T. Verdin Company (Cincinnati, Ohio) in 1975. A time-evolving spectrum of the A4 bell is shown in Fig. 6.I-11 for comparison. From the spectrum and Table 6.I-3 we discover several interesting facts. First, contrary to the usual assertions that the pitch of a bell is determined by the strike tone (and prime mode), we see that the perceived pitch of these bells is in fact the pitch of the hum tone. Secondly, since there is no perceived pitch change as the bell is struck, the strike tone for these bells must be about the same as the hum tone. Since the second, fourth, and fifth modes have frequency ratios of almost exactly 2, 3, and 4 times the frequency of the fundamental (hum tone), the perceived pitch caused by the difference tone will again be that of the hum tone. By comparing frequencies from the table, the three bells are seen to be quite accurately tuned to A3 = 220, A4 = 440, and A5 = 880 Hz, respectively.

Table 6.I-3. Frequencies and frequency ratios for first five modes of each of three carillon bells. (Courtesy of I. G. Bassett.)

Mode Name	Frequency			Frequency Ratio		
	A3	A4	A5	A3	A4	A5
1 hum	220.0	439.2	881.0	1.00	1.00	1.00
2 prime	440.3	878.5	1763.1	2.00	2.00	2.00
3 third	523.6	1045.0	2095.6	2.38	2.38	2.38
4 fifth	662.3	1318.5	2649.2	3.01	3.00	3.01
5 nominal	880.0	1755.9	3524.1	4.00	3.99	4.00

Often, a fifth of a second or so after a bell is struck a characteristic pulsating tone is heard. This variation in intensity is caused by beats due to slight differences in pitch between vibrating segments of the bell, probably resulting from some nonuniformities in the mass distribution. This effect can be seen in Fig. 6.I-11, along with the frequencies and decay rates of the various modes.

Computer Analysis Bell 21 (A₄)

Fig. 6.I-11. Time evolving bell spectrum. (Courtesy of I. G. Bassett.)

A carillon is a set of at least two octaves of bells, very carefully tuned to chromatic intervals, playable from a keyboard which permits control of expression through variation of touch. Although the carillon console resembles an organ console (with hand and foot operated levers rather than a keyboard and a pedalboard), the action is more closely related to that of the piano. The console, although it appears to be clumsy, has been designed to permit the performer (without any assistance from electrical or pneumatic devices) to articulate with sensitivity and virtuosity on bells weighing many tons. Because of the extremely high cost involved in manufacturing carillon bells and because of the large space required for a carillon tower, chimés or electroacoustic devices have often been employed to simulate bells. In section 6.J we will consider the electroacoustic method in more detail.

Exercises

1. When a membrane is struck in the center, which modes will not be present? When a pate is clamped in the center which modes are not possible?

2. Sketch the 5th vibration mode for a vibrating bar.

3. Describe how the restoring forces associated with a stiff bar compare to the restoring forces in a piano string. Describe the relative importance of tension and stiffness in each case. What bearing does this have on the sketching of the partials in each case?

4. What observations can be made about the frequencies of the higher modes relative to the frequency of the lowest mode for nonpercussive instruments? For percussive string instruments? For vibrating bars? For vibrating membranes?

5. How do inharmonic partials affect waveshapes as compared to waveshapes created by harmonic partials? How do sound qualities differ?

6. Are there more or fewer modes present within a given frequency range for the tone of a percussive string as compared to a tone with harmonic partials? For a bar? For a membrane? How big are the differences?

7. Through comparison of chime and bell spectra (Figs. 6.I-8 and 6.I-11), make arguments as to why chimes can often be used to simulate bells.

8. Determine the frequencies and amplitudes of the first six modes for the chime tone shown in Fig. 6.I-8 right after the tone began. Determine the amplitudes of the modes at several later times. Which mode decays first? Which lasts the longest?

9. Repeat exercise 8 using the time-evolving spectra of other instruments.

*10. a. Decide on the gross features of a percussive instrument that you wish to construct.
   b. By consulting appropriate sources determine the physical properties of your instrument (such as material, length of vibrators, vibrator mounting, etc.) so that it will produce the desired frequencies.
   c. Construct the instrument.
   d. Take the instrument into the laboratory and make measurements to determine how well you have achieved your design goals.

Further Reading

Backus, chapter 14

Benade, chapters 5, 9

Culver, chapter 14

Kinsler and Frey, chapters 3, 4

Olson, chapters 5, 6

Grutzmacher, M., W. Kallenbach, and E. Nellessen. 1965-66. "Acoustical Investigations on Church Bells," Acustica 16, 34-45.

Hardy, H. C., and J. E. Ancell. 1961. "Comparison of the Acoustical Performance of Calfskin and Plastic Drumheads," J. Acoust. Soc. Am. 33, 1391-95.

Jones, A. T. 1937. "The Strike Tone of Bells," J. Acoust. Soc. Am. 8, 199-203.

Rossing, T. D. 1976. "Acoustics of Percussion Instruments—Part I," The Physics Teacher 14, 546-55.

Rossing, T. D. 1977. "Acoustics of Percussion Instruments—Part II," The Physics Teacher 15, 278-88.

Schad, C., and H. Warlimont. 1973. "Acoustic Investigation of the Influence of the Material on the Sound of Bells," Acustica 29, 1-14.

Audiovisual

1. *Vibrations of a Drum* (Ealing Film Loop 80-3924)
2. *Vibrations of a Metal Plate* (Ealing Film Loop 80-3932)

Demonstrations

1. Effect of striking drum head at various points

6.J. ELECTRONIC INSTRUMENTS

The musical instruments discussed in the preceding sections used mechanical vibrators to excite acoustical resonators, which in turn radiate the sound energy directly into the air. In this section we will consider electronic instruments in which the oscillations exist in the form of electrical signals at some point and are electronically amplified and radiated into the air by means of a loudspeaker. The basic elements of an electronic musical instrument appear in Fig. 6.J-1. (The extensive use which has been made of electronics for the reproduction of sound is considered in chapter 7.) There are two basic types of electronic instruments: one in which a mechanical oscillation is converted into an electrical oscillation and electronically modified and another in which the original oscillation is electrical. These two types are considered in this section.

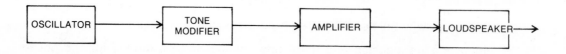

Fig. 6.J-1. Elements of an electronic musical instrument.

Tone Modifiers

The instruments belonging to this group are characterized by a mechanical oscillation that is converted into an electrical waveform by means of an appropriate transducer. The tone modifier consists of electronic circuitry designed to modify the waveforms by filtering, frequency division, addition of noise, etc. Instruments included in this group are the electropiano, electric guitar, electric carillon, and "electrical wind instruments."

An early application of electronics to the piano was to remove the soundboard and use electronic pickups to capture and subsequently amplify the sound of the piano strings. (With the soundboard missing the strings by themselves would be almost inaudible.) The sound produced by this method, however, had only the barest resemblance to the tone of a piano, one problem being that the strings vibrate for a longer time when the soundboard is not present to dissipate the energy. A later development in electronic pianos utilized a vibrating reed, struck by a hammer, with electronic pickup. The reed vibration decays in about the same time as a normal piano string, but the harmonic structure of the vibrating reed is quite different from that of a string, and thus the resulting sound resembles a piano only in its decay characteristics. More recently an electropiano has been developed which largely retains the conventional piano tone quality, while enabling the piano to have a wider dynamic range, as well as new tonal capabilities. This instrument was briefly described in section 6.H.

The electric guitar consists of six strings mounted on a heavy body that radiates little sound energy. The vibrations of the strings are transmitted to the bridge, where they are picked up by mechanical-to-electrical transducers and converted into electrical signals. Because of the amplification available (see Fig. 6.J-1), the instrument can compete in loudness with an ensemble of conventional instruments. (This same approach is applicable to other stringed instruments, such as the violin.)

The electrical carillon consists of small mechanical vibrators, such as tuned bars or rods, which have partials similar to those of a bell that are controlled from a keyboard. Mechanical-to-electrical transducers are used to convert the vibrations into electrical signals which are filtered (to eliminate undesirable harmonics) and then presented over loudspeakers.

An "electrical" clarinet is illustrated in Fig. 6.J-2. It consists of a conventional clarinet with a small microphone mounted so as to sense the pressure inside the clarinet mouthpiece and convert it into an electrical signal. The tone modifier part of this system consists of a frequency divider and a set of filters. The frequency divider makes it possible to obtain a tone one or more octaves lower than the one played, so that a performer could accompany himself with a simulated bass clarinet tone an octave lower. The filters provide a means for modifying the spectrum of the tone and thus enable the production of different tonal qualities. If the frequency divider is not used, the system can function as an amplifier for the clarinet tone. The resulting sound output will be that coming from the instrument directly plus that coming from the loudspeaker. This same approach is applicable to other wind instruments, such as flute or trumpet.

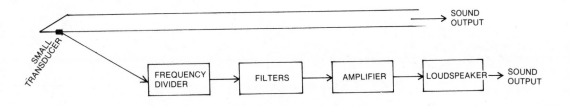

Fig. 6.J-2. Tone modifier for clarinet.

One kind of electronic music is created by recording natural sounds, such as speech, bird songs, or city noise, and then passing them through a tone modification system to achieve the desired results.

Electronic Organs

At the present time the most extensive application of electronics to the direct production of sound is in the electronic organ. Almost all of the models being manufactured today use various circuits which generate the electrical oscillations directly in one of two ways: additive synthesis or subtractive synthesis. *Additive synthesis* constructs a complex tone by adding together various amounts of simple tones having different frequencies. By a proper selection of frequencies and their relative amplitudes various tone colors can be achieved. (Recall that any complex musical tone can be constructed with frequency components of various amplitudes.) The additive synthesis process is represented in Fig. 6.J-3.

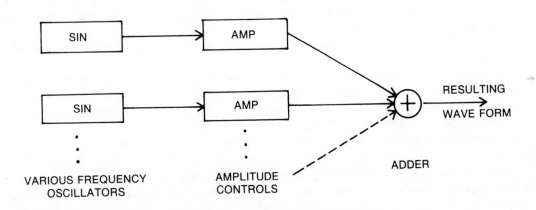

Fig. 6.J-3. Elements of additive synthesis.

Subtractive synthesis starts with a tone which is rich in harmonics and deletes or attenuates these harmonics to produce the desired tone quality. The attenuations are produced by various filters, and the original waveform (before filtering) is usually one which is easy to generate electronically, such as a sawtooth or a square wave. The subtractive synthesis process is illustrated in Fig. 6.J-4.

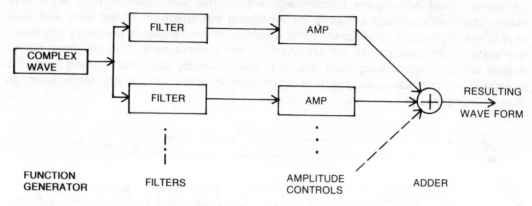

Fig. 6.J-4. Elements of subtractive synthesis.

An example of a commercially available instrument which utilizes additive synthesis is the Hammond organ. With this instrument, the performer has control over the relative amplitudes of nine independent harmonics, each of a different frequency but related to the fundamental. To a first-order approximation this enables the performer to reproduce the steady-state spectrum of any musical instrument tone, including pipe organ. (The fact that the Hammond is unable to accurately reproduce such a tone is evidence of the importance of factors other than steady-state spectrum. Factors such as attack time, decay time, and ensemble must all be considered if a realistic representation of an actual musical instrument is desired.)

Subtractive synthesis is used in organs such as the Baldwin, Conn, and Schober. The following basic components may be utilized in schemes of this sort (see Fig. 6.J-5):

1. A sine wave oscillator (SIN) produces a pure tone (sinusoid) of any desired frequency.
2. A saw-tooth oscillator (SAW) produces waves resembling a saw blade.
3. A pulse train (PUL) is a series of sharp pulses which occur at any desired repetition rate.
4. A square wave oscillator (SQU) produces a waveform that changes from steady positive values to steady negative values.
5. A triangular wave oscillator (TRI) produces waves having the appearance of repeated triangles.
6. A noise generator (NOI) produces a waveform whose consecutive values are random.
7. A low-pass filter (LPF) passes low frequencies, but attenuates high frequencies.
8. A high-pass filter (HPF) passes high frequencies, but attenuates low frequencies.
9. A band-pass filter (BPF) attenuates both high and low frequencies, but passes a central band of frequencies.
10. An amplitude control (AMP) controls the relative amount of energy.

Suppose, for example, that one desired an oboelike sound on an electronic organ. It could be produced by following Fig. 6.J-4 and choosing a saw-tooth wave of the desired frequency (the frequency is controlled by the note one depresses) as the complex wave, then passing the saw-tooth wave through two band-pass filters, one peaking at 1000 Hz and one at 3000 Hz (see section 6.D), and adding the outputs.

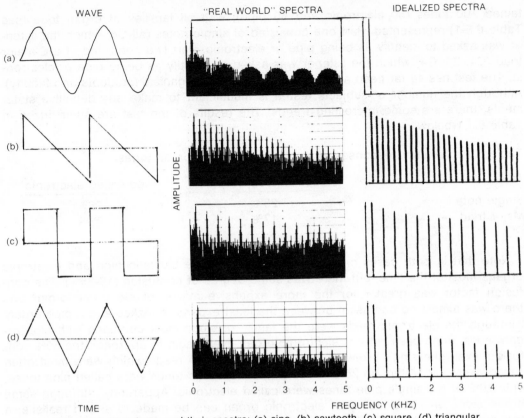

WAVE   "REAL WORLD" SPECTRA   IDEALIZED SPECTRA

AMPLITUDE

TIME   FREQUENCY (KHZ)

Fig. 6.J-5. Waveforms and their spectra: (a) sine, (b) sawtooth, (c) square, (d) triangular.

Although the tone quality of electronic instruments which use subtractive synthesis is closer to that of a pipe organ than the additive synthesis devices, they still leave much to be desired. Even though some work has been done toward achieving realistic attack and decay times, there is (in most instruments) no ensemble effect since only 12 tone generators are employed for the entire instrument. This results in all "pipes" being mathematically in tune so that inadequate amounts of beating occur (see section 6.A). Comparing this to even a small pipe organ having hundreds of individual tonal sources, it is no wonder that the imitation is a rather poor one. Recently the Allen Organ Company has marketed a "digital computer organ" which uses an enclosed computer to generate sounds resembling those of a pipe organ. When the performer selects a desired tone quality the sound is constructed from stored information available to the computer. An instrument such as this goes a long way toward removing some of the limitations on electronic simulation of pipe organ quality.

We have recently completed a preliminary study on whether organists can differentiate between electronic organ tones and pipe organ tones when presented via a tape recording. (We realize that a pipe organ achieves a spatial effect which no number of well-placed speakers can duplicate.) The tests involved tones from seven of the major manufacturers of electronic organs and nine different makes of pipe organ. The electronic instruments varied from new dealer demonstration models to instruments more than ten years old. The pipe organs varied from recently installed instruments to fifty-year old instruments. Only concert (or church) model electronic instruments were recorded. Small electronic entertainment models were by-passed so that there would be more chance for comparison; i.e., small entertainment organs do not make any attempt to duplicate the sounds of a pipe organ and therefore are of no interest to our study. Some of the electronic instruments were situated in rooms with very little reverberation, as were some of the pipe organs (practice rooms). Likewise, many of the pipe organs and a number of their electronic counterparts were located in fairly reverberant churches. The test consisted of two parts; each part contained 100 tones (50 electronic and 50 pipe), with all families of organ tone (see Table 6.F-1) represented. Part one consisted of single notes (all C4) which the listener was asked to identify as being pipe or electronic. Part two consisted of a C-major triad (C4, E4, G4) which the listener was asked to identify as being pipe or electronic. The test has so far been administered to a dozen organists (students and faculty). Although the number of subjects tested is insufficient to make any definitive statements, there are some interesting trends. The results of the test are summarized in Table 6.J-1 below.

Table 6.J-1. Listener responses to electronic and pipe organ tones.

Test	Electronic called pipe	Pipe called electronic
Single note	25%	16%
Major triad	17%	23%

Note that in both cases organists could differentiate between pipe and electronic organs most of the time, but there was some degree of confusion (16-25%). The confusion factor was greater for the more expensive makes of electronic organ and there was almost no confusion between the least expensive makes and a pipe organ. Although the electronic reeds were the family of stops most confused with pipe organ reeds, in no case did any electronic organ even begin to approach the 50% confusion level (guessing). Likewise, it was the pipe organ reeds which were most often called electronic. Although 25% of the single electronic tones were called pipe tones, only 16% of the single pipe tones were called electronic. Apparently, although some organ stops on some makes of electronic organ can be made to sound realistic, a real pipe does not sound like an imitation quite as often. For the major triad, the results were reversed: only 17% of the electronic tones were called pipe, while 23% of the pipe tones were confused with electronic tones. Even the more expensive makes were confused much less often on this test; probably the lack of real ensemble in the electronic organ tones (even when the manufacturer has made heroic attempts to provide ensemble at a reasonable price) is apparent to the organist's trained ear. Why, then, were pipe organ tones, when played as a chord, confused with electronic tones more often?

While we cannot provide a definitive answer to the question at this time, we are prepared to venture a guess. We believe that one of the primary clues used to differentiate pipe organ tones from electronic organ tones is the initial attack. When one listens to individual tones the attack is very perceptible and, as we discovered in section 6.A, the attack time gives important information about the perceived tone. When a chord is played, there are three notes and up to two dozen individual pipes sounding. Obviously, it is impossible to distinguish an individual ensemble in such a case. We therefore conclude that while the ensemble effect so characteristic of pipe organs helps to make many of the electronic instruments sound "dead" in comparison, the same ensemble muddles an important acoustic clue (the attack of individual pipes) so that more of the pipe organs are confused with their electronic imitators.

Synthesizers

Synthesizers consist basically of some control mechanism (punched paper tape, keyboard, etc.) attached to various electronic elements, such as oscillators and filters. The control mechanism permits the "performer" to control such things as frequency, vibrato, tremolo, and spectrum change of the electronic elements used for synthesis.

The RCA synthesizer utilizes preprogrammed punched paper tape to control a rather elaborate electronic synthesis device. The output of the synthesizer is stored on tape and rather complex musical structures can be achieved by means of multiple recording.

Other synthesizers, such as the Moog and Arp, employ a keyboard as the control mechanism. The performer can control spectrum, frequency, attack and decay time, and other musically useful parameters.

Synthesizers were originally devised to produce music a line at a time, record it on one track of a multitrack tape, and then mix the results as many times as required to produce the complexity desired in the musical structure. The first synthesizers were limited to the control of one oscillator at a time from the keyboard. However, recent synthesizers permit the control of several oscillators at a time and ensembles of synthesizers can sometimes provide complete live performances. (Usually part of the "live" program is prerecorded onto magnetic tape.)

To be musically successful a synthesizer should be designed to meet the following requirements:

1. The synthesizer should perform all of the basic generating and modifying operations of the classical studio and provide resources consistent with the state of the art.
2. The design should contain no unnecessary inherent limitation.
3. The operation of the synthesizer should be straightforward, easy for a musician to understand, and should not require extensive technical knowledge.
4. Features which speed up the composition process and give the composer direct control should be stressed.
5. The instruments should be stable and precisely calibrated for repeatability and ease in working with a score.
6. The complete synthesizer should be as lightweight and portable as possible, and should present an aesthetically pleasing appearance.
7. Reliability should be achieved through the use of conservative design, quality components, and rugged construction techniques.
8. By using suitable control devices, the synthesizer should be readily adaptable to both live performance and programmed control.

(Whether the synthesizers currently available meet all of these requirements is open to question.)

The electronic functions listed under the previous heading are generally available on a synthesizer. In addition, modified square waves and saw-tooth waves can provide additional tone colors. The various waveforms can then be filtered, modulated, and combined in many unique ways. Finally, reverberation may be added. Fig. 6.J-6 depicts a typical synthesizer arrangement.

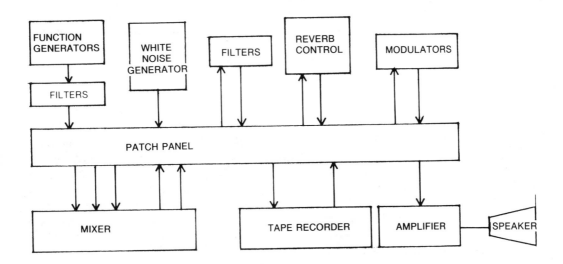

Fig. 6.J-6. Typical synthesizer arrangement.

313

Digital Computers

Computers have been used to study and simulate a wide range of musical tone phenomena. They have proven very useful in the analysis and synthesis of musical tones. In addition they can be used to simulate synthesizers. For example, the computer has been used to analyze oboe and bassoon tones over a wide range of frequencies and dynamic markings. The results indicate that the spectral envelope concept, as shown in Fig. 6.D-11, is basically valid and that the envelopes are remarkably stable over the entire range of the instrument. Oboe and bassoon tones were then synthesized using the following information: the fundamental frequency, and the frequencies and amplitudes of the two peaks in the spectral envelope (Plitnik, 1975). The two peaks in the spectral envelope are modeled with two parallel filters, each having its own amplitude controls, as shown in Fig. 6.J-4. The input is a pulse train having a fundamental frequency equal to that of the desired note. A pulse train has a spectrum with all harmonics having equal amplitude. When the pulse train passes through the two parallel filters, frequencies near the resonant peaks of the filters are passed through while frequencies not near the peaks are attenuated, which results in a spectrum similar to that shown in Fig. 6.D-11. The different dynamic markings are obtained by increasing or decreasing the amplitude of each resonant peak via the amplitude controls. The resulting tone qualities are clearly oboelike and bassoonlike, although not all notes sound completely natural.

The above scheme was used to produce the first known arrangement of Bach's Two-Part Invention in C Major for synthetic oboe and bassoon. The synthetic oboe was programmed to play the upper voice, while the synthetic bassoon played the lower voice. The result can best be described as "interesting." On the plus side, it was possible to produce the music with minimal effort on the part of the programmer. On each data card the following information was specified: the note to be played, its time value, the metronome marking, the dynamic marking, and the phrasing. After a separate recording of each voice was made, they were added together via tape recorder. On the negative side, there were slight round-off errors in the frequencies and time values of the notes, which resulted in the two voices being slightly out of tune and synchronization. The results of this experiment are available on section 6.J of the audiotape "Descriptive Acoustics Demonstrations."

Music compilers have been developed that permit a "composer" to specify musical instrument characteristics and note characteristics and sequences. The instrument characteristics are supplied to the compiler, which simulates the instruments and "plays" the specified notes on them. The resulting sequences of numbers are then D/A converted and presented via a loudspeaker. Computers offer great flexibility for simulation of systems, but suffer from the disadvantage of not performing in real time. The time required to calculate the musical waveforms is usually many times longer than the duration of the waveform created, so they must be calculated and stored and then played out all at once when the calculations are complete.

Computers have been used recently to control electronic and pipe organs. There are several advantages in having a computer attached to an organ. First of all, the computer can play music that humans are not capable of performing. Second, the computer can be used as an aid (e.g., for registration changes) for a live performer. Finally, the computer can be used to record (for later playback) a live performance.

Computers have also been used to compose music. Typically this process involves the specification of rules which are stated in the form of the likelihood that something will occur such as one note following another. These rules can apply to melody, harmony, note sequences, note durations, etc. Random numbers are generated by the computer and put through the rules of the system to generate note strings. In this fashion many different musical structures can be generated, but all of the structures will have the same rules in common.

Exercises

1. Why is the purely electronic generation of piano tone (no mechanical vibrators used) difficult? (See section 6.H.)

2. Suppose one wanted to simulate bassoon sounds on a synthesizer. Which of the elements mentioned in this section would be used? Draw a diagram to indicate how they would be combined.

3. Draw a block diagram showing the elements and the combination which would be most useful for synthesizing flute sounds. Hint: The flute has relatively few harmonics (see section 6.E).

4. What device must be used with a digital computer in order to hear the computer sounds? Why is it necessary?

5. Describe how you might use the various functions available on a synthesizer to create flute sounds or clarinet sounds. Illustrate each case with a diagram.

6. Figure 6.J-2 is a diagram of a tone modifier system for a clarinet.
   a. How could the frequency divider be used to change the sound?
   b. How could the filters be used to change the sound?

7. Take a fairly simple melody (a Stephen Foster tune, for example) and determine the following:
   a. the number of times each note occurs;
   b. the number of times particular two-note sequences occur;
   c. the number of times particular three-note sequences occur.

8. Construct a new "Stephen Foster-like" melody by using a set of two dice as your random number generator for determining two-note sequences. The following probabilities exist for getting each number from a throw of the dice.

Number	Probability	Number	Probability
1	0/36	7	6/36
2	1/36	8	5/36
3	2/36	9	4/36
4	3/36	10	3/36
5	4/36	11	2/36
6	5/36	12	1/36

Start with some arbitrary note such as C. If there is a 1/2 chance of having an E follow the C and a 1/2 chance of having a G follow the C you might choose to have an E anytime you throw a 2, 4, 5, 6, or 10 with the dice (1/36 + 2/36 + 3/36 + 4/36 + 5/36 + 6/36 = 1/2) and a G for any other number thrown. Then once you have "thrown" an E or a G you can determine the next note by means of the probabilities from exercise 7 in a similar fashion.

9. Suppose you were given the task of creating the "ideal" keyboard instrument. How would you provide control of frequency, vibrato, tremolo, tone color, transients, ensemble effects, etc., to the performer?

*10. Experiment with a synthesizer if you have access to one. Try to simulate lip reed, mechanical reed, air reed, string, and percussive tones.

*11. Visit several music stores and observe the tonal characteristics of the various electronic organs.

Further Reading
Backus, chapter 15
Culver, chapters 15, 16

Olson, chapters 5, 10

Winckel, chapter 9

Dorf, R. 1968. *Electronic Musical Instruments,* 3rd ed. (Radiofile, N.Y.).

Douglas, A. 1957. *The Electronic Musical Instrument Manual,* 3rd ed. (Pitman, N.Y.).

Dudley, J. D., and W. J. Strong. 1975. "Analysis and Synthesis of Oboe Tones by the Linear Prediction Method," J. Acoust. Soc. Am. *58,* 5131.

Hartman, W. M. 1975. "The Electronic Music Synthesizer and the Physics of Music," Am. J. Physics, *43,* 755-763.

Hiller, L. A., and L. M. Isaacson. 1959. *Experimental Music* (McGraw-Hill, N.Y.).

Mathews, M. V. 1969. *The Technology of Computer Music* (MIT Press, Cambridge, Mass.).

Moog, R. 1967. "Electronic Music—Its Composition and Performance," Electronics World.

Moorer, J. A. 1977. "Signal Processing Aspects of Computer Music—A Survey," Computer Music Journal *1,* no. 1, 4-37.

Olson, H. F., H. Belar, and J. Timmens. 1960. "Electronic Music Synthesis," J. Acoust. Soc. Am. *32,* 311-19.

Plitnik, G. R. 1975. "A Digital Computer Technique for the Production of Synthetic Oboe Tones," Frostburg J. of Math.

Plitnik, G. R., and W. J. Strong. 1970. "Digital Technique for Synthesis of Bassoon Tones," J. Acoust. Soc. Am. *47,* 131(A).

Plitnik, G. R., and W. J. Strong. 1969. "Digital Technique for Synthesis of Oboe Tones," J. Acout. Soc. Am. *45,* 313(A).

Pritts, R. A., and J. R. Ashley. 1975. "Live Electronic Music in Large Auditoriums," J. Audio Eng. Soc. *23,* 374-80.

Audiovisual

1. *Descriptive Acoustics Demonstrations,* section 6.J (GRP)
2. *Percussion, Mechanical, and Electronic Instruments* (3 3/4 ips, 2-track, 30 min, UMAVEC)
3. *Sounds—The Raw Materials of Music* (1 7/8 ips cassette, SILBUR)
4. *Music from Mathematics* (Bell Labs Record #122227)
5. Carlos, *Switched On Bach II* (Columbia Record KM 32659)
6. Tomita, *Pictures at an Exhibition* (RCA Records)
7. Kraft and Alexander, *1812 Overture* (London, Phase 4)

Chapter 7

Electronic Reproduction of Music

7.A. RECORDED MUSIC AND SOUND REPRODUCTION

The basic principle involved in all sound recording is to capture minute pressure variations in the air and to store them in a convenient form so they can be released again at will. The goal of high-fidelity audio equipment manufacturers is to make this reconstituted sound as indistinguishable from the original as is technologically feasible. The fact that engineers have been successful in achieving this result is apparent from numerous recent listening tests where audiences were unable to differentiate between live music and a stereo reproduction both played behind a screen. In this section we consider the nature of high fidelity and stereo and then the components of a modern stereophonic system. This is followed by a brief survey of high fidelity component specifications and simulated environments.

What Is High Fidelity?

Although the term *high fidelity* (which is usually abbreviated *hi-fi*) has been in general use since the Second World War, it has never been satisfactorily defined. A definition often encountered is "close resemblance to an original sound when reproduced." Manufacturers have not hesitated to exploit such vague implications for commercial profit by labeling many inferior products with the ubiquitous term *hi-fi*. Today the term has, for all practical purposes, become synonomous with *sound system*. Further confusion has been generated by the general public's acceptance of the mere label as somehow magically improving the quality of generally mediocre equipment. Even so, the term *hi-fi* remains the only expression available to convey the ideal of all such systems: a reproduction which is completely faithful to the original sound. Furthermore, the foisting of the term *hi-fi* upon a gullible public should not obscure the very impressive progress, made by the audio industry in recent years, toward achieving such a goal. In fact, it is safe to say that the current "state of the art" is such that the desired degree of faithfulness for reproduced music is limited only by your pocketbook. A practical definition of *hi-fi* for the potential enthusiast is "the best sound reproduction equipment available which is within your budget."

There are three different types of hi-fi enthusiast. The type of equipment deemed necessary for a particular enthusiast depends, to a great extent, into which group he fits. The first type is primarily a music lover who wishes to hear his favorite performance with a minimum of distortion. Since any reasonable quality of reproduction will satisfy this desire, slight distortions of the system will go unnoticed. The second type is a technical perfectionist who is more concerned with the responses of his equipment than with the music itself. He may derive great satisfaction from listening to loud sound with the least possible amount of distortion. The vast majority of hi-fi owners, however, do not fall into either camp. Rather, they represent the compromise position: they are not only interested in obtaining distortion-free reproductions, but they also enjoy the music itself.

True faithfulness in a reproduced sound requires not only a distortion-free reproduction, but also a realistic reproduced sound. During the years that the science of high fidelity was developing many attempts were made to achieve realistic monophonic reproductions. These attempts failed because there is no way that acoustic space can be perceived with a single-channel recording. The only way to achieve realism in a reproduction is to use two (or more) separately recorded and reproduced signals, and this is the essence of stereophonics. The term *stereophonic* is derived from two Greek words: *stereos,* meaning "solid," and *phone,* meaing "sound." The composite word thus literally means "solid sound" or "three-dimensional sound." The term, *stereophonic,* coined by Alexander Graham Bell in 1880, was actually a misnomer for *binaural* (or "two-eared") sound reproduction. The aim of binaural reproduction is to take the listener to the source of sound; stereophonic reproduction attempts to bring the source of sound to the listener. When music is recorded binaurally, two closely spaced directional microphones are mounted on the ear positions of a dummy head. The sound is then presented to a listener by earphones. Stereophonic recording utilizes two (or more) widely spaced microphones, and the playback is through two loudspeakers. As we have seen in section 3.C, we perceive sounds binaurally and use this "two-channel" information to localize sound or to listen selectively. When Mr. Bell experimented with binaural sound reproduction, he compared the spatial effect of the binaural reproduction with the apparent visual solidity seen in the stereopticon popular at that time. (The stereopticon uses two views of the same object photographed from slightly different positions.) The comparison, unfortunately, was not a good one. The eye is capable of quite remarkable accuracy in depth perception and in judging distance. The ear, however, has no comparable process which can be used to elicit depth sensations or to determine the distance of a sound source (although we do often judge the distance of a source by its relative loudness). Our visual perception of depth is largely determined by the degree to which an observer's eyes converge on the object in order to bring it into focus. Since human ears cannot move to any appreciable degree, aural depth perception is nearly impossible. (However, some animals do possess movable ears: they may very well have the ability to determine the distance of a sound source.)

Hi-fi and Stereo Components

A modern high-fidelity reproduction system always embodies three basic components: (1) a transducer to change the information stored on record or tape into an electrical signal, (2) a device to amplify and otherwise modify the electrical signals to a form suitable for (3) a second transducer, which transforms the electrical signal into sound. The first transducer may be a record player, a tape deck, or even a radio tuner. The second component usually is associated with pre-amplifiers, equalizers, tone control, and a power amplifier. The final transducer may be earphones or a loudspeaker. A diagram of these components, which we now consider in more detail, is given in Fig. 7.A-1.

The input transducer is usually a record player, a tape deck, or a radio tuner. A record player consists of three components: a turntable, a tone arm, and a cartridge. The turntable rotates a record at a constant speed of 33 1/3 or 45 rpm (revolutions per minute). The tone arm pivots so as to guide the cartridge across the record.

When a record is recorded, however, the recording head moves radially across the disc, as shown by the dotted line of Fig. 7.A-2. When a record is played, the tone arm swings in an arc about a pivot point: the stylus thus follows the curved solid line shown in Fig. 7.A-2. Over most of the record, then, the stylus is angled slightly to the actual recorded groove. The greatest angle of error (usually expressed in degrees) is

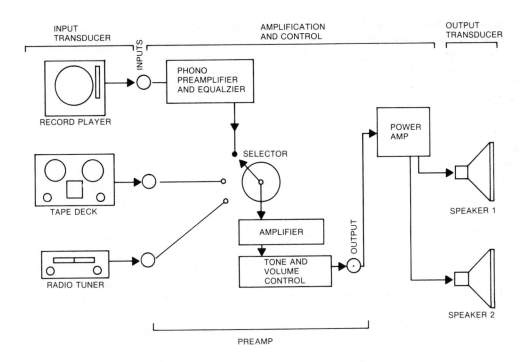

Fig. 7.A-1. Diagram of a complete high fidelity stereo sound system.

known as the *tracking error*. Tracking error can be reduced by using a long tone arm and by positioning the pivot more advantageously. For example, an 8-inch arm can be positioned so as to give a maximum tracking error of only 2.5 degrees (the use of a 16-inch arm would reduce the error to 1.5 degrees). A poorly mounted 8-inch arm, however, can given an error of some 18 degrees—obviously causing considerable distortion of the reproduced sound. The *tracking force* is the downward force required (usually expressed in grams) to keep the stylus in the record groove and follow the undulations. Lighter tracking forces mean less wear on the record and the stylus, but if the tracking force is too low the record may be damaged by groove skipping. The cartridge, a small unit located at the end of the tone arm, is the actual transducer which changes mechanical motion into an electrical signal. The transduction is accomplished by a stylus (needle) which is vibrated by the changing pattern on the wall of the groove in the rotating record. (We will consider the nature of the interaction between record and stylus in more detail in section 7.C.)

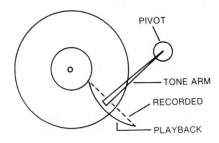

Fig. 7.A-2. Tracking error.

The stylus itself is usually sapphire or diamond. Despite the greater initial cost, diamond needles last approximately 20 times longer than sapphire needles and thus help to increase the longevity of records.

319

There are two methods by which the mechanical motion of the stylus can be converted into an electrical signal: by piezoelectric or by electromagnetic means. The piezoelectric effect, defined in section 1.B, causes a voltage to be generated across the surface of a crystal which is proportional to the force supplied by the motion of the stylus. The voltage is generally sufficient to be conducted directly to the amplifier without the need for preamplifiers or equalizing circuitry. Because piezoelectric cartridges are quite sensitive to excess heat, and because their high-frequency response tends to drop off, piezoelectric cartridges are used only for simple "lo-fi" systems. Electromagnetic cartridges contain a magnetic field in which a coil of wire attached to the stylus shaft vibrates because of the stylus motion caused by the record groove. When a conductor moves in relation to a magnetic field, a voltage is generated in the coil in accordance with Faraday's Law (see section 1.B). The voltage generated is several orders of magnitude smaller than that generated with piezoelectric cartridges. Because of the small voltage, a wide-frequency range can be reproduced without excessive distortion. However, the output requires boosting in the low-frequency range and attenuation of the high-frequency end of the spectrum. (This process, termed "equalization," will be discussed in more detail later.) Also, because of the very small cartridge voltages generated, at least one stage of pre-amplification is necessary.

A tape deck retrieves magnetic patterns from a tape and converts them into electrical signals. (More detail on magnetic tapes will be presented in section 7.C.) Because of the small signal output from the tape head, and because of the need for full equalization, the tape deck includes its own preamp and equalizer circuit.

A radio tuner picks up radio signals (via an antenna) and converts them into electrical signals which can be amplified in the stereo system. The two types of input signals are AM (amplitude modulation) and FM (frequency modulation). Since circuits for both types of reception can be provided for about the same price as either one alone, most inexpensive units sold are AM/FM combination tuners. The difference between AM and FM transmission is that in AM the intensity of the brodcast signal is constantly varied. For FM the intensity is constant but the frequency is modulated. To the average person who purchases a tuner, the listening results are more salient than the method used to receive the signals. Let us now summarize the main differences in these systems. AM reception can be over longer distances than FM, and much more background noise is perceptible with AM. Although there are 20 times as many AM stations to choose from, only a restricted range of frequencies is broadcast and true high fidelity is thus impossible. Furthermore, in many localities it is difficult to accurately tune the desired station, and often the signal is garbled by competing stations (particularly at night). On the other hand, monaural FM transmits frequencies up to 15,000 Hz (almost to the limit of hearing), allowing high-fidelity reproduction. Also, the signals are low in noise and a station (once tuned accurately) remains in tune. The automatic frequency control (AFC) is a further refinement which holds the required station against "drift" effects. Finally, the programs broadcast are generally of higher caliber than those found on AM stations. FM signals can also be used to transmit stereophonic signals effectively. Often FM radio has been subject to the criticism that it is very wasteful to transmit only a relatively narrow range of audio frequencies over a bandwidth which could transmit a far greater frequency range. By means of multiplex techniques this ordinarily wasted capacity can be used to transmit two-channel stereo on a one-frequency channel. The simplest way to do this, if the frequency band available for transmission is from 0 to 53,000 Hz, would be to have one channel carried on the 0-15,000 Hz part of the band while the second channel would occupy the 23,000-53,000 Hz part.

Although this method is simple, it would never have been approved by the FCC because someone who did not have a stereo FM tuner would receive only one of the two channels (the 0-15 kHz channel). A practical FM stereo broadcasting system must be compatible with existing monaural tuners. This is accomplished in the following manner. First, the left (L) and right (R) signals are added to give a sum (L + R). The sum signal is broadcast on the 0-15 kHz channel, thus giving the "monaural equivalent" of a stereo program. The R signal is also subtracted from the L signal to give an L–R "difference" signal. This difference signal is used to modulate a 38-kHz carrier signal; so the information is contained in the 23-53 kHz region, well beyond the range of audible frequencies. The stereo tuner detects the L + R and the L–R signals, then adds and subtracts them from each other. In other words, we generate two new signals from alegbraically manipulating the received channels. The operations yield the following: (L + R) + (L–R) = 2L and (L + R) – (L–R) = 2R. Note that this operation enables us to recover the original L and R channels as independent entities. Although this seems like a rather complex way to proceed, it is necessitated by the problem of having FM stereo broadcasting be compatible with existing monaural equipment.

The preamp is electronic circuitry which takes the very small varying voltage from the magnetic pickup and makes it strong enough for the amplifier to use. Usually the preamp (when separate from the power amp) also contains all the control units and the appropriate equalization circuits. The preamp also contains the selector switch so that the tape deck and tuner can be controlled. The purpose of the equalizer is to obtain a flat frequency response from records. When records are made, the amplitude of low frequencies is reduced while the amplitude for high frequencies is increased. (The reasons for this will be discussed in section 7.C.) When the record is played, however, we want to compensate for the above condition; therefore, electronic circuitry which is somewhat unresponsive to high frequencies and very responsive to low frequencies is included. The effect on input signals of constant amplitude at all frequencies would be the curve shown in Fig. 7.A-3; this is called the RIAA (Radio Industries Association of America) equalization curve. Some of the basic controls usually associated with preamps are: volume controls for each channel and bass and treble boost controls for each channel. Often one will also find balance controls (which increase the intensity of one channel while simultaneously reducing the intensity of the other) and several types of switches. The switches control various filters. For instance, a scratch filter will remove high frequencies while a rumble filter attenuates the low frequencies. One further control is always necessary to prevent the ever-present danger of uncomfortably loud volume levels—self control!

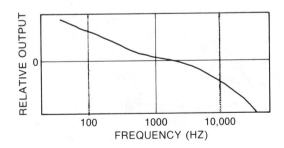

Fig. 7.A-3. RIAA playback equalization curve.

The power amplifier is the device which takes the signals from the preamp (or from a tape deck or tuner) and amplifies them to a level sufficient to drive the loudspeakers. This might seem to be a simple task, but it must be performed at an even level over the system's full frequency range with no audible distortion or noise. The *RMS power* rating of an amplifier is the maximum power (in watts) that the amplifier can deliver continuously without exceeding its rated distortion. To be at all meaningful, an amplifier power rating must include (1) the frequency range over which it can deliver this power, (2) the distortion present at its rated power output, (3) the operating conditions (either one channel or both), and (4) impedance (resistance) of the speakers used. In considering the power required for a home hi-fi system, it is important to remember that doubling the power of an amplifier will not double the resulting sound. Rather, because of the way the ear perceives changes of loudness, one would have to buy an amplifier with ten times the power rating in order to approximately double the maximum sound levels from the speakers!

The final component of any hi-fi system is the loudspeakers or, optionally, headphones. Because of the importance of loudspeakers to the audio field, we will consider them in greater detail in section 7.D.

Component Specifications

The components in any sound reproducing system are not ideal and introduce deficiencies and distortions of their own. The sound output from a reproducing system will be no better than the weakest link in the system will permit, and in many cases it will be worse if the components interact and compound the distortion. One common deficiency of reproducing systems is that they have limited bandwidths (i.e., they do not respond to all audio frequencies). Table 7.A-1 contains some illustrative examples of sound reproduction bandwidths.

Table 7.A-1. Bandwidths of some common sound reproduction systems. Bandwidths are shown in terms of approximate lower and upper frequencies.

System	Bandwidth-Hz
Normal hearing	16–16,000
Telephone	300– 3000
Hearing aid	200– 4000
Small tape recorder	100– 8000
Professional tape recorder	15–25,000
Hi-fi sound system	20–20,000

"Hi-fi" implies a uniform intensity response over a wide frequency range (often 20-20,000 Hz). It is not clear, however, that a listener always prefers this system to one having a more limited response. Some experiments indicate that the opposite is true. For older people, hi-fi is a less serious concern because their hearing at high frequencies is often reduced and they cannot hear the high-frequency sounds even if they are present. It is also very likely that the high-frequency information on records is partly destroyed after a few hundred playings.

Other deficiencies and means of rating the components in a sound reproducing system are detailed below.

1. *Intermodulation distortion* (IM), which results in the creation of higher harmonics and sum and difference frequencies of frequencies present in the original signal. Just perceptible IM distortions have about 1% or more of the power of the original signal.
2. *Total harmonic distortion* (THD) is the amplitude of each individual harmonic generated within the system measured relative to the fundamental frequency. The THD is expressed in percent of the fundamental.
3. Additive noise (or static) is a problem common to all systems. *Signal-to-noise ratio* is the ratio of signal (wanted) to noise (unwanted). S/N ratios of 60 dB or more are desirable.
4. *Hum* results when some of the house current (60 Hz) "leaks" into the audio channels, thus producing a humming sound.
5. Directivity of loudspeaker arrays.
6. Loudspeaker resonances resulting in abnormally high response at certain frequencies.

322

7. Limited dynamic range.
8. Peak clipping resulting from overdriving.
9. Phase delay, in which some frequency components in the signal are delayed in relation to other frequencies.
10. *Transient distortion,* in which some element of the system (typically a loudspeaker) fails to reproduce the transients in the waveform adequately.
11. *Rumble* due to mechanical vibrations of a record player being picked up by the needle.
12. *Wow* and *flutter* due to variation in record or tape speeds which results in a frequency modulation. (Wow is a speed variation below 20 Hz; flutter is above 20 Hz.) Distortions of any kind perturb the sound quality.
13. The *frequency response* of a system shows (usually by a graph) how the different frequencies are related to each other in amplitude. A flat response indicates that all frequencies are treated about the same. If expressed as ± so many dB, the indication is that the response curve does not deviate by more than this amount over the desired frequency range.
14. *Crosstalk* (or stereo separation) is the ability of a transducer to separate the two channels of information, or to prevent one channel signal from leaking into the other channel. Expressed in dB, the higher the separation value, the better.

Some of the desirable component responses for good-quality hi-fi equipment are given in table 7.A-2 below.

Table 7.A-2. Component specification (medium quality).

Component	Frequency response	S/N ratio	Crosstalk	THD + IM
Phono cartridge	20-20,000 Hz ± 3 dB	Rumble of –35 dB (unweighted)	20 dB	1%-4%
Tape deck (open reel)	30–18,000 Hz ± 3 dB at 7½ ips	50 dB	30 dB	—
FM tuner	30–15,000 Hz ± 1 dB	60 dB at full output	—	1% at full output
Amplifier	20–20,000 Hz ± 1 dB	60-70 dB for each input	—	1% or less at rated power

Let us consider briefly several other important considerations for hi-fi components. The *compliance* of a phono cartridge indicates how freely the stylus moves in the groove. It is the amount of deflection (in cm) that a force of 1 dyne will cause. The larger the compliance, the better the response (e.g., it should be at least 10^7 cm/dyne). The tracking force of a cartridge should be between 0.75 and 3.0 gm for optimum results; the tracking error of the tone arm should be less than 2 or 3 degrees. The wow and flutter of a turntable or a tape deck should be less than 0.2%.

Other important specifications of an FM tuner include the sensitivity and the capture ratio. The *sensitivity* is the minimum antenna signal which the tuner can transform into a satisfactory sound signal. It is expressed in microvolts (10^{-6} volt) of signal strength at which the tuner can still suppress the noise by 30 dB at the output. A reasonable specification would be 3 microvolts for 30-dB quieting. The *capture ratio* describes the ability of a tuner to suppress the weaker of two signals on the same frequency while receiving only the stronger. A good specification would be a 3.0-dB, or smaller, capture ratio. With the above information, a wary buyer will be able to read hi-fi specification sheets to be certain that he is buying components that match; hopefully he will then avoid some mismatch in his system which will be the weak link limiting the quality of the entire system.

Alternate channel selectivity is possibly even more important, particularly in areas with many FM stations. This selectivity describes the ability of the tuner to suppress strong neighboring stations without affecting the desired received station.

Simulated Environments

Any reproduced sound ends up in an environment different from that in which it originated. However, several systems have been devised that make it possible to capture much of the flavor of the original recording environment. These have included high fidelity, stereophonic, and quadraphonic reproduction. Contributions of hi-fi were largely considered in the preceding section. Basically these include system components with sufficiently wide frequency and dynamic ranges to capture the essence of the original sound without the introduction of undue annoyances.

Stereophonic reproduction represents an attempt to simulate the spatial characteristics of an original sound. In monophonic recording the outputs of all microphones are mixed and recorded on one channel. In stereophonic recording the outputs of two or more microphones are recorded on two channels so that the two loudspeakers (or headphones) used in reproduction will produce different signals and thus reproduce in part the original spatial setting. (See also section 3.C for comments on headphone vs. loudspeaker stereo listening.) More recently quadraphonic sound has come into the picture, with four channels for capturing spatial features.

The diagram in Fig. 7.A-4 is a speculative illustration of a type of environment control for listening to recorded sounds. This control would permit tailoring of the sound so that listening in a living room could be perceived as listening in a concert hall, a cathedral, etc. The perception of sound depends on the reverberation time of the room and on relative proportions of direct and reflected sounds and their corresponding times of arrival. Suppose a living room is made as nearly anechoic as possible by covering the various surfaces with acoustic tile, drapery, and carpet. Four speakers are used, one mounted in each corner of the "anechoic living room." The character of the perceived listening room (whether concert hall, etc.) is determined by imposing various amounts of delay time and various losses in the separate channels. (Experiments indicate that more than four speakers would provide more realistic simulations of varied environments.)

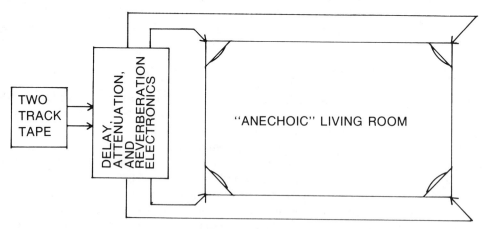

Fig. 7.A-4. Anechoic living room with loudspeaker in each corner and electronics to simulate various listening environments.

With sound signals available in electrical form it is possible to perform various modifications that will have significant perceptual effects. Better balance among various parts of an orchestra or between a soloist and the orchestra can be achieved in the electronic mixing process by varying the gains on the different channels to achieve the desired balance. Artificial reverberation can be added to the signal by electronically delaying it and attenuating it and then adding it to a slightly later portion of the recording. Selective filtering of the signal can be used to eliminate undesirable noise components or to spectrally change the signal in some desired manner.

Exercises

1. High-fidelity sound systems may have a frequency range which extends beyond the range of human hearing. Is this extended range of any value to the reproduced sound? (Hint: See section 2.C on Fourier analysis and synthesis.)

2. What characteristics of the ear make bass and treble boost controls essential on modern hi-fi equipment? (Assume we want to listen to music which is balanced across the audible frequency range at both high-intensity and low-intensity levels. Hint: See section 3.B.)

3. If the volume of a preamp is boosted so that output power is doubled, will the music sound twice as loud? Explain.

4. Explain why a single-channel hi-fi recording can never (no matter how many speakers are used) give a realistic reproduction of the original sound.

5. How far apart should stereo loudspeakers be placed for the best reproduction of the original sound field? What happens if the speakers are too close? Too far apart?

6. Which would sound more realistic as you move around a room, stereo loudspeakers or binaural earphones? Explain.

7. Why can more than two recording microphones often be used advantageously for stereo recording while more than two mikes for a binaural system would be superfluous? Explain.

8. When high-quality tape recordings of music are made, the speed of the tape is usually 7.5 ips. Why would a lower speed not suffice?

9. Can better separation of stereo channels be achieved from a record or a tape? Why?

10. Explain why an equalization circuit is an essential part of modern hi-fi preamps.

11. Why is stereo AM radio broadcasting not practical, or even particularly desirable?

12. The following specs are for a Big Moose Hi-Fi Receiver. Evaluate the receiver from the specs given. Would you buy this equipment?
    Frequency response: 30-30,000 Hz $\pm$ 25 dB
    S/N ratio: 1 to 1
    Power output: 12 watts
    Extras: single channel mixing

13. How should the delay, attenuation, and reverberation controls be adjusted for the anechoic living room to simulate the effect of listening in an anechoic room? A cathedral? A concert hall?

Further Reading

Aldred, chapters 9, 10
Official Guide, chapters 1, 2, 4, 5, 6, 10
Olson, chapter 9
Foster, E. J. 1977. "How to Judge Record-playing Equipment," High Fidelity, April 1977, 60-68.
Hope, A. 1976. "Will Ambisound Shatter the Peace and Quiet of the Stereo Market?" New Scientist 69, 222-24.

Maxwell, J. 1977. "Phono Cartridge Noise," Audio, March 1977, 40-42.

Nakayama, T., and T. Minura. 1971. "Subjective Assessment of Multichannel Reproduction," J. Audio Eng. Soc. *19*, 744-51.

Audiovisual

Frequency Response in Audio Amplifiers (30 min, color, AIM)

Introduction to Radio Receivers (30 min, color, AIM)

Waves, Modulation and Communication (19 min, B&W, 1969, HRAW)

7.B. HISTORY OF HIGH FIDELITY

One hundred years ago the almost simultaneous invention of the telephone and the phonograph began the historic development which was to culminate in the complex world of high fidelity that we know today. The present section will trace the development of devices for recording and reproducing music from Edison's original invention to recently developed quadraphonic systems. We will end with a critique of quadraphonics.

From Edison to High Fidelity

In 1877 Thomas Edison invented a device which was patented as the "phonograph." This first model was little more than a hand-rotated cylinder covered with tinfoil. Although the machine could only be used to repeat several spoken syllables, it did demonstrate that sound reproduction was a definite possibility. The sound track was a groove, with vertical hills and dales embossed on the foil at the bottom of the groove. Although the "record" could not be removed, stored, or duplicated, Edison soon lost interest in his experimental device and left it to others to develop the first commercial model. The first commercial model, devised by Chichester Bell and Sumner Tainter, was called (in reverse of Edison) the "graphophone." The graphophone, which appeared in 1887, substituted wax cylinders for Edison's tinfoil, thus improving both the longevity of the disc and the quality of the reproduced sound. The basic elements of this system are diagrammed in Fig. 7.B-1. When a sound enters the horn, a diaphragm at the small end of the horn vibrates the cutter, thus scratching a groove in the rotating wax cylinder, with hills and valleys representing the wave motion. When the cylinder was used to reproduce the sound, a playback needle followed the squiggles of the wax cylinder. The needle, however, was attached to a diaphragm which was forced to vibrate, thus sending the reproduced sound out the large end of the horn. Although this system could handle only a limited range of frequencies, it was able to reproduce intelligible speech. Since the wax on the cylinder was relatively soft and deteriorated rapidly when played, the record had a somewhat limited lifetime. As there was not any reliable method for making duplicate cylinders, they had to be cut one at a time. The unfortunate artist who wished to record his skills for posterity was forced to perform into a dozen recording horns simultaneously, sixty or seventy times a day!

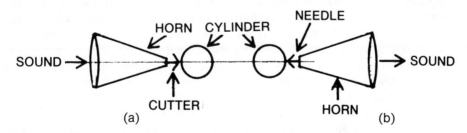

Fig. 7.B-1. Edison recording/reproducing system: (a) recording mode, (b) playback mode.

In 1888 a German-American, Emile Berliner, invented a whole new approach to reproduced sound. He substituted a flat disk, rotated on a turntable, for the cylinder used by Edison and Bell/Tainter. More importantly, the groove was changed from an up-and-down type to a lateral type, which gave an improved production. In other words, the recording needle would vibrate from side to side, rather than up and down, producing a wiggly groove of constant depth on the disc of wax-coated zinc.

327

Because of patents already in effect, however, Berliner had to devise a new way of creating the record groove. This he did by a process, developed after much laborious experimentation, of etching the zinc disk with acid after the recording needle had left its trace on the wax coating. The basis of this system is diagrammed in Fig. 7.B-2. Sound entering the horn causes the diaphragm to oscillate, which vibrates the cutter perpendicular to the groove. During the playback process the needle follows the squiggles in the groove, thus vibrating the diaphragm and generating sound. Although the etching process left a grainy pattern which caused a great deal of surface noise, Berliner's device had one immediate practical advantage over its rival: the playback needle automatically followed the helical groove, while the needle on the cylinder player had to be cranked sideways while playing. Berliner's greatest triumph, however, came several years later. After more laborious experimentation he was able to perfect a method for mass-producing records—the same method which is still used today. The process involves plating the original disk with copper and nickel, which, when removed, makes a reverse master disc (the grooves project as ridges). When the master is pressed against a heated plastic disk, a replica of the original disk is created.

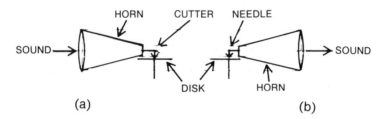

Fig. 7.B-2. Berliner's disk recording device: (a) recording, (b) playback.

The final ingredient needed for Berliner to perfect his invention (which he patented as the gramophone) was a motor which would move the turntable at a constant speed. When the motor, devised by Eldridge Johnson, was finally added to the rest of Berliner's system, the Victor Talking Machine Co. was founded, and its trademark, the "Victrola," became widely known and used.

While the disk system was gaining popularity, the earlier cylinder devices were also making impressive gains. By the turn of the century the rivalry between disk and cylinder manufacturers was intense. Furthermore, each lacked some vital ingredient necessary to perfect its product, which the other had protected by patents. Berliner had exclusive rights to the disk record and the lateral groove, but he could not use the superior wax recording medium used by graphophone people. On the other hand, the graphophone manufacturers still had no reliable way to reproduce their superior sounding wax cylinders, which were still being sold a handful at a time. In 1901 the matter came to a head and the recording industry reached a temporary standstill due to these patent problems. Fortunately, each major interest had something to give and also needed something from the other, so in 1902 the impasse was resolved by an epoch-making patent pool. By making the basic patents available to all, the advantages of both systems were combined to the benefit of all. Berliner immediately abandoned his long-time zinc process and began recording on wax, while the graphophone companies switched from cylinders to the much more convenient disks. Thus began the golden age of reproduced sound: an age which was to last until the advent of radio brought the next major advance in sound reproduction systems.

Until 1925 the recording and reproduction of sound was an entirely mechanical process. The invention of the vacuum tube and electronic amplifiers, however, paved the way for the inevitable wedding of mechanical sound reproduction with electronics. (That the marriage has been a happy one is evidenced by the plethora of electronically-based offspring.)

Although electronics led to tremendous improvements in sound reproduction, the basic principles remain unaltered. A sound signal is received by a microphone and amplified before driving the cutter. The electronic amplification has made possible finer grooves, smaller needles, and lighter pickup arms. Nevertheless, until the rapid advancement of the electronics industry which occurred during the Second World War, sound reproducing systems were still basically "lo-fi." Recorded performances did not resemble closely the original sound: the frequency response was not uniform, the bass was deficient, and the treble was grossly attenuated. After the war, true hi-fi systems became a possibility. Not only was the frequency range extended, but defects in the frequency response were compensated for by filters and other electronic means. Nevertheless, as we have seen in section 7.A, high fidelity could never realistically reproduce recorded sound without stereo.

Quadraphonic Sound

Each major step in the one-hundred-year development of sound reproduction systems brought a new degree of realism. In the early years of this century the quality of reproduced sound was not important. Rather, people were enchanted by the idea of having a "magic talking box." When the first electrical recordings appeared in 1925 people raved over the "lifelike quality" of the reproduction, in spite of the missing bass and treble. When hi-fi systems became popular after World War Two, people were very impressed with the quality of sound, even though their systems sound unbelievably crude to our ears today. With the advent of hi-fi stereophonic systems in the late 1950s, rather faithful reproductions of concert hall environments were at last made possible. Although stereo added a feeling of breadth to the sound, some felt that the effect still lacked depth: there was no feeling of being completely surrounded by sound. The next logical step in the development of total realism was the recent introduction of 4-channel sound: quadraphonics (from the Latin word *quadri,* meaning "four"). For quadraphonic sound reproduction, four independent loudspeakers, each with its own power amplifier, are used. The speakers are arranged with two in front of the listener and two behind him. When the storage medium is tape, four independent tracks are used for recorded music (one track for each loudspeaker). Stereo records, however, have only two channels available (as we will see in section 7.C), and therefore the four channels required for quadraphonic sound had to be encoded into the two available channels. This has been done in two basically different ways: matrix encoding and the discrete (or CD-4) system. The word *matrix* refers to a "mathematical mixing" of signals: the four program channels of a live performance or of a quad tape are combined into two, more complicated, electronic signals by a *matrix encoder.* After being encoded, the signals can be treated like ordinary two-channel stereo (for FM broadcasting or record creation) until the signals emerge from the preamp. At this point a *matrix decoder* converts two channels back into four channels, which are independently amplified and sent to the four loudspeakers. Fig. 7.B-3 presents a diagrammatic representation of this process. (The two main systems of matrix recording in use are the SQ and QS systems.) Since four signals have been mathematically combined into two signals, the decoding process, no matter how good it is, will not be able to recover completely four independent channels. Furthermore, all matrix systems tend to reduce the front speaker separation (stereo effect) to gain front-rear separation.

Matrix records are completely compatible with existing stereophonic systems. Furthermore, since all stereo records contain some out of phase information on the two channels, it is possible to use the decoder to obtain quadraphonic effects from stereo records. (Matrix-encoded records make use of this out-of-phase principle to derive front-rear separation.)

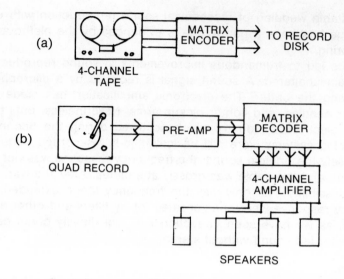

Fig. 7.B-3. Matrix quadraphonic system: a) encoding; (b) decoding.

The discrete (CD-4) record is compatible with existing stereo systems if a special phono cartridge is used. The reason for the special cartridge will become apparent after we discuss how the system operates. Each channel input from the record contains two signals having frequencies ranging from 30 Hz to 15 kHz. The left channel contains the sum of the left-front (LF) and the left-back (LB) channels; the right channel contains the sum of the right-front (RF) and right-back (RB) channels. The difference between CD-4 recording and an ordinary record disc, however, is the addition of a 30-kHz carrier signal on each channel. This high-frequency signal is modulated by the difference between each respective front and back signal. In other words, the left channel carries LF-LB while the right carries RF-RB. This, of course, is exactly the same method used to transmit FM stereo (as described in section 7.A). To recover the four channels requires a demodulator. This device first recovers the high frequencies (from 20 kHz to 45 kHz), which modulate the 30-kHz carrier frequency. After these high frequencies have been converted back to audio signals, the four channels which then exist are combined so as to separate out the four original channels. The combinations are as follows:

$$(LF + LB) + (LF - LB) = 2\,LF$$
$$(LF + LB) - (LF - LB) = 2\,LB$$
$$(RF + RB) + (RF - RB) = 2\,RF$$
$$(RF + RB) - (RF - RB) = 2\,RB$$

Because each original channel is, in principle, completely recovered, the CD-4 records are termed discrete.

Quadrifizzle?

The term *quadrifizzle* was coined by J. R. Ashley of the University of Colorado Electrical Engineering Department to describe his opinion of quadraphonic sound systems. We have already mentioned that *quadri* means "four." "Fizzle" implies a good start with high promise and a sputtering finish. According to Ashley, there are four broad areas of quadraphonic home entertainment which have fizzled out, which we will summarize in the remainder of this section.

The *first fizzle* concerns the idea that if two channels are better than one, four channels must be better than two. This fallacious reasoning resulted from the observation that a stereo system of 1960 sounded much better than a monaural system of 1955. The stereophonic effect was undoubtedly one of the major reasons for this, but there were also other engineering developments (such as magnetic cartridges and good preamps) which improved the fidelity of the stereo system. Furthermore, certain other engineering revolutions (such as the development of acoustic suspension loudspeakers) brought the price of stereo systems down to that of a comparable monophonic system of five years earlier.

When we consider going from two to four channels we must ask ourselves two basic questions: (1) What do we get for two more channels? (2) What do the additional channels cost? For two additional channels we get a slight front-back effect (under good listening conditions), sometimes at the expense of stereo separation. Since there are no major technological breakthroughs which will reduce component costs on the horizon, adding two more channels increases the cost of your system by a factor of 1.5 to 1.8.

The *second fizzle* is the utter failure of four-channel sound to better reproduce the listening environment of a concert hall in the average living room. The two reasons often given for the supposed superiority of four-channel sound over two-channel sound are (1) enhanced directional effects, and (2) added ambience. If we picture a listener facing two loudspeakers, with two identical speakers behind him, we think of being in a concert hall and hearing direct sound from the front speakers and reflected sounds from the rear speakers. This, however, is not how the music is usually recorded. When four independent channels, made from four microphones located in the listening areas of a concert hall, are played back via independent channels (such as four-track tape) the result is described by listeners as "confusionphonic" (Tager, 1967). Furthermore, Tager comes to the conclusion that a complete two-channel system (from mike through loudspeaker) while not perfect, is optimum. When more channels are added, the results get worse.

Ambience is, roughly speaking, equivalent to the reverberant sound field in an auditorium. Can four-channel sound systems bring reverberant sound into your home better than two channels? If we record a concert with two microphones facing front and two aimed toward the rear of the auditorium, the "rear" mikes will record mostly reverberant sound. Since the volume of the reverberant sound is usually substantially lower than that of the direct sound, it is superfluous to require two extra channels to carry this limited amount of information. Why not just add the ambience information into the front channels to get a realistic reverberation effect?

The *third fizzle* has to do with the fact that the sale of electronic equipment is coupled to the availability of program material. Sales of four-channel equipment have been generally slow, in spite of the blitz of advertising material which has appeared in the past few years. The mono-to-stereo revolution came about because the public could hear clearly the superiority of stereo recordings over monaural records. Stereophonic records and FM radio were developed in response to public demand for stereophonic sound. Quadraphonic systems, on the other hand, were developed before there was much public interest. Even now, the selection of quadraphonic disks available is small compared to the number of stereo disks available.

The *fourth fizzle* is the failure to develop a quadraphonic record that has both the required channel separation and low distortion. The two quad record systems we have discussed, matrix and CD-4, show these problems, respectively (Scheiber, 1971). The CD-4 system, because of the modulation process and wide bandwidth required, results in both bandwidth problems and distortion troubles. With the matrix systems the two 20-kHz channels are retained without distortion, but a significant amount of crosstalk is created between channels. Thus, in choosing a system for quadraphonic records you must choose between the lesser of two evils: crosstalk or distortion.

In conclusion, we can state that four-channel equipment costs approximately twice as much as stereo equipment, whereas the results (if noticeable at all) are often detrimental to the overall quality of the reproduction.

Exercises

1. Why was the graphophone an improvement over Edison's phonograph? What feature of the original phonograph made it unusable as a device which could be exploited commercially?

2. Why were the original phonographs better for reproducing music than speech? Explain.

3. By referring to the appropriate material of section 2.E, explain how complex sounds, such as the human voice, can be represented by one "squiggle" on a phonograph recording.

4. Compare the graphophone and the gramophone. How were they similar? Different?

5. Why did the introduction of electronic amplifiers in sound reproduction systems mean that smaller needles and lighter tone arms could be used? Explain.

6. Why was it easier to adapt tape recorders to quadraphonic sound reproduction than it was to adapt records?

7. Why must there be significant amounts of crosstalk in the matrix-encoded quad system?

8. Why is a special cartridge necessary for the CD-4 quadraphonic system?

9. Compare and contrast the process used to broadcast FM stereo with the CD-4 record system.

10. How would you devise a quadraphonic FM system which could broadcast four discrete channels on one station? What problems might be encountered with such a system?

11. Why are the matrix-encoding methods presently compatible with FM stereo, while the CD-4 system is not?

12. Explain why a stereo system in 1960 did not cost substantially more than a comparable monaural system in 1955.

13. Explain why ambience effects are better handled with two channels (although four recording microphones may be used) than with four.

14. Explain why the CD-4 system is bound to introduce distortion into a quadraphonic reproduction.

15. What problems can you foresee in balancing the sound from four speakers during a quad reproduction? Is there any way to avoid these problems?

Further Reading

Official Guide, chapter 9

Olson, chapter 9

Ashley, J. R. 1976. "Is Four Channel a Quadrifizzle?" IEEE Newsletter (Acoustics, Speech, and Signal Processing Society) 38, 1-6.

Bauer, B., D. Gravereaux, and A. Gust. 1971. "A Compatible Stereo-Quadraphonic (SQ) Record System," J. Audio. Eng. Soc. 19, 636-46.

Eargle, J. M. 1971. "Multichannel Stereo Matrix Systems: An Overview," J. Audio Eng. Soc. 19, 552-59.

Hodges, R. 1977. "Will Quadraphonics Rise Again?" Stereo Review, April 1977, 33.

Inoue, T., N. Takahashi, and I. Owahi. 1971. "A Discrete Four-channel Disc and its Reproducing System (CD-4 System)," J. Audio Eng. Soc. 19, 576-83.

Scheiber, P. 1971. "Four Channels and Compatibility," J. Audio Eng. Soc. 19, 267-79.

Tager, P. G. 1967. "Some Features of Physical Structure of Acoustic Fields of Stereophonic Systems," Journal SMPTE 76 (2), 105-10.

7.C. RECORDING MEDIA

When we wish to save a certain set of musical sounds for future recall, or for later enjoyment, we must turn the sound into a form which can be stored conveniently. Perhaps the oldest way of doing so is by means of sheet music. Although the storage medium is very simple, an extremely complicated reproducing mechanism (the human performer) is required. Other media used extensively in the past to store music have included the holes on a player piano roll and spiked drums in music boxes. Although these methods are simple and work amazingly well, they are not particularly convenient or readily accessible. Today three different media are commonly used to store almost any desired sound. The methods are mechanical, magnetic, and optical. The mechanical method is that used to produce plastic records; the magnetic media refers to tape recorder tape; while the optical is used to store the sound track of movie film. In this section we will consider the first two media in some detail.

Disk Recording

To successfully create a record requires four important links, each of which must maintain extremely high quality. The links are (1) the recording machine, (2) the master disc, (3) the copying apparatus, and (4) the final product. We will consider these devices, as well as the magnetic pick-up cartridge in more detail.

The recording machine is a sophisticated, high-quality lathe which engraves a spiral track on an aluminum disk coated with cellulose acetate. The engraved disk. which must be absolutely free of any blemish, is known as the *master*. The recording head includes a steel stylus attached to a coil of wire located in a magnetic field. When a varying voltage (the signal to be recorded) is introduced in the coil the stylus vibrates, thus tracing out a mechanical version of the time-varying input signal. The cutting edge of the stylus is kept very clean and sharp so that the groove walls will be smooth. As the cutting process proceeds, a highly inflammable waste material, known as *swarf,* is formed. The swarf must be removed immediately; often this is accomplished by air suction. The recording process is diagrammed in Fig. 7.C-1.

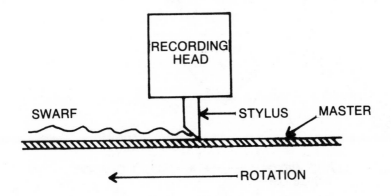

Fig. 7.C-1. The disk recording process.

The depth of the cut made by the stylus must be very carefully controlled: too deep a cut will yield an incorrect groove width, while too shallow a cut will not enable the reproducing head to track correctly. The correct depth is maintained by a feedback device which monitors the cutting process. In order to produce the fine groove spacing required of 33 1/3-rpm records, the cutting stylus is made extremely small. The radius of its tip is 0.00064 cm (see Fig. 7.C-2), and the resulting groove width is .0064 cm with a depth of .0031 cm. The number of grooves per inch averages out to about 250; for soft sounds the grooves are placed closer together, while for loud sounds they are more widely spaced. The space between grooves, called the land width, is the area available for carrying the signal. Obviously, the groove excursion cannot exceed the land width, but we do want the maximum excursion to be as great as possible to mask the surface noise. If the excursion is too great, however, the resulting stress will deform the material of the master. Since almost all of the large-amplitude components in music occur in the low-frequency range, the low frequencies are artificially attenuated in the recording process. Also, since most record playback noise occurs in the high frequencies, and since high-frequency tones tend to be undermodulated on a disc recording, the high frequencies are artificially boosted. The amount by which the bass is attenuated and the treble is boosted for disk recording has been standardized as the RIAA curves discussed in section 7.A. Obviously the playback preamp stage must perform the reverse process in order to reproduce the original sound.

Two further sounds of distortion in disk recording are radial distortion and the "pinch effect." *Radial distortion* occurs because the actual speed of the stylus on the disk depends on how far the stylus is from the center: as the stylus moves closer to the center it moves more and more slowly along the groove. This causes a progressive reduction in the ability of the reproducing stylus to trace the high frequencies, even though they are present on the disk. The *pinch effect* arises because the cutting edge of the recording stylus is chisel-shaped and not spherical like the reproducing stylus. The width of the recorded groove consequently depends on the angle of the groove to a line perpendicular to the disk radius (see Fig. 7.C-3). If the width of the groove decreases too much, the recording stylus will be "pinched upward" and perhaps even forced out of the groove (hence the name "pinch effect"). To compensate for this problem, there must be a certain amount of vertical "give" in the reproducing stylus.

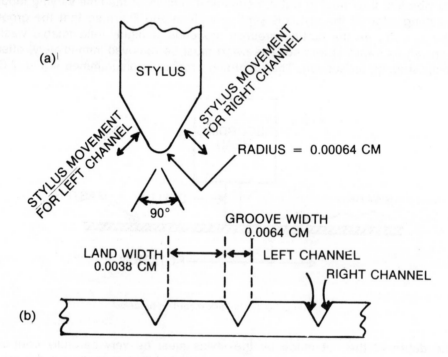

Fig. 7.C-2. (a) Cutting stylus, and (b) master groove.

After the master acetate disk has been cut, it is cleaned and dipped into a bath of silver nitrate, which causes a very thin layer of silver to be deposited. The silver makes the disk electrically conductive so that a nickel or copper plate can be made. A hammer and chisel are then used to remove the metal negative (called the *father*) from the master, which is destroyed in the process. Since the father is too valuable to be used for stamping out records, a *mother* is next made from the father in the

334

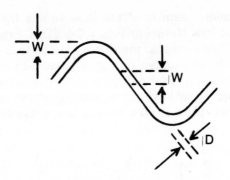

Fig. 7.C-3. The pinch effect: w is width of cutting stylus; d is width of the groove; note d is less than w.

same way that the father was produced from the master, except that the father is not destroyed in this process. Since the mother is a playable disk, flaws can be detected and corrected at this stage of the process. The mother is then used to generate further negatives, appropriately called sons, which are used to stamp out the vinyl records.

Since stereo records require two independent channels, the cutter must vibrate in two perpendicular directions, as is indicated in Fig. 7.C-2. Each of the two walls of the groove (at 45 degrees to the perpendicular) has one waveform impressed upon it; the left wall carries the left channel information, while the right wall carries the right channel. During playback, the stylus mimics the original motions of the cutter. The complex motion of the stylus can then be resolved into two separate signals, which become the two channels of stereophonic sound.

Tape Recording

Although the idea for recording sounds by means of magnetization was first expounded at the turn of the century, it was not until the Second World War that German scientists made the idea practical. The original models recorded magnetic patterns onto a steel wire, but the Germans developed the plastic tape coated with iron oxide particles which is in universal use today. The magnetic recording device consists of a drive mechanism to move the tape at a constant speed past a recording head. The recording head is nothing more than a soft iron core in a C-shape with a coil of wire wound around it, as shown in Fig. 7.C-4. When a varying voltage, corresponding to the input signal, is introduced into the coil, a varying magnetic field is produced. The soft iron core enhances the strength of this field so that at the gap it is intense enough to permanently magnetize the iron oxide on the magnetic tape passing by. In principle the magnetization produced on the tape is proportional to the input signal, so that when the tape is passed over the playback head, the original electrical signals are reconstructed by the reverse process. In practice this entire process is extremely nonlinear, and the resulting reproduction is of very poor quality. Because of the nonlinear magnetic properties of iron oxide, the magnetism induced in the recording head is not directly proportional to the signal strength. If we plot a

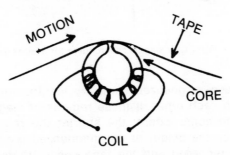

Fig. 7.C-4. A magnetic tape head.

graph of the magnetization intensity of the tape versus the strength of the magnetizing field, we get the curve shown in Fig. 7.C-5. This extremely nonlinear curve is the cause of distortion. Furthermore, there is a serious loss of high-frequency energy and the bass tends to be attenuated during playback. In order to eliminate the distortion, a very high-frequency sinusoid (with a frequency of 50 kHz) is used as a biasing signal. The audio signal is superimposed upon the biasing signal, with the result that the tape magnetization is now directly proportional to the signal strength. If

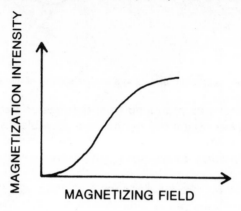

Fig. 7.C-5. Magnetization intensity vs. magnetizing field for a tape recorder.

we were to construct the complete magnetization curve for iron oxide we would get the loop (called a hysteresis loop) shown in Fig. 7.C-6. When a rapidly alternating signal is presented to the tape, the magnetization follows this loop. The part from the origin to the curve is followed only when we first present the signal to an unmagnetized tape. If the amplitude of this signal were gradually reduced, the size of the loop would decrease until the residual magnetism was zero. Since the bias frequency is oscillating at two or three times the highest audio frequency, the loop is being transversed very rapidly. With the audio signal present, the magnetization does not spiral down to zero. Instead, it stops at a level which is proportional to the signal at that instant.

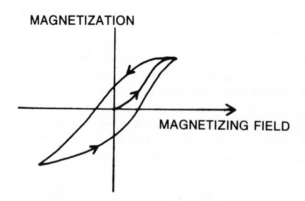

Fig. 7.C-6. Magnetization curve for iron oxide.

When a tape has been magnetized along its length, adjacent magnetic particles may be in opposition to each other. If these regions of magnetization are short they neutralize each other to some extent; the shorter the regions, the greater the demagnetization. But, since the length of the magnetized regions is determined by the audio frequency being recorded and the tape speed, there is an upper limit to the frequency which can be recorded at a given tape speed (the higher the speed, the higher this frequency will be). The length of the magnetized bars on the tape decreases as the audio frequency increases. When the frequency gets so high that the wavelength is approximately equal to the width of the gap of the recording head, demagnetization occurs. The upper frequency is thus limited by the gap in the tape head and the speed of the tape. Fig. 7.C-7 shows typical magnetic recording and playback curves for a speed of 3 3/4 ips. In order to obtain a good S/N ratio as well as to increase the bass, the treble is enhanced during recording while the bass level is increased during playback. The turnover point (marked X) moves up the scale as the tape speed is increased, as can be seen from Table 7.C-1. For this reason high-quality tape recordings of music are recorded at 7 1/2 ips while a higher-quality commercial tape made for producing a master disk is recorded at 15 ips.

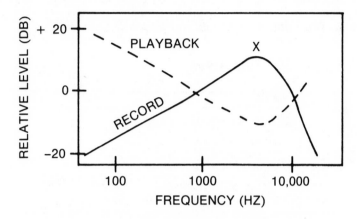

Fig. 7.C-7. Magnetic recording and playback curves.

When a magnetic tape is tightly wound on its spool a strongly magnetized section of the tape may slightly magnetize the tape on an adjacent layer. This phenomenon, known as "print-through," causes a "ghost sound" in the reproduction. Print-through can be reduced by using a heavier tape (with resulting shorter playing time) and by winding the tape less tightly. It is a known fact that the rewind mechanism of tape recorders winds the tape much more tightly than when the tape is played. Hence, tapes will suffer less print-through when stored on the take-up-reel and rewound just before playing.

Exercises

1. Why do low-frequency sounds have a greater amplitude than high-frequency sounds which appear equal in loudness?

2. Explain how the RIAA characteristics discussed in 7.A help to reduce surface noise and groove excursion.

3. Why does the erase head of a tape recorder have a wider gap than the record/playback head?

4. Why is there an upper frequency limit inherent in tape-recorded music? Why does raising the tape speed increase this frequency limit?

5. Why is it not practical to reduce the size of the gap in the tape recording head to even smaller spacings than are in use today?

6. Why are discrete four-channel tapes the only practical way to achieve quadraphonic sound via tape recordings?

7. If you own a tape where print-through is a particular problem, why would it be wise not to rewind the tape after you finish playing it? Explain.

8. Compare and contrast records and tapes as a sound storage medium. What are the advantages and disadvantages of each?

9. Name and describe all the transducers used in the process of tape recording a sound and then playing it back.

10. Name and describe all the transducers involved in making a stereophonic record and then playing it back.

Further Reading

Official Guide, chapter 7

Aldred, chapters 6, 8

Weiler, H. D. 1956. *Tape Recorders and Tape Recording* (Radio Mag., Inc., Mineola, New York).

Woram, J. M. 1976. *The Recording Studio Handbook* (Sagamore).

Audio-visual

The Soundman (15 min, color, 1970, UEVA)

338

7.D. ELECTROACOUSTIC TRANSDUCERS

It is well known that in 1876 Alexander Graham Bell invented the telephone. Perhaps it is realized less widely that in order to transmit a telephone message he had to invent electroacoustic transducers—devices which change sound into electrical energy and vice versa. In this section we will be concerned with two basic types of electroacoustic transduction devices: the microphone and the loudspeaker.

Microphones

As we have seen in chapter one, microphones are transducers which transform acoustic energy into electrical energy. Since sound waves consist of tiny pressure variations in air, the molecules of air move from regions of greater density to regions of reduced density. The movement of air molecules as the pressure rises and falls is known as the *particle velocity*. We can think of a sound wave as consisting of two time-varying components: the particle velocity and the pressure variation. Ideally, a microphone should respond to both of these components: in practice, most microphones respond only to the pressure variations. In order to respond accurately to the velocity component of a sound wave the mechanical element of the microphone must be so light that the air "does not know" it is present. In practice this is impossible to achieve. We will now consider several representative types of pessure microphones and one type of velocity microphone.

For pressure microphones, a sound wave impinges on a diaphragm, causing the diaphragm to vibrate. The mechanical motion of the diaphragm produces the varying voltage in one of several possible ways. In section 1.E we discussed the dynamic microphone as one way. We now consider three additional types of pressure microphones: the carbon microphone, the crystal microphone, and the condenser microphone.

The *carbon microphone* consists of a diaphragm, the central portion of which is attached to a small enclosure filled with carbon granules as shown in Fig. 7.D-1. A constant voltage is placed across the carbon granules. As the diaphragm is displaced by an incoming sound wave the carbon granules are subjected to varying pressure. This change in pressure on the carbon causes the resistance of the carbon granules to change in an amount proportional to the pressure of the sound wave. The varying resistance then causes a varying voltage proportional to the pressure variations of the original sound wave. The carbon microphone has a high sensitivity, high durability, and low cost, and has therefore been found useful for telephone and radio communication purposes. The microphone does not, however, have a wide range of frequency response, and the distortion of the original sound wave is often significant.

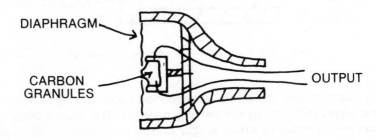

Fig. 7.D-1. Elements of a carbon microphone.

A *crystal microphone* generates a varying voltage which is proportional to the pressure of the original sound wave by means of the deformation of a piezoelectric crystal. A piezoelectric crystal (such as Rochelle salt) generates a voltage when it is deformed. The diaphragm is connected to one corner of the crystal, as shown in Fig.

7.D-2. Microphones of this type are small, are fairly inexpensive, and have a fairly uniform response from 20 to 10,000 Hz. The crystal microphone is widely used in P.A. systems and in hearing aids. The principal disadvantage of these microphones is that relatively small output voltages are obtained for normal acoustic pressures and consequently more amplification is needed.

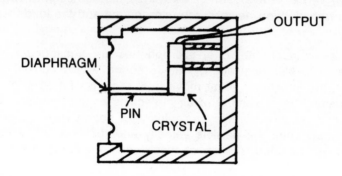

Fig. 7.D-2. Elements of a crystal microphone.

A *condenser microphone* depends for its operation on the variation in capacitance (defined as the ratio of electric charge to voltage for a charged conductor) between a fixed plate and a tightly stretched metal diaphragm; as the distance between the diaphragm and the back plate varies, the capacitance varies accordingly. Since the electrical output varies directly with the changing capacitance, it corresponds to the motion of the diaphragm. A diagram of this type of microphone is shown in Fig. 7.D-3. The condenser microphone has a very high-quality response, but a high voltage (200 to 400 volts) must be supplied in the vicinity of the mike. Also, a preamplifier is required close to the mike to boost the power output. This mike is used extensively in hi-fidelity recording and for research purposes, regardless of the disadvantages.

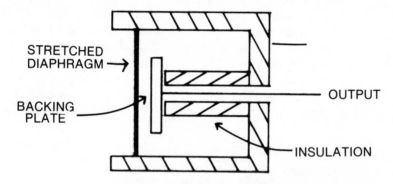

Fig. 7.D-3. Elements of a condenser microphone.

For *velocity microphones* the electrical response corresponds to the particle velocity in a sound wave, rather than the varying pessure. The most common type is the velocity ribbon microphone, in which a light corrugated metallic ribbon is suspended in a magnetic field with both sides freely accessible to surrounding air. The ribbon is driven by the difference in sound pressure between the two sides, which corresponds to the velocity. The motion of the ribbon in a magnetic field then produces a varying voltage which corresponds to the original sound. This type of microphone has good low-frequency response which is quite uniform, but the upper useful limit is usually about 9000 Hz. The response pattern of the microphone is bidirectional as compared to the omnidirectional response of pressure mikes. Since the microphone responds equally to sounds on both sides, it is useful for discriminating against undesirable sounds, balancing orchestral instruments, and in picking up dialogue.

Loudspeakers

A *loudspeaker* is a transducer actuated by electrical energy that produces an acoustical waveform essentially proportional to the time-varying electrical input. The most common type of loudspeaker is the direct radiator dynamic loudspeaker; it is the inverse of the dynamic mike, previously described. The varying input voltage is fed to the voice coil, which is situated in a magnetic field produced by a permanent magnet. The varying voltage in the magnetic field causes the coil to experience a force and thus to move. But the coil is attached to a large diaphragm; so the motion of the coil drives the diaphragm, causing the air to vibrate and thus producing a sound. Because of the simplicity of construction, the small space requirements, and the fairly uniform frequency response, this type of speaker is used almost universally in radio, phonograph, and intercom systems.

The type of response achieved with a speaker, however, depends to a large extent on the type of mounting system used. If the speaker were unmounted, a condensation produced at the front would be partially canceled by the equivalent rarefaction wave from the rear. This effect is most severe at low frequencies; at high frequencies the cone is vibrating so rapidly that condensation rarefactions do not have time to travel around from front to rear. If the speaker is mounted in an enclosure of some type, this effect can be negated to some extent. Loudspeakers can be mounted in many different ways; here are brief descriptions of five of the most common.

A *baffle* is any structure used to regulate or control the sound output from a loudspeaker. A flat baffle mounting, shown in Fig. 7.D-4a, requires a relatively large baffle to obtain an adequate low-frequency response. For instance, to get a good bass response down to 70 Hz the baffle would have to be about 150 cm in diameter.

An open-back cabinet mounting, shown in Fig. 7.D-4b, is widely used for radios, console phonographs, and television sets. The cabinet resonance (usually occurring at 100 to 200 Hz) is used to enhance a weak bass, but this gives the characteristic "boomy" sound usually associated with speakers of this nature. Furthermore, the low-frequency response *below* the resonance is decreased.

The phase-inverter or bass-reflex cabinet is completely enclosed, as shown in Fig. 7.D-4c, with a small port, or opening, cut in the cabinet under the speaker. The port inverts the phase of the waves; otherwise they would cancel. When this type of cabinet is designed properly it not only extends the low-frequency response, but it gives a more uniform output without the undesirable boomy quality of the open-back cabinet.

The acoustic labyrinth principle (see Fig. 7.D-4d) is a variation of the bass-reflex cabinet, except that the path followed by the compressions and rarefactions from the rear of the speaker absorbs energy to prevent cancelation with energy from the front rather than to reinforce the bass. Because of the "folded up" path for the back wave, the total enclosure can be smaller in size than a completely enclosed system.

The completely enclosed cabinet, or infinite baffle mounting, as shown in Fig. 7.D-4e, has no openings which communicate to the outside air. Since the cabinet is completely enclosed, the stiffness on the speaker is increased, which raises the resonant frequency. These considerations must be kept in mind when speakers of this type are designed, and it may be necessary to use a large cabinet in order to extend the bass range.

A popular speaker system for home use is the *acoustic-suspension* design—a form of infinite baffle. Acoustic suspension systems, however, are different because they require very flexible speaker cones mounted with a very loose rim construction. In free air the cone can undergo very large excursions with very little applied power. When these speakers are mounted in air-tight enclosures, the entrapped air "stif-

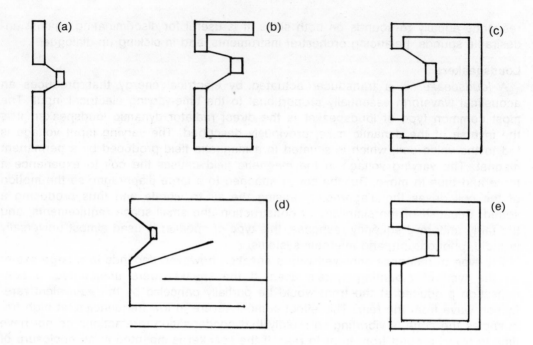

Fig. 7.D-4. Common types of loudspeaker mounting and baffling systems: (a) flat baffle; (b) open-back cabinet; (c) bass reflex cabinet; (d) acoustic labyrinth; (e) infinite baffle cabinet.

fens" the speaker suspension system. An acoustic suspension system enables us to achieve a quite good bass response in a relatively small cabinet. These favorable characteristics, however, are achieved at the expense of efficiency; most are less than 1% efficient. When enclosed speaker cabinets are properly designed, appreciable outputs can be obtained even at 40 Hz for quite small (50,000 cm^3) cabinets. This type of speaker, however, would have a reduced high-frequency output, which would necessitate the use of a multispeaker system.

A *horn loudspeaker* consists of an electrically driven diaphragm mounted to a horn, as shown in Fig. 7.D-5. The horn provides the coupling between the diaphragm and the air and thus increases the low-frequency output. Direct radiator loudspeakers usually have quite low efficiencies, on the order of three to five percent, and are therefore not suitable for installations requiring large amounts of power, such as for public address systems. Horn loudspeakers can obtain efficiencies of 25 to 50 percent and consequently require much less power to drive than a direct radiator speaker would for the same output power.

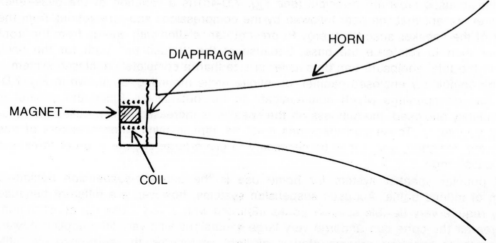

Fig. 7.D-5. Elements of a horn loudspeaker.

A single loudspeaker cannot adequately reproduce tones ranging from very low frequencies (under 100 Hz) to very high frequencies (up to 20,000 Hz). Most well-designed speaker systems, therefore, use two or more separate speakers of various sizes. The bass tones are reproduced by the woofer, and treble by a tweeter. A circuit, called the crossover network, is used to divide the frequencies and channel them to the appropriate speaker.

Loudspeaker Specifications

While most loudspeaker manufacturers seem to agree that the frequency response of their units is an important specification, there is no agreement as to how these measurements should be made or how they should be presented. If an average listening room is used to make the measurements, the normal modes of the room (as discussed in section 2.F) will greatly influence the results. If an anechoic chamber is used for the measurements, however, the speakers are being measured in an unnatural environment and we learn little about how they will perform in an actual listening situation. The published frequency response curves for loudspeakers are therefore, in all probability, not very meaningful.

There are two other speaker specifications which are very important: the impedance and the efficiency. The impedance of a speaker is a quantity related to electrical resistance. To prevent impedance mismatches (which prevent the maximum transfer of power from amplifier to speakers) it is important to connect 8-ohm speakers to the 8-ohm amplifier output, etc. Since speakers range in efficiency from less than 1% to greater than 10%, the speaker efficiency must be matched to the amplifier. A 10-watt amplifier connected to a 10% efficient speaker will produce just as much output as a 100-watt amplifier feeding a 1% efficient loudspeaker.

Exercises

1. Describe how a dynamic microphone works (refer to section 1.E).

2. Which type of microphone would be most useful for each of the following applications: a. An accurate calibration of another microphone? b. A hearing aid? c. A tape recorder used to record speech? d. A high-fidelity recording of an orchestra? e. A mike to record dialogue on a radio program? f. A mike for use with dramatic or musical productions? Explain.

3. Does the size and style and baffle used to enclose a loudspeaker affect the bass or treble? Why?

4. Why must a woofer be relatively large? Why is a tweeter always small?

5. Why are tweeters always mounted on the front surface of a multispeaker enclosure?

6. When you hear a neighbor's hi-fi system playing, why do you hear mostly bass sounds?

7. Why do some multispeaker units have several identical tweeters, each angled in a slightly different direction?

8. Contrast and compare the various speaker enclosures of Fig. 7.D-4. Some items to consider are: expense, efficiency, and bass response.

Further Reading

Official Guide, chapter 3
Olson, chapter 9
Aldred, chapters 2, 4
DiLavore, P., ed. 1975. *The Loudspeaker* (McGraw-Hill, New York).
Olson, H. F. 1976. "A History of High-Quality Studio Microphones," J. Audio Engin. Society *24,* 798-807.

Appendices

APPENDIX 1. REVIEW OF ELEMENTARY MATH

Definitions
1. Constants
    a. Numerical constants: a quantity that always retains the same value. Example: speed of sound in air at T = 20°C.
    b. Arbitrary constants: numerical values assigned to a particular situation. Example: Assuming a constant speed of 20 mph, a man travels ten miles in ½-hour time period.
2. Variables: quantities to which an unlimited number of values can be assigned.
    a. Independent variable: variable for which the numerical value is chosen arbitrarily.
    b. Dependent variable: variable for which the numerical value depends on the value chosen for the dependent variable. Example: Given that a car is traveling at a constant speed of 20 mph, two variables are still involved, time and distance. Taking time as the independent variable, choose any arbitrary time interval, such as two hours. Then the dependent variable (distance) is determined to be 40 miles.
3. Ratio: a quotient or indicated division, often expressed as a common fraction.

Review of Elementary Concepts of Math
The math used in physics can be thought of as a set of symbolic relations used to show how "real things" relate to or depend on one another. For example, the distance (or length) traveled by an auto can be expressed with word symbols, as

$$\text{distance traveled (or length) is equal to speed}$$
$$\text{of travel (or velocity) multiplied by time traveled,}$$

or with mixed symbols, as

$$\text{length} = (\text{velocity}) (\text{time}),$$

or with other symbols, as

$$\ell = (v)(t).$$

We often deal with *equations,* which show some set of symbols equal to some other set. In many cases all symbols but one have known values. Then it becomes our task to calculate a value for the unknown symbol. To solve algebraic equations, a simple rule is to treat each side of the equation as one of a pair of identical twins. Each time something is done to one side of the equation (or to the twin) the same thing must be done to the other side of the equation (or twin). The object is to get the unknown quantity on one side of the equation and the known quantities on the other. Consider the following examples:

(1) $f = 1/T$ $T = 0.1$

$$f = \frac{1}{0.1} = \frac{(1)(10)}{(.1)(10)} = \frac{10}{1} = 10.$$

(2) $l = (v)(t)$ $l = 1000$ $t = 5$ $v = ?$

$$(1000) = (v)(5).$$

Divide each side by 5.

$$\frac{(1000)}{5} = \frac{(v)(5)}{5}$$

$$200 = v$$

$$v = 200.$$

There are in addition to equations some special symbols that represent operations or special actions. "SQRT" is the square root symbol. It means to find a number which when multiplied by itself gives the number under the square root symbol. Study the following examples:

(1) $\text{SQRT}(4) = 2$ $2 \times 2 = 4.$
(2) $\text{SQRT}(5) \simeq 2.24$ $(2.24)(2.24) = 5.0176 \simeq 5$

The logarithm (or "log") of a number is the power to which 10 must be raised to produce that number. The log of the product of two numbers is the sum of the logs of the two numbers.

$\log 10 = 1$	$10^1 = 10$
$\log 1 = 0$	$10^0 = 1$
$\log 1 = 2$	$10^2 = (10)(10) = 100$
$\log 0.1 = -1$	$10^{-1} = 1/10 = 0.1$
$\log (10)(10) = \log 10 + \log 10 = 2$	
$\log (2)(10) = \log 2 + \log 10 = 1.3$	

The "sine" of an angle (where the angle is expressed in fractions of a cycle) is simply a number. The number can represent "real things," such as displacement, force, etc. The reason for using the sine function is that it can conveniently represent different kinds of waves. The sine for different fractions of a cycle can be generated by taking a stick L units in length and pinning one end at the point where the horizontal and vertical axes cross each other. The sine of the fractional rotation (θ) of the stick is defined as the distance (D) of the other end of the stick from the horizontal axis divided by L. Compare the values determined in this manner with those listed in the tables. Once a table of sine values has been put together we can look in the table for the sine value we want, rather than generating it with our rotating stick.

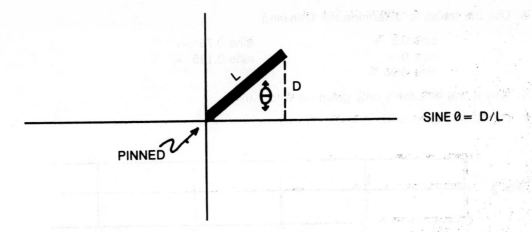

SINE θ = D/L

Exercises

Work the problems below. Use the tables where necessary.

1. If $v = f\lambda$,

given:		find:
(a) $\lambda = 3$	$v = 21$	$f =$
(b) $f = 6$	$\lambda = 3$	$v =$
(c) $v = 16$	$\lambda = 5$	$f =$
(d) $f = 8$	$v = 12$	$\lambda =$

2. If $f = 1/T$,

given:	find:
(a) $T = 1$	$f =$
(b) $T = 0.2$	$f =$
(c) $f = 30$	$T =$
(d) $f = 10$	$T =$

3. If $d = vt$,

given:		find:
(a) $d = 30$	$v = 10$	$t =$
(b) $v = 2$	$t = 5$	$d =$
(c) $t = 100$	$d = 2$	$v =$

4. If $f = \dfrac{1}{6} \text{SQRT} \left(\dfrac{s}{m} \right)$

given:		find:
(a) $s = 90$	$m = 10$	$f =$
(b) $s = 360$	$m = 10$	$f =$

347

5. Use the tables to determine the following:

sine 0.5 =	sine 0.75 =
sine 0 =	sine 0.125 =
sine 0.25 =	

6. Why is the sine table only given for 0 to 1.0?

7. Plot sine Θ for Θ = 0, .125, .25, .375, .5, ...

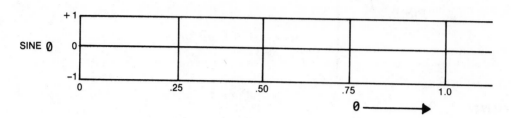

8. Solve for the following:
   (a) log 100 =
   (b) log 1000 =
   (d) log 2 =
   (d) log 4 =
   (e) log 10 =
   (f) log 1 =

9. Solve the following:
   (a) 10^0 =
   (b) 10^1 =
   (c) $10^{.6}$ =
   (d) $10^{.3}$ =
   (e) 10^3 =
   (f) 10^2 =

10. Solve for the following:
    (a) log 200 =
    (b) log 40 =
    (c) log 60000 =
    (d) log 4000 =
    (e) log 9000 =

APPENDIX 2. GRAPHS AND GRAPHICAL ANALYSIS

If is frequently useful to observe how one part of a physical system changes with respect to some other part; that is, when one variable (the independent variable) is changed, how does another variable (the dependent variable) change? If a quantitative relationship exists it may be expressed in one of three ways: (1) in a table, (2) by an equation, or (3) by a graph.

Let us analyze some typical relationships found in the physical world. For each case the information will be expressed in each of the above three forms and the relationship which exists will be explicitly pointed out.

Example One

A moose remains at rest on the ground. How does the displacement of the moose change with time?

This is a static situation: the displacement of the moose does *not* change with time. In tabular form we would have:

Time (sec)	Distance (m)
0	0
1	0
2	0
3	0
4	0
5	0

The equation would be: displacement = constant = 0. The graph would look like:

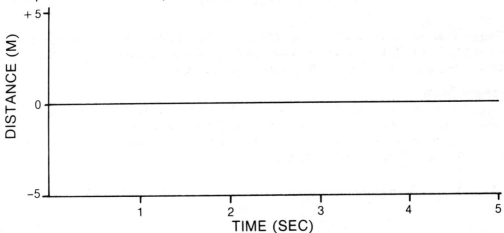

Example Two

The moose now rises and begins to walk in a straight line at a constant speed of 10 meters per second. If we call the moose's original position 0 m and the time when he started walking 0 min, the table would be:

Time (sec)	Distance (m)
0	0
1	10
2	20
3	30
4	40
5	50
6	60
7	70
8	80
9	90
10	100

The equation would be: displacement = speed × time

$$= 10 \text{ (m/sec)} \cdot t$$

The graph would be:

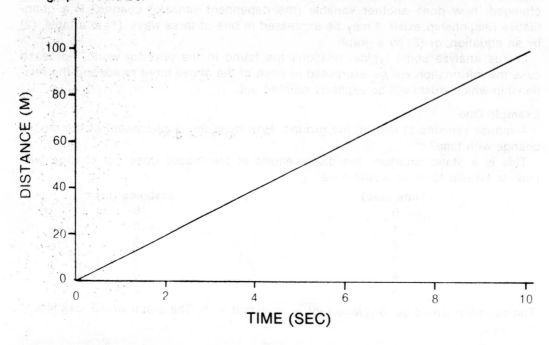

Notice that the distance the moose walks from his original position increases *in direct proportion* to the time he has been walking. When variables are related in this manner they are said to be *directly proportional* variables.

Example Three

Suppose the moose now gets in a motorized cart and at zero time he starts accelerating from rest. The cart accelerates in a straight line at a constant rate of 2 m/sec/sec, in other words, every second the cart goes at a speed which is 2 m/sec faster. The table of data now appears as:

Time (sec)	Distance (m)
0	0
1	1
2	4
3	9
4	16
5	25
6	36
7	49
8	64
9	81
10	100

The equation relating the variables is now:

displacement = ½ (acceleration) × time²

$$= ½ \text{ (2m/sec/sec)} \times t^2$$

$$= t^2.$$

350

The graph appears as:

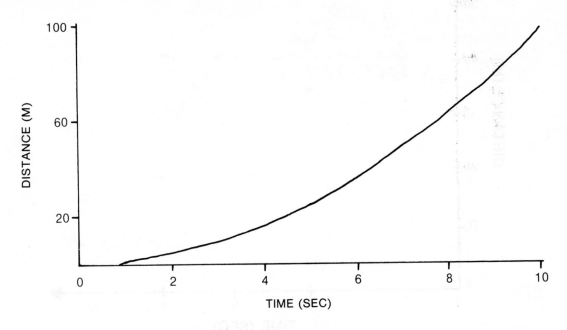

Note that the relationship is no longer directly proportional.

Example Four

A moose awakens from a nap and at time = 0 starts walking due north at a constant rate of 10 meters per second for 5 seconds. He then stops for 2 seconds to eat an eggplant in a farmer's field. The farmer (who was out standing in his field) sees the moose devour the eggplant and immediately begins to shoot. The shots scare the moose and he begins accelerating (due north) at 2 m/sec/sec for the next 3 seconds. The table below gives the displacement of the moose from his original position as time varies. Plot this in the given space.

Time (sec)	Distance (m)
0	0
1	10
2	20
3	30
4	40
5	50
6	50
7	50
8	51
9	54
10	59

Example Five

Suppose you are given five bundles consisting of $100.00 each and asked to divide it equally among five different groups of students. Group 1 has two students, group 2 has five students, group 3 has ten students, group 4 has twenty students, and group 5 has fifty students. Calculate the money each student will receive in each group and fill in the table below.

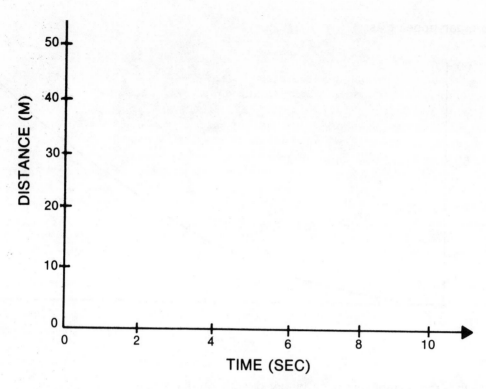

Group	Number of Students	Money per Student
1	2	
2	5	
3	10	
4	20	
5	50	

Plot this data in the graph provided. What kind of relationship exists between the "number of students" and the "money per student"?

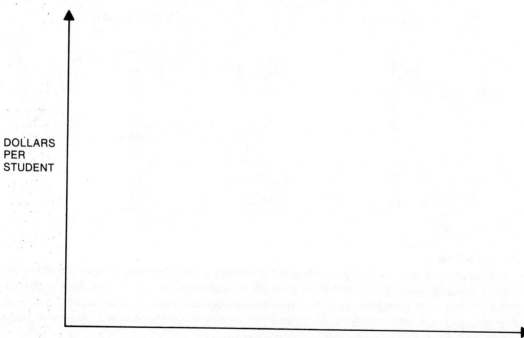

APPENDIX 3. SYMBOLS AND UNITS

In the hypotheses and abstractions of science it is necessary to state relationships between natural phenomena. To express these relationships special symbols are typically employed, because they are a shorthand way to represent various physical quantities and hence allow us a great deal of economy compared to writing everything out in words. Special symbols are not peculiar to science; they are used extensively in music, speech, and other facets of our lives. It is very helpful to use standardized special symbols in science so that we do not have to learn new symbol sets every time we study a different person's writings. However, in science as practiced, although much standardization exists, uniformity of symbol usage is not complete.

When carrying out the measurement step in a scientific method it is necessary to quantify the results of the various measurements. Standard units of measure should be employed so that one set of measurement results can be easily communicated to and interpreted by others. Other scientists making similar observations can easily compare their experimental results with one another when standard units of measure are employed. (Think for a moment of the confusion that might develop in comparing various measurements of length of a particular object if each of ten different observers used as a unit of length the length of his/her shoe or index finger. In this case, you can see that the same object would be described in terms of ten length measures, each having a different numerical value.)

The definition of standard symbols and standard units of measure is quite arbitrary in many respects. Once the definitions have been made, however, a consistent use of the standards makes their utility value very significant.

Symbol	Quantity	Units
a	acceleration	meter/second^2
d	displacement	meter
f	frequency	cycles/second
ℓ	length	meter
m	mass	kilogram
n	integer	——
p	pressure	newton/meter^2
s	stiffness	newton/meter
t	time	second
v	velocity	meter/second
x	any unknown quantity	——
A	area	meter^2
E	energy	joule
F	force	newton
I	intensity	joule/second—meter^2
T	period	second
V	volume	meter^3
Δ	symbol meaning a small part	——
δ	mass density	kilogram/length or area or volume
λ	wavelength	meter
Θ	angle	fraction of cycle
ϕ	phase	fraction of cycle

APPENDIX 4. LIST OF FORMULAS

You should be able to use the following formulas in a meaningful way:

$d = vt$	which relates displacement, velocity, and time
$v = f\lambda$	which relates velocity, frequency, and wavelength
$f = 1/T$	which relates frequency and period
$F = ma$	which relates force, mass, and acceleration
$f = .16\ SQRT(s/m)$	which relates frequency, stiffness, and mass for a simple vibrator
$f_l = f_s (v + v_l)/(v - v_s)$	which relates the frequency perceived by a listener to the frequency produced by a source
$v = SQRT(F/\delta)$	which relates wave velocity to the stretching force and density of a string
$v = SQRT(1.4P/\delta)$	which relates wave velocity to the pressure and density of a gas
$E = Fd$	which relates energy, force, and displacement
Power = Energy/Time	which relates power, energy, and time
I = Power/Area	which relates intensity, power, and surface area
$dB = 20\ \log (P/.00002)$	which relates pressure level to pressure
$dB = 10\ \log (I/10^{-16})$	which relates intensity level to intensity
$d = d_0\ sine\ (ft + \phi)$	which relates displacement to displacement amplitude, frequency, time, and phase
$p = p_0\ since\ (ft + \phi)$	which relates pressure to pressure amplitude, frequency, time, and phase
Area $= 4\pi d^2$	which gives the surface area of a spherical shell in terms of the distance from the center of the sphere.
$v = 3400\ cm/sec$	the approximate speed of sound in air

APPENDIX 5. TABLES OF FUNCTIONS

Table of Square Roots		Table of Logarithms		Table of Sine Values	
n	SQRT(n)	n	log n	θ (in fraction of cycle)	Sine θ
1	1.00	0	——	0	.00
2	1.41	1	.00	.05	.31
3	1.73	2	.30	.10	.59
4	2.00	3	.48	.15	.81
5	2.24	4	.60	.20	.95
6	2.45	5	.70	.25	1.00
7	2.65	6	.78	.30	.95
8	2.83	7	.85	.35	.81
9	3.00	8	.90	.40	.59
10	3.16	9	.95	.45	.31
11	3.32	10	1.00	.50	.00
12	3.46	20	1.30	.55	−.31
13	3.61	50	1.70	.60	−.59
14	3.74	100	2.00	.65	−.81
15	3.87	1000	3.00	.70	−.05
16	4.00			.75	−1.00
17	4.12			.80	−.95
18	4.24			.85	−.81
19	4.36			.90	−.59
20	4.47			.95	−.31
21	4.58			1.00	
22	4.69			(or 0)	0
23	4.80			1.05	
24	4.90			(or .05)	.31
25	5.00				

Bibliography

PUBLICATIONS

AAPT. 1972. *Apparatus for Teaching Physics* (American Association of Physics Teachers, Washington, D.C.).

Aldred, J. 1963. *Manual of Sound Recording* (Fountain Press, London).

Ashford, T. 1967. *The Physical Sciences,* 2nd ed. (Holt, Rinehart, Winston, New York).

Backus, J. 1977. *The Acoustical Foundations of Music,* 2nd ed. (Norton and Co., New York).

Baines, A. 1969. *Musical Instruments through the Ages* (Penguin Books, Baltimore, Maryland).

Ballif, J. R., and W. E. Dibble. 1972. *Physics: Fundamentals and Frontiers* (John Wiley and Sons, New York).

Baron, R. A. 1970. *The Tyranny of Noise* (St. Martin's Press).

Bartholomew, W. T. 1965. *Acoustics of Music* (Prentice-Hall, Englewood Cliffs, New Jersey).

Benade, A. H. 1976. *Fundamentals of Musical Acoustics* (Oxford university Press, New York).

Beranek, L. L. 1962. *Music, Acoustics and Architecture* (Wiley, New York).

Bragdon, C. R. 1970. *Noise Pollution: The Unquiet Crisis* (University of Pennsylvania Press).

Bragg, S. W. 1968. *The World of Sound* (Dover, New York).

Broch, J. T. 1971. *Acoustic Noise Measurements* (Bruel and Kjaer Instruments, Cleveland).

Chedd, G. 1970. *Sound* (Doubleday, New York).

Crawford, F. S. 1968. *Waves, Berkeley Physics Course—Volume 3* (McGraw-Hill, New York).

Culver, C. A. 1956. *Musical Acoustics,* 4th ed. (McGraw-Hill, New York).

Doelle, L. L. 1972. *Environmental Acoustics* (McGraw-Hill, New York).

Denes, P. B., and E. N. Pinson. 1973. *The Speech Chain* (Doubleday Anchor Books, Garden City, New York).

Edge, R. D., ed. 1974. *Experiments on Physics and the Arts* (University of South Carolina).

Fant, G. M. 1960. *Acoustics Theory of Speech Production* (Mouton).

Flanagan, J. L. 1972. *Speech Analysis Synthesis and Perception,* 2nd ed. (Springer-Verlag).

Flanagan, J. L., and L. R. Rabiner. 1973. *Speech Synthesis* (Dowden, Hutchinson, and Ross).

Fletcher, H. 1953. Speech and Hearing in Communication (Van Nostrand).

French, A. P. 1971. *Vibrations and Waves, MIT Introductory Physics Seriss* (Norton and Co., New York).

Gerber, S. E. 1974. *Introductory Hearing Science* (W. B. Sanders Co., Philadelphia).

Helmholtz, H. L. F. 1954. *On the Sensations of Tone* (Dover, New York).

Hutchins, C. M., ed. 1975. *Musical Acoustics, Part I: Violin Family Components* (Dowden, Hutchinson, and Ross, Stroudsburg, Pennsylvania).

Hutchins, C. M., ed. 1975. *Musical Acoustics, Part II: Violin Family Functions* (Dowden, Hutchinson, and Ross, Stroudsburg, Pennsylvania).

Institute of High Fidelity. 1974. *Official Guide to High Fidelity* (Howard W. Sams, Indianapolis).

Jeans, J. 1968. *Science and Music* (Dover, New York).

Johnson, K., and W. Walker. 1977. *The Science of Hi-Fidelity* (Kendall-Hunt, Dubuque, Iowa).

Josephs, J. J. 1973. *The Physics of Musical Sound* (Van Nostrand Momentum Book, New York).

Kent, E. E., ed. 1977. *Musical Acoustics: Piano and Wind Instruments* (Dowden, Hutchinson, and Ross, Stroudsburg, Pennsylvania).

Kinsler, L. E., and A. R. Frey. 1962. *Fundamentals of Acoustics,* 2nd ed. (John Wiley and Sons, New York).

Knudsen, V. O., and C. M. Harris. 1950. *Acoustical Designing in Architecture* (Wiley, New York).

Kock, W. E. 1971. *Seeing Sound* (Wiley Interscience, New York).

Kock, W. E. 1965. *Sound Waves and Light Waves* (Doubleday, New York).

Krauskopf, K., and A. Beiser. 1971. *Fundamentals of Physical Science,* 6th ed. (McGraw-Hill, New York).

Kryter, K. D. 1970. *The Effects of Noise on Man* (Academic Press).

Ladefoged, P. 1975. *A Course in Phonetics* (Harcourt Brace Jovanovich, New York).

Lass, N. I., ed. 1976. *Contemporary Issues in Experimental Phonetics* (Academic Press, New York).

Meiners, H. F., ed. 1970. *Physics Demonstration Experiments, Volume I, Mechanics and Wave Motion* (The Ronald Press Co., New York).

Miller, D. C. 1935. *Anecdotal History of the Science of Sound* (Macmillan, New York).

Miller, G. A. 1951. *Language and Communication* (McGraw Hill, New York).

Olson, H. F. 1967. *Music, Physics, and Engineering* (Dover, New York).

Peterson, A. P. G., and E. E. Gross, Jr. 1972. *Handbook of Noise Measurement,* 7th ed. (General Radio Co.).

Pierce, J. R., and E. E. David, Jr. 1958. *Man's World of Sound* (Doubleday, New York).

Plomp, R., and G. F. Smoorenburg, ed. 1970. *Frequency Analysis and Periodicity Detection in Hearing* (A. W. Sythoff, Lieden).

Roederer, J. G. 1975. *Introduction to the Physics and Psychophysics of Music,* 2nd ed. (Springer-Verlag, New York).

Sabine, W. C. 1964. *Collected Papers on Acoustics* (Dover, New York).

Sanders, D. A. 1977. *Auditory Perception of Speech* (Prentice-Hall, Englewood Cliffs, New Jersey).

Sears, F. W., M. W. Zemansky, and Young. 1975. *College Physics* (Addison-Wesley, Cambridge, Massachusetts).

Seashore, C. E. 1967. *Psychology of Music* (Dover, New York).

Singh, S. S., and K. S. Singh. 1976. *Phonetics: Principles and Practices* (University Park Press, Baltimore).

Stevens, S. S., and H. Davis. 1938. *Hearing* (Wiley, New York).

Stevens, S. S., ed. 1951. *Handbook of Experimental Psychology* (Wiley, New York).

Stevens, S. S., and F. Warshofsky. 1965. *Sound and Hearing* (Time Science Series, New York).

Taylor, C. A. 1965. *The Physics of Musical Sounds* (American Elsevier, New York).

United States Gypsum. 1972. *Sound Control Construction,* 2nd ed. (United States Gypsum, Chicago).

Van Bergiik, W. A., J. R. Pierce, and E. E. David. 1960. *Waves and the Ear* (Doubleday, New York).

Von Bekesy, G. 1960. *Experiments in Hearing* (McGraw-Hill, New York).

White, F. A. 1975. *Our Acoustic Environment* (Wiley, New York).

Winckel, F. 1967. *Music, Sound and Sensation* (Dover, New York).

Wood, A. 1974. *The Physics of Music* (Halsted Press, New York).

AUDIOVISUAL RESOURCES

AIM	Associated Instructional Media
BELL	Bell Telephone Companies
CF (COR)	Coronet Instructional Films
EBE (EBEC)	Encyclopedia Britannica Educational Corporation
EFL	Ealing Film Loop
EHF	Educational Horizons Films
EMCC	EMC Corporation
ES	Edmund Scientific Company
FLMFR	Filmfair Communications
FRSC (FRS)	Folkways Records and Services Corporation
GRP	G. P. Tape Service, Department of Physics, Frostburg State College, Maryland.
GSF	Garden State/Novo Incorporated
HRAW	Holt, Rinehart and Winston
ILAVD	Institute of Laryngology and Voice Disorders
IOWA	Iowa State University
IPS	Introductory Physics Series
IU	Indiana University
JACBMC	Jacobs Manufacturing Company
MGH (MGHF)	McGraw-Hill Films
MLA	Modern Learning Aids Division of Ward's Natural Science
OPRINT	Out of print, but available in many film libraries
PAROX	Paramount-Oxford Films
RANK	J. Arthur Rank Organ
SILBUR	Silver Burdett Company
SF	Sterling Educational Films
UEVA	Universal Educational and Visual Arts
UIOWA	University of Iowa
UKANMC	University of Kansas Medical Center
UMAVEC	University of Michigan Audio Visual Education Center
UMICH	University of Michigan
USNAC	U.S. National Audiovisual Center
WD	Walt Disney

ABOUT THE AUTHORS

William J. Strong was born in Idaho. He received his B.S. and M.S. degrees at Brigham Young University, and his Ph.D. at the Massachusetts Institute of Technology. He is currently professor of physics at BYU.

Dr. Strong's particular areas of interest are varied and include the analysis and synthesis of musical tones, machine recognition and synthesis of speech, aids for the deaf, physical properties of musical instruments, and computer modeling of musical instruments, speech systems, and architectural systems. He is a member of Sigma Xi, IEEE, A. G. Bell Association, and the Acoustical Society of America, where he has served on a number of committees, chairing the Musical Acoustics Committee in 1971-73.

Dr. Strong and his family live in Provo, Utah.

George R. Plitnik, a native of New Jersey, received his B.A. and B.S. degrees from Lebanon Valley College in Pennsylvania, his M.A. from Wake Forest University, and his Ph.D. from Brigham Young University. He holds memberships in the Acoustical Society of America, the American Association of Physics Teachers, and the American Guild of Organists. He is currently associate professor of physics at Frostburg State College, Frostburg, Maryland.

In addition to physics, Dr. Plitnik has great enthusiasm for the pipe organ. He is an accomplished organist and currently serves as organist at the LaVale (Maryland) United Methodist Church. He spends much of his spare time rebuilding pipe organs, and he has a rebuilt instrument in his home.

Dr. Plitnik and his family reside in rural Mt. Savage, Maryland, on a small farm.